WIRELESS
SENSOR
NETWORKS

WIRELESS
SENSOR
NETWORKS

Architecture • Applications • Advancements

S. R. Vijayalakshmi, PhD
&
S. Muruganand, PhD

MERCURY LEARNING AND INFORMATION
Dulles, Virginia
Boston, Massachusetts
New Delhi

Publisher: David Pallai
MERCURY LEARNING AND INFORMATION
22841 Quicksilver Drive
Dulles, VA 20166
info@merclearning.com
www.merclearning.com
(800) 232-0223

S. R. Vijayalakshmi, PhD & S. Muruganand, PhD. *Wireless Sensor Networks*.
ISBN: 978-1-68392-225-4

Library of Congress Control Number: 2018938985

181920321   This book is printed on acid-free paper in the United States of America.

CONTENTS

PREFACE

Wireless sensor networks (WSN) are more and more frequently seen as the solution to large scale tracking and monitoring applications. WSN provide a bridge between the real physical and virtual worlds. They allow the ability to observe the previously unobservable at a fine resolution over large scales. The deployment of a large number of small, wireless sensors that can sample, process, and deliver information to external systems (such as the satellite network or the Internet), opens many novel application domains. They have a wide range of potential applications to industry, science, transportation, civil infrastructure, and security. The Internet of Things (IoT), which is technically supported by WSN and other relevant technologies, is discussed. Wireless Sensor Networks is an essential guide for anyone interested in wireless communications for sensor networks, home networking, or device hacking.

This book covers a large number of topics encountered in the architecture, application, and advancements of a wireless sensor network. It covers the basic idea and advanced technologies in the field of sensor networking. It also covers the research ideas behind any application of a sensor network.

Chapter 1 discusses the Wireless Sensor Network. Introductory points, usage, applications, difficulties, basic requirements, and history of wireless networks are covered. Wireless sensor network architecture, communication protocols, and WSN sensors are introduced in this chapter.

Chapter 2 discusses different types of the node hardware architecture. The components of a wireless sensor node architecture, modular sensor node architecture, PIC node architecture, IMote node architecture, XYZ node architecture, and Hogthrob node architecture are discussed.

Chapter 3 discusses the software architectures of a wireless sensor network. The operating system requirements, characteristics of WSN, components of the software architecture, a cluster-based service-oriented architecture, software development for sensor nodes, Tiny OS, and ZigBee are covered.

Chapter 4 discusses the wireless body sensor network. The architecture of wireless body sensor networks, bio signal monitoring using wireless sensor networks, differences between a wireless sensor network and a body sensor network, methodology for development of biomedical signals acquisition and monitoring using WSN, health monitoring, wearable computing, simulators for WSN and research ideas in medical field are discussed in this chapter.

Chapter 5 discusses the ubiquitous sensor network. Applications of USNs, monitoring volcanic eruptions with a ubiquitous sensor network, wireless sensor networks on regional environmental protection, design and the development of USNs for a paddy rice cropping monitoring application, WSNs in the smart grid, smart water sensor networks, intelligent transportation using WSNs, traffic infrastructures, WSNs in smart homes, monitor system for structure and IP-WSN based integration systems are discussed in this chapter.

Chapter 6 discusses underwater wireless sensor networks. WSNs for oceanographic monitoring, aerial wireless sensor networks for oceanographic monitoring, underwater acoustic sensor networks, underwater wireless sensor networks, challenges in network protocol for ocean monitoring, distinctions between UWSNs and ground-based sensor networks, networking architectures for UWSNs, water quality monitoring, limitations of UWSNs, and research challenges in UWSNs are discussed in this chapter.

Chapter 7 covers the integration of the Internet of Things with WSNs. This chapter introduces the Internet of Things (IoT), context awareness, integrating a WSN to the Internet, architecture of IoT for WSN, gateway-based integration, IoT and WSN design principles, big data and IoT in a WSN, challenges in IoT WSN, and finally, simple versus embedded control gateways.

Chapter 8 discusses Wireless Multimedia Sensor Networks (WMSN). This chapter discusses introduction of WMSNs, factors influencing the design of multimedia sensor networks, network architecture of WMSN, WMSNs as distributed computer vision systems, the application layer, transport layer, network layer, MAC layer, physical layer, cross-layer design, wireless video sensor networks, three tier architecture of video sensor networks and wireless video sensor network for autonomous coastal sensing.

Chapter 9 discusses the Mobile Ad hoc Networks (MANETs). Wireless Ad hoc Sensor Networks, Mobile Ad hoc Networks, Classification of routing protocols for MANETs, security in ad hoc networks, Ad Hoc Networks and Internet Connectivity, Mobile Ad hoc networking for the military, Vehicular ad hoc and Sensor Networks (VANET), applications of VANET, routing for VANET, security in VANET, VSN architecture for micro-climate monitoring are all discussed in this chapter.

Chapter 10 discusses routing and security in the WSN. Algorithms for wireless sensor networks, routing protocols, security in a Wireless Sensor Network, obstacles of sensor security, security requirements, attacks, intrusion detection, and finally, defensive measures are discussed in this chapter.

S. R. Vijayalakshmi
S. Muruganand

CHAPTER 1

WIRELESS SENSOR NETWORKS

This chapter deals with the introduction of wireless sensor networks. By the end of the chapter, one can learn the answers to simple questions like: What is a wireless sensor network? What are the uses and advantages of these network systems? Where could these devices be implemented?

1.1 Introduction to Wireless Sensor Networks

A sensor is a small, lightweight device which measures in the environment physical parameters such as temperature, pressure, relative humidity, and many more. Sensor networks are highly distributed networks of wireless sensor nodes, deployed in large numbers to monitor an environment or system. Sensor nodes are used for sensing, processing, and communicating. The nodes then sense environmental changes and report them to other nodes over flexible network architecture. Sensor nodes are great for deployment in hostile environments or over large geographical areas.

A wireless sensor network (WSN) is a collection of nodes organized into a cooperative network. Each node consists of processing capability (one or more microcontrollers, CPUs, or DSP chips), may contain multiple types of memory (program, data, and flash memories), has an RF transceiver (usually with a single omni-directional antenna), has a power source (e.g., batteries and solar cells), and can accommodate various sensors and actuators. The nodes communicate wirelessly and often self-organize after being deployed in an ad hoc fashion. Systems of thousands or even 10,000 nodes are anticipated. Such systems can revolutionize the way we live and work.

Currently, wireless sensor networks are beginning to be deployed at an accelerated pace. It is not unreasonable to expect that in 10–15 years the world will be covered with wireless sensor networks that can be accessed via the Internet. This can be considered as the Internet becoming a physical network, as shown in Figure 1.1. This new technology is exciting, with unlimited potential for numerous application areas including environmental, medical, military, transportation, entertainment, crisis management, homeland defense, and smart spaces. The information collected by the sensor nodes is transmitted to the base station. The base station is connected to the Internet. So anyone can access the information anywhere at any time.

1.2 Usage of Sensor Networks

Sensor networks have been useful in a variety of domains. The primary domains in which sensors are deployed are as follows:

Environmental Observation

Sensor networks can be used to monitor environmental changes. An example could be water pollution detection in a lake that is located near a factory that uses chemical substances. Sensor nodes could be randomly deployed in unknown and hostile areas and relay the exact origin of a pollutant to a centralized authority to take appropriate measures to limit the spreading of pollution. Other examples include forest fire detection, air pollution, and rainfall observation in agriculture.

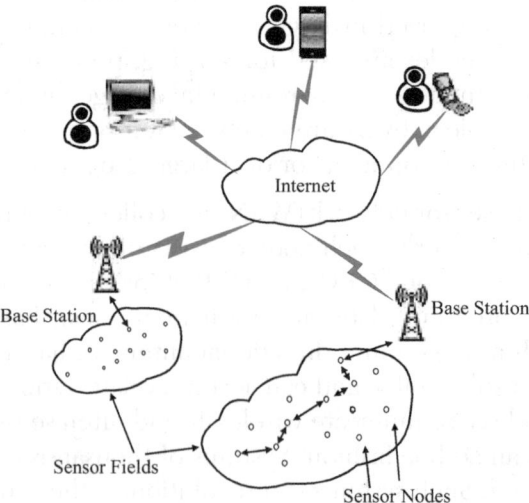

FIGURE 1.1 Accessing WSNs through the Internet.

Military Monitoring

The military uses sensor networks for battlefield surveillance; sensors could monitor vehicular traffic, track the position of the enemy, or even safeguard the equipment of the side deploying sensors.

Building Monitoring

Sensors can also be used in large buildings or factories to monitor climate changes. Thermostats and temperature sensor nodes are deployed all over the building's area. In addition, sensors could be used to monitor vibrations that could damage the structure of a building.

Healthcare

Sensors can be used in biomedical applications to improve the quality of the provided care. Sensors are implanted in the human body to monitor medical problems like cancer and help patients maintain their health.

1.3 Applications of WSNs

The following are the some of the applications of wireless sensor networks. But they are not limited to these applications. Researchers in this field will introduce new application areas in the future.

- Constant monitoring and detection of specific events
- Military, battlefield surveillance
- Forest fire and flood detection
- Habitat exploration of animals
- Patient monitoring
- Home appliances
- Disaster relief operations
- Intelligent buildings (or bridges)
- Reducing energy wastage by proper humidity, ventilation, air conditioning, heating, ventilation, and air-conditioning (HVAC) control
- Measurements about room occupancy, temperature, air flow, etc.
- Monitoring mechanical stress after earthquakes

- Machine surveillance and preventive maintenance
- Embedding sensing/control functions into places no cable has gone before, e.g., tire pressure monitoring
- Precision agriculture
- Bringing out fertilizer/pesticides/irrigation only where needed
- Medicine and health care
- Post-operative or intensive care
- Long-term surveillance of chronically ill patients or the elderly

The following are brief discussions of some of the application areas.

Military Applications

Sensor networks can provide variety of services to the military and air forces like information collection, battlefield surveillance, intrusion detection, and attack detection. Application sensor networks have quite an advantage over other networks, because enemy attacks can damage or destroy some of the nodes, but node failure in WSN doesn't affect the whole network. Possible uses of WSN in the military are:

1. *Intrusion Detection:* Sensor networks can be used as a two-phase intrusion detection system. Instead of using mines, intrusion can be detected by establishing sensor networks in that area. Mines are dangerous to civilians, so instead sensor nodes sense the intrusion and alarm the army. The response to prevent intrusion can now be decided by the military.

2. *Enemy Tracking and Target Classification:* Moving objects with significant metallic content can be detected using specially designed sensors, so enemies can be tracked and civilians are ignored. This system especially helps in detecting armed soldiers and vehicles.

3. *Battlefield Surveillance:* Critical areas and borders can be closely monitored using sensor networks to obtain information about any enemy activity in that area. The quick gathering of information provides time for a quick response.

4. *Battlefield Damage Assessment:* Sensor networks can be deployed after the battle or attacks to gather information about damage assessment.

5. *Detection of NBC Attacks:* Sensor networks can be used as a Nuclear, Biological, and Chemical warning system. If any nuclear, biological, or chemical agents can be detected by sensors, an embedded alert system can now send a warning message. It provides the military critical response time to check the situation and prevent possible attacks, which can save lives of many.

6. *Targeting System:* Sensors can be embedded in weapons. Exact information about the target, like distance and angle, can be collected and sent to the shooter so sensors can collaborate with weapons for better target assessment.

Applications for Automobiles

Sensors can even be employed in vehicles, which provides an advance tracking mechanism for vehicles as well as allowing police to track stolen vehicles. Applications of WSN for vehicles are:

1. *Detecting and Monitoring Car Thefts:* Sensor nodes are being deployed to detect and identify threats within a geographic region and report these threats to remote end users by the Internet for analysis.

2. *Vehicle Tracking and Detection:* Sensor nodes connected to WSNs can be embedded in vehicle designs, so the vehicle can be tracked with the help of sensor networks.

3. *Traffic Control:* Sensor networks have been used for vehicle traffic monitoring and control for quite a while. Most traffic intersections have either overhead or buried sensors to detect vehicles and control traffic lights. Furthermore, video cameras are frequently used to monitor road segments with heavy traffic, with the video sent to human operators at central locations.

Industrial Applications

Commercial industry has long been interested in taking advantage of sensing as a means of lowering cost and improving machine performance and maintainability.

1. *Machine Health Monitoring:* Wireless sensor networks have been developed as a condition-based maintenance (CBM) solution for machinery. In wired systems, the cost of wiring limits the installation of enough sensors. Machine "health" is monitored through the determination of vibration or wear and lubrication levels. WSNs enable the insertion of sensors into regions that are inaccessible by humans, like rotating machinery, or hazardous or restricted areas.

2. *Monitoring Industrial Environments:* Wireless sensors are used for environmental condition monitoring in industries, such as monitoring the level of water in overflow tanks in nuclear power plants and the temperature inside refrigerators.

3. *Managing Inventory Control:* Each item in a warehouse may have a sensor node attached to it. The end users can find out the exact location of the item with the help of sensor and tally the number of items in the same category stored in the database.

4. *Interactive Museums:* In the future, places like museums will become alive with the help of sensor networks. Children will be able to interact with objects in museums to learn more about them and the objects will be able to respond to their touch and speech with the help of sensors.

Based on the specific requirements of industrial production, the IWSN applications can be classified into three groups:

1. *Environmental Sensing:* This group generally represents the widest field of WSN applications nowadays. IWSN applications for environmental sensing cover the problems of air and water (together with waste water) pollution, but cover production material pollution monitoring as well. Furthermore, in hazardous environments, there are numerous needs for fire, flood, or landslide sensing. Finally, security issues arise in markets with competing products and with service providers, where IWSNs are used for point of interest, area, and barrier monitoring.

2. *Condition Monitoring:* This group generally covers the problems of structural condition monitoring, providing both structure health information (the condition of the buildings, construction, bridges, supply routes, etc.) and machine condition monitoring including possible automatic maintenance. Therefore, this group of IWSN applications is vital for production in all branches of industry.

3. *Process Automation:* The last group of applications provides the users with information regarding resources for the production and service provision (including the materials, current stock, and supply chain status, as well as the manpower included in the industrial process).

Finally, one of the most important issues from the user perspective is the production performance monitoring, evaluation, and improvement that are achieved through IWSNs.

1.4 Difficulties in WSN Research

Since a wireless sensor network is a distributed real-time system, a natural question is how many solutions from distributed and real-time systems can be used in these new systems? Unfortunately, very little prior work can be applied, and new solutions are necessary in all areas of the system. The main reason is that the set of assumptions underlying previous work has changed dramatically. Most past distributed systems research has assumed that the systems are wired, have unlimited power, are not real-time, have user interfaces such as screens and mice, have a fixed set of resources, treat each node in the system as very important, and are location independent. In contrast, for wireless sensor networks, the systems are wireless, have scarce power, are real-time, utilize sensors and actuators as interfaces, have dynamically changing sets of resources, aggregate behavior, and are location dependent. Many wireless sensor networks also utilize minimal capacity devices, which places a further strain on the ability to use past solutions. The following are wireless network concerns.

1. *Interference Issues*

   Radio frequency (RF) interference can lead to disastrous problems with wireless deployments. With 2.4 GHz wireless, there are several sources of interfering signals, including microwave ovens, cordless phones, Bluetooth enabled devices, FHSS wireless LANs, and neighboring wireless LANs.

2. *Power Management*

   Electricity in batteries is a limited resource. It is also a concern in the case of wireless networks.

3. *System Interoperability*

   The ability of two or more systems or components to exchange information and to use the information that has been exchanged is also a difficult process.

4. *Security Concerns*

Wireless security is the prevention of unauthorized access or damage to computers using wireless networks.

5. *Security Threats*

Radio waves can easily penetrate walls. One can passively retrieve the radio signal without being noticed (Figure 1.2). Someone could maliciously jam a wireless network and create electronic damage.

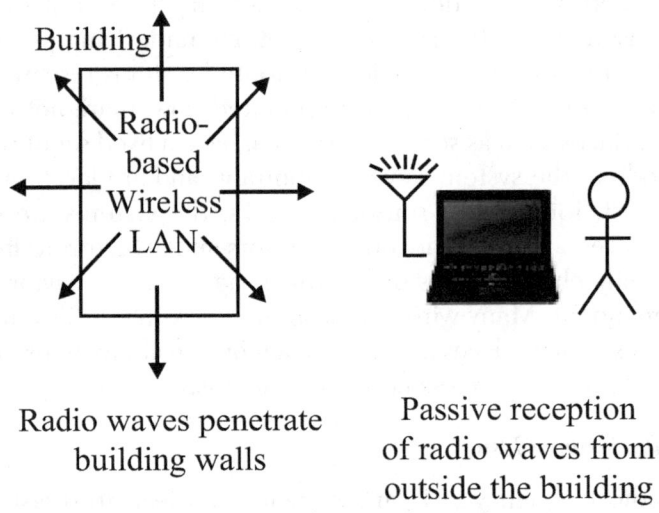

Radio waves penetrate building walls

Passive reception of radio waves from outside the building

FIGURE 1.2 The passive reception of wireless network data.

6. *Installation Issues*

Wireless coverage as a contour is shown in Figure 1.3. So, two different interferences are Intra-system interference (e.g., between its own access points) and Inter-system interference (e.g., from external Bluetooth, which is also on 2.4 GHz).

7. *Health Risks*

So far, there is no conclusive answer!! Radio is safer than cellular phones!! A wireless network is even safer, as it operates at 50~100 milliwatts, compared to 600mw~3w of cellular phones.

8. *Wireless Standards*

The two wireless standards used by WSNs are 802.15.4 and Zigbee. The characteristics of these protocols are as follows:

- They are low-power protocols
- Performance is an issue
- Maximum distance is around 100 m

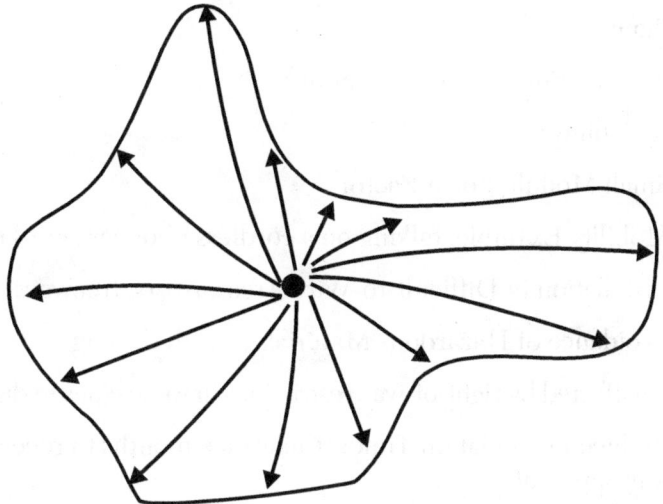

FIGURE 1.3 The resulting radiation pattern of an omni-directional antenna within an office building.

The following are the channels and their bit rates. Channels are:

- 868.0-868.6MHz → 1 channel (Europe)
- 902.0-928.0MHz → 10 channels (USA)
- 2.40-2.48GHz → 16 channels (worldwide)

Bit Rates are:

- 868.0-868.6MHz → 20/100/250 Kb/s
- 902.0-928.0MHz → 40/250 Kb/s
- 2.40-2.48GHz → 250 Kb/s

1.5 Basic Requirements for WSNs

The following are the basic requirements for wireless sensor networks:

- Low Power Consumption
- Ease of Use
- Scalability
- Responsiveness
- Range
- Bi-Directional Communication
- Reliability
- Small Module Form Factor
- Mobility Example, talking on a cordless phone vs. cord phone
- Installation in Difficult-to-Wire Areas: rivers, freeways, old buildings
- Avoidance of Hazardous Materials as with drilling
- Unaffected by right-of-way restrictions in some cities to dig in the ground
- Reduced Installation Time: It may take months to receive right-of-way approvals
- Increased Reliability: cable vs. cable-less
- Long-Term Savings: never need re-cabling

1.6 History of Wireless Networks

Indians used fire and smoke in ancient days. Later, a messenger on horseback in the sixteenth century carried the information. Telephone lines in the nineteenth century were used for voice communication. Now networks are used. Traditional networks (Local Area Network, Metropolitan Area Network, and Wide Area Network) have provided great convenience in offices, hotel rooms, or homes. But one cannot utilize the service unless the people are physically connected to a LAN or a telephone line. In 1980s, amateur radio hobbyists built TNCs (terminal node controllers) to interface "ham" radio equipment and their computers (Figure 1.4).

Radio Waves
Pocket Data transmission

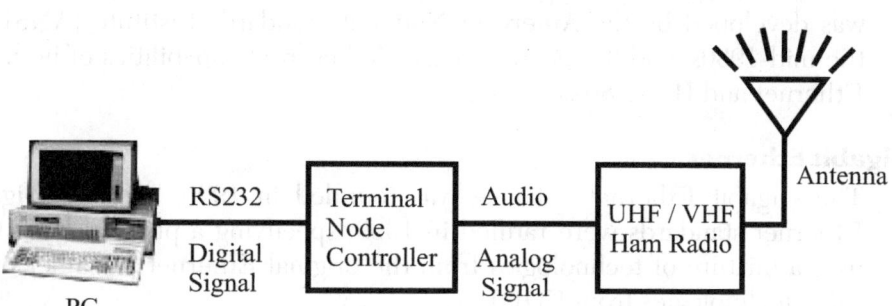

FIGURE 1.4 Ham radio terminal node controllers.

In 1985, the Federal Communications Commission (FCC) authorized the use of ISM bands for Industrial, Scientific, and Medical for commercial development. ISM bands are equal to 902MHz and 5.85 GHz. ISM is very attractive to vendors because no FCC license is required. In 1980s, small-size computers started to appear. Laptops, palmtops, Personal Digital Assistants, and Wireless LAN products appeared. In wireless LAN, the IEEE 802.11 standard was finalized in July 1997. IEEE 802.11a, b, e, g, i, r, etc. wireless WAN Packet radio networks (e.g., RAM) were implemented.

Historical Development and Standards

Much of this information contains a thorough summary of communication network standards, topologies, and components.

Ethernet

The Ethernet was developed in the mid 1970s by Xerox, DEC, and Intel, and was standardized in 1979. The Institute of Electrical and Electronics Engineers (IEEE) released the official Ethernet standard IEEE 802.3 in 1983. The Fast Ethernet operates at ten times the speed of the regular Ethernet, and was officially adopted in 1995. It introduces new features such as full-duplex operation and auto-negotiation. Both these standards use IEEE 802.3 variable-length frames having between 64 and 1514-byte packets.

Token Ring

In 1984 IBM introduced the 4Mbit/s token ring network. The system was of high quality and robust, but its cost caused it to fall behind the Ethernet in popularity. IEEE standardized the token ring with the IEEE 802.5

specification. The Fiber Distributed Data Interface (FDDI) specifies a 100Mbit/s token-passing, dual-ring LAN that uses fiber optic cable. It was developed by the American National Standards Institute (ANSI) in the mid-1980s, and its speed far exceeded current capabilities of both the Ethernet and IEEE 802.5.

Gigabit Ethernet

The Gigabit Ethernet Alliance was founded in 1996, and the Gigabit Ethernet standards were ratified in 1999, specifying a physical layer that uses a mixture of technologies from the original Ethernet and fiber optic cable technologies from FDDI.

Client-Server

Client-Server networks became popular in the late 1980s with the replacement of large mainframe computers by networks of personal computers. Application programs for distributed computing environments are essentially divided into two parts: the client or front end, and the server or back end. The user's PC is the client and the more powerful server machines interface to the network.

Peer-to-Peer Networking

Peer-to-peer networking architectures have all machines with equivalent capabilities and responsibilities. There is no server, and computers connect to each other, usually using a bus topology, to share files, printers, Internet access, and other resources.

Peer-to-Peer Computing

Peer-to-peer computing is a significant evolutionary step over P2P networking. Here, computing tasks are split between multiple computers, with the result being assembled for further consumption. P2P computing has sparked a revolution for the Internet Age and has obtained considerable success in a very short time. The Napster MP3 music file sharing application went live in September 1999, and attracted more than 20 million users by mid-2000.

802.11 Wireless Local Area Network

IEEE ratified the IEEE 802.11 specification in 1997 as a standard for WLAN. Current versions of 802.11 (i.e., 802.11b) support transmission up to 11Mbit/s. WiFi, as it is known, is useful for fast and easy networking of

PCs, printers, and other devices in a local environment, that is, the home. Current PCs and laptops as purchased have the hardware to support WiFi. Purchasing and installing a WiFi router and receivers is within the budget and capability of home PC enthusiasts.

Bluetooth

Bluetooth was initiated in 1998 and standardized by the IEEE as Wireless Personal Area Network (WPAN) specification IEEE 802.15. Bluetooth is a short range RF technology aimed at facilitating communication of electronic devices between each other and with the Internet, allowing for data synchronization that is transparent to the user. Supported devices include PCs, laptops, printers, joysticks, keyboards, mice, cell phones, PDAs, and consumer products. Mobile devices are also supported. Discovery protocols allow new devices to be hooked up easily to the network. Bluetooth uses the unlicensed 2.4 GHz band and can transmit data up to 1Mbit/s, can penetrate solid non-metal barriers, and has a nominal range of 10m that can be extended to 100m. A master station can service up to 7 simultaneous slave links. Forming a network of these networks, for example, a piconet, can allow one master to service up to 200 slaves. Currently, Bluetooth development kits can be purchased from a variety of suppliers, but the systems generally require a great deal of time, effort, and knowledge for programming and debugging. Forming piconets has not yet been streamlined and is unduly difficult.

Home RF

Home RF was initiated in 1998 and has similar goals to Bluetooth for WPAN. Its goal is shared data/voice transmission. It interfaces with the Internet as well as the Public Switched Telephone Network. It uses the 2.4 GHz band and has a range of 50 m, suitable for home and yard. A maximum of 127 nodes can be accommodated in a single network.

IrDA (Infrared Data Association)

IrDA (Infrared Data Association) is a WPAN technology that has a short-range, narrowtransmission-angle beam suitable for aiming and selective reception of signals.

ZigBee

ZigBee takes full advantage of a powerful physical radio specified by IEEE 802.15.4. ZigBee adds logical network, security, and application software.

ZigBee continues to work closely with the IEEE to ensure an integrated and complete solution for the market. The features are

- Low power consumption

- Low cost

- Low offered message throughput

- Supports large network orders (<= 65k nodes)

- Low to no QoS (Quality of Service) guarantees

- Flexible protocol design suitable for many applications

Wireless WAN vs. LAN

The following table (Table 1.1) compares Wireless WAN with LAN.

Table 1.1 Comparison between Wireless WAN and Wireless LAN

Wireless WAN	Wireless LAN
transmission speed: 10K-1M ubiquitous coverage roaming speed: vehicular real-time voice supported: circuit-switching	transmission speed: more than 1Mbps coverage: a few hundred meters roaming speed: pedestrian packet-switching

1.7 Electromagnetic Spectrum

The spectrum is a range of electromagnetic radiation, as shown in Figure 1.5, and bands are spectrum parts, as given in Table 1.2.

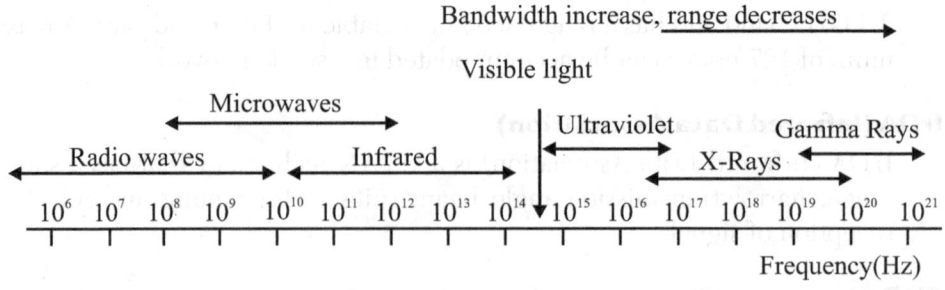

FIGURE 1.5 The electromagnetic spectrum.

Table 1.2 The Various Radio Bands and Their Common Use

Frequency	Band name	Applications
<3 KHz	Extremely Low Frequency (ELF)	Submarine
3 KHz - 30 KHz	Very Low frequency (VLF)	communications
		Marine
30 KHz - 300 KHz	Low Frequency (LF) or Long Wave (LW)	communications
300 KHz - 3 MHz	Medium Frequency (MF) or Medium	AM radio
3 MHZ - 30 MHz	Wave (MW)	AM radio
30 MHz - 300 MHz	High Frequency (HF) or Short Wave	AM radio
300 MHz - 3 GHz	(SW) Very High Frequency (VHF)	FM Radio-TV
	Ultra High Frequency (UHF)	TV - cellular
		telephony
3 GHz - 30 GHZ	Super High Frequency (SHF) Extra	Satellites
30 GHZ - 300GHZ	High Frequency (EHF)	Satellites - radars

HF band enables worldwide transmission. HF signals are reflected off the ionosphere and thus can travel very large distances. Radio spectrum for wireless technology is shown in Figure 1.6. The separate frequencies are "extremely low," "very low," "medium," "high," "very high," "ultra high," "microwave region," "infrared region," "visible light region," and "X-ray."

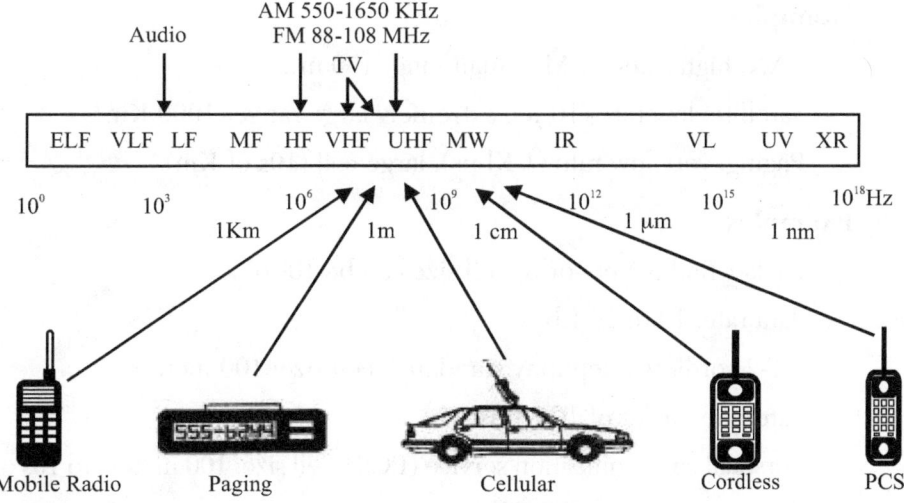

FIGURE 1.6 The radio spectrum.

Audio frequency range is 20 Hz to 20 KHz. The AM radio stations use 1 MHz. The FM and TV use 100 MHz. Cell sizes can vary from tens of meters to thousands of kilometers. Data rates may range from 0.1 K to 50 Mbps (Figure 1.7).

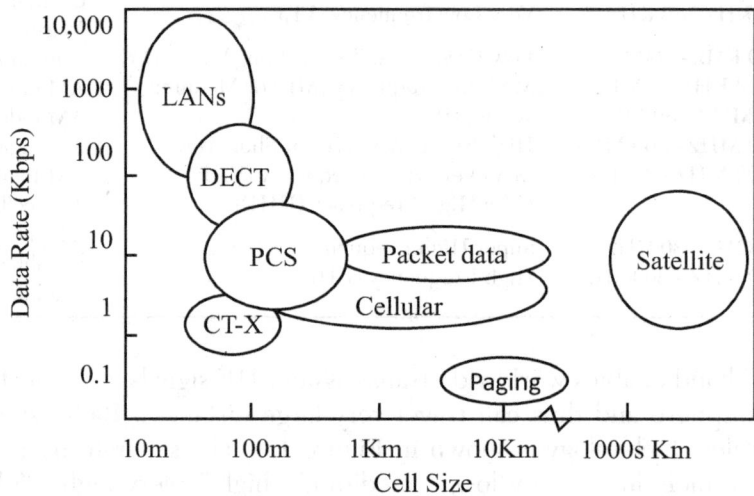

FIGURE 1.7 Relationship of networks.

Examples:

> LAN: high rate (11 M), small range (50 m).
>
> Satellite: low rate (10 K), extremely large range (1000 Km).
>
> Paging: very low rate (1 Kbps), large cell (10s of Km).

Examples:

> Packet Radio Networks: cell size can be 10s of km
>
> data rate: 10 to 20 Kbps
>
> CT-2 cordless telephony standard): cell size: 100 meters
>
> date rate: order of 10 Kbps
>
> Personal communication service (PCS): cell size: 100 meters to 10s of km
>
> data rate: order of 100 Kbps

Smaller cell size implies higher data rate, less power consumption, more handovers, and more frequency reuse.

There are 3 approaches.

1. ISM band

2. Narrow band

3. Spread spectrum

1. ISM Bands

In 1985, the FCC modified part 15 to stimulate the use of wireless networks. ISM stands for Industrial, Scientific, and Medical. It is unlicensed and one can use freely install and move. Only 2.4 GHz is the world-accepted ISM band. 902 MHz is easier in manufacturing. Figure 1.8 shows the Industrial, Scientific and Medical (ISM) frequency bands.

902 MHz
2.4 GHz   2.4 GHz
5.7 GHz

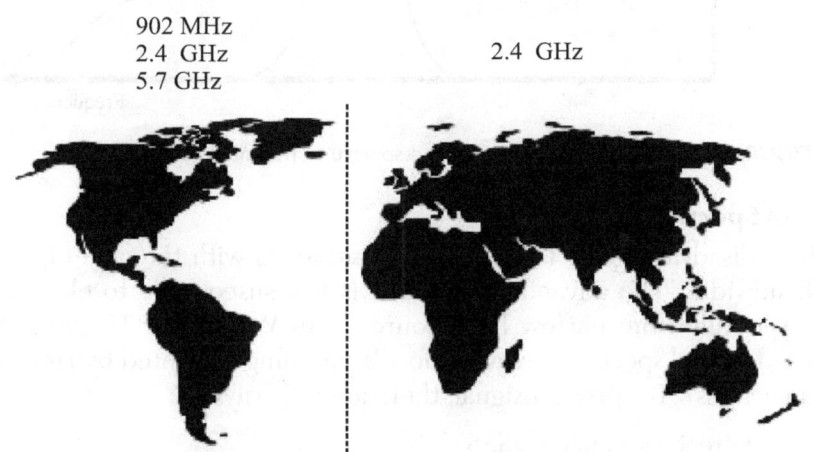

FIGURE 1.8 ISM spectrum availability.

2. Narrow Band Modulation

It concentrates all transmission power in a narrow range of frequencies. Efficient use of the radio spectrum will save bandwidth. Ex: For example, television, AM, and FM.

Advantage

For long-distance transmission (e.g., metropolitan area).

Disadvantages

Noise can easily corrupt the signals. It is necessary to obtain FCC (Federal Communications Commission) licenses to coordinate use.

3. Spread Spectrum Modulation

Spread spectrum modulation is defined as "spreading" a signal's power over a wider band of frequency. Figure 1.9 shows narrow band versus spread spectrum modulation.

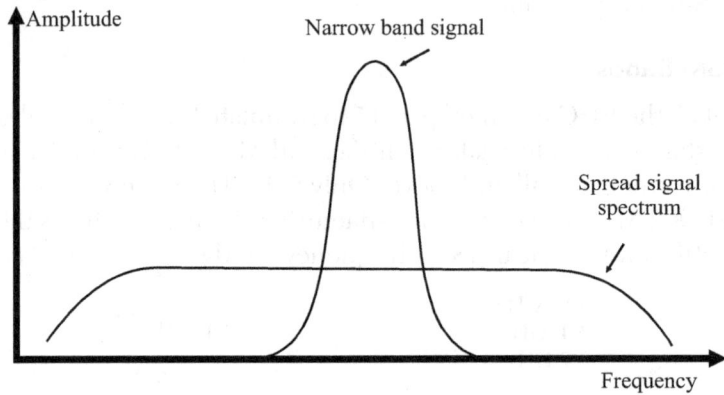

FIGURE 1.9 Narrow band versus spread spectrum modulation.

Spread Spectrum

The disadvantage of this is that it contradicts with the goal of conserving bandwidth. The advantage is that it is less susceptible to electrical noise (especially from narrow band sources). In World War II, the U.S. Army used spread spectrum to avoid hostile jamming (invented by Hedy Lamarr, an actress). To spread a signal, there are two ways:

1. Direct sequence (DSSS)

2. Frequency hopping (FHSS)

Direct Sequence Spread Spectrum (DSSS)

It uses a bit sequence to represent "zero" and "one" as given in Figure 1.10. It is also referred to as "chipping code." Longer chipping codes are more resilient to noise. The minimum length is equal to 10 (by FCC). IEEE 802.11 uses 11 chips per data bit.

DSSS is a transmission technology used in Local Area Wireless Network transmissions where a data signal at the sending station is combined with a higher data rate bit sequence, or chipping code, that divides the user data according to a spreading ratio. The chipping code is a redundant bit pattern for each bit that is transmitted, which increases the signal's resistance to interference. If one or more bits in the pattern are damaged during transmission, the original data can be recovered due to the redundancy of the transmission.

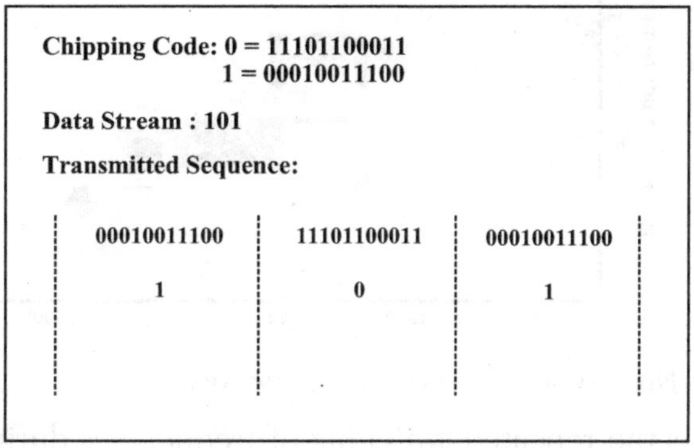

Chipping Code: 0 = 11101100011
1 = 00010011100
Data Stream : 101
Transmitted Sequence:
00010011100
11101100011
00010011100
1
0
1

FIGURE 1.10 The operation of the direct sequence spread spectrum.

Frequency Hopping Spread Spectrum (FHSS)

Data is modulated by carrier signals that hop from frequency to frequency as a function of time, over a wide band of frequencies. Hopping Code is used to determine the order of hopping frequencies. The receiver must "listen" to incoming signals at the right time at the right frequency. FCC regulation is at least 75 frequencies, with a maximum dwell time of 400ms. The advantage is it is very resilient to noise. Orthogonal hopping codes are sets of hopping codes that never use the same frequencies at the same time, as in Figure 1.11 (they can be adjusted online by software). They allow multiple wireless LANs to coexist.

Time
Hopping Pattern CABED
D
E
B
A
C
2.40
2.41
2.42
2.43
2.44
2.45
Frequency
(GHz)

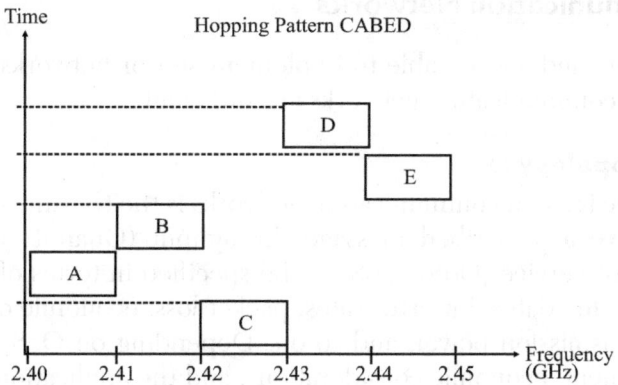

FIGURE 1.11 A frequency hopping spread spectrum.

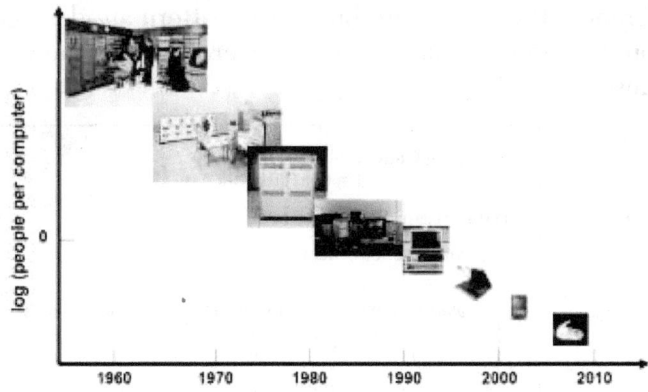

FIGURE 1.12 Number of people per computer vs. year.

FHSS is the repeated switching of frequencies during radio transmission, often to minimize the effectiveness of "electronic warfare"; that is, the unauthorized interception or jamming of telecommunications. Spread spectrum enables a signal to be transmitted across a frequency band that is much wider than the minimum bandwidth required by the information signal. The transmitter "spreads" the energy, originally concentrated in narrowband, across a number of frequency band channels on a wider electromagnetic spectrum. Benefits include improved privacy, decreased narrowband interference, and increased signal capacity.

Figure 1.12 shows the number of people per computer. The number of people using the computer is consistently increasing.

1.8 Communication Networks

To understand and be able to implement sensor networks, the basic concepts of communication networks are sufficient.

Network Topology

The basic issue in communication networks is the transmission of messages to achieve a prescribed message throughput (Quantity of Service) and Quality of Service (QoS). QoS can be specified in terms of message delay, message due dates, bit error rates, packet loss, economic cost of transmission, transmission power, and so on. Depending on QoS, the installation environment, economic considerations, and the application, one of several basic network topologies may be used.

A communication network is composed of nodes, each of which has computing power and can transmit and receive messages over communication links, wireless or cabled. The basic network topologies are shown in Figure 1.13 and include fully connected, mesh, star, ring, tree, and bus. A single network may consist of several interconnected subnets of different topologies. Networks are further classified as Local Area Networks (LAN), for example, inside one building, or Wide Area Networks (WAN), for example, between buildings.

Fully Connected Networks

Fully connected networks suffer from problems of NP-complexity, because as additional nodes are added, the number of links increases exponentially. Therefore, for large networks, the routing problem is computationally intractable even with the availability of large amounts of computing power.

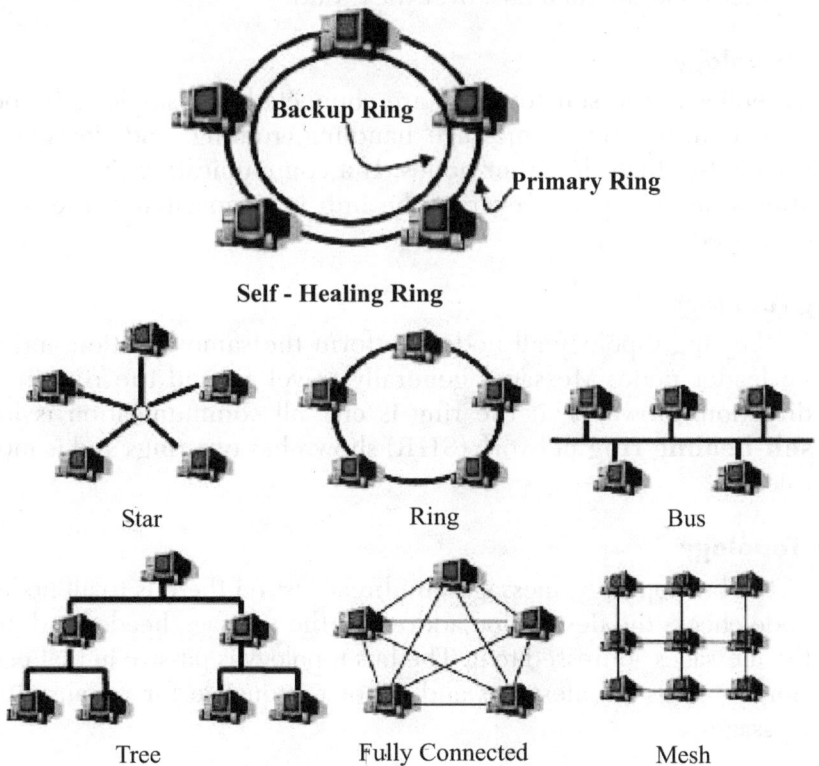

FIGURE 1.13 Basic network topologies.

Mesh Networks

Mesh networks are regularly distributed networks that generally allow transmission only to a node's nearest neighbors. The nodes in these networks are generally identical, so that mesh nets are also referred to as peer-to-peer (Figure 1.13) nets. Mesh nets can be good models for large-scale networks of wireless sensors that are distributed over a geographic region, for example, personnel or vehicle security surveillance systems. Note that the regular structure reflects the communications topology; the actual geographic distribution of the nodes need not be a regular mesh. Since there are generally multiple routing paths between nodes, these nets are robust to failure of individual nodes or links. An advantage of mesh nets is that, although all nodes may be identical and have the same computing and transmission capabilities, certain nodes can be designated as "group leaders" that take on additional functions. If a group leader is disabled, another node can then take over these duties.

Star Topology

All nodes of the star topology are connected to a single hub node. The hub requires greater message handling, routing, and decision-making capabilities than the other nodes. If a communication link is cut, it only affects one node. However, if the hub is incapacitated, the network is destroyed.

Ring Topology

In the ring topology all nodes perform the same function and there is no leader node. Messages generally travel around the ring in a single direction. However, if the ring is cut, all communication is lost. The **self-healing ring** network (SHR) shown has two rings and is more fault tolerant.

Bus Topology

In the bus topology, messages are broadcast on the bus to all nodes. Each node checks the destination address in the message header and processes the messages addressed to it. The bus topology is passive in that each node simply listens for messages and is not responsible for retransmitting any messages.

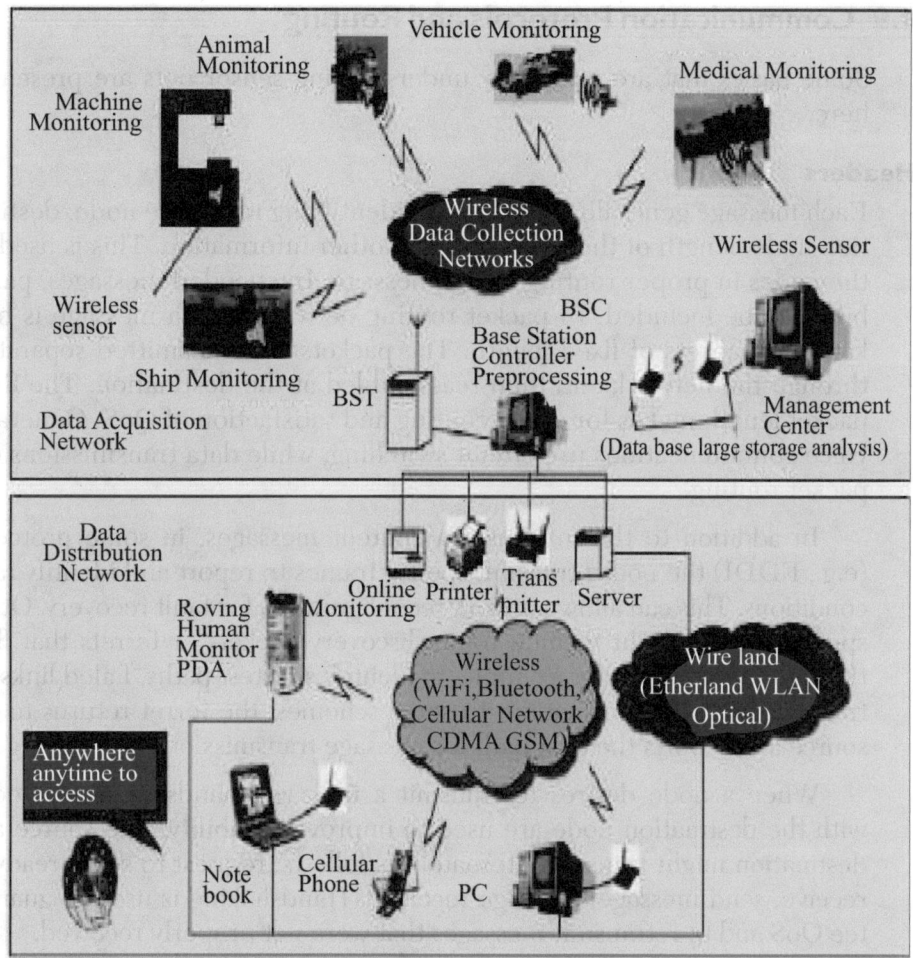

FIGURE 1.14 Complexity of wireless sensor networks.

Figure 1.14 shows the complexity of wireless sensor networks, which generally consist of a data acquisition network and a data distribution network, monitored and controlled by a management center. The plethora of available technologies makes even the selection of components difficult, let alone the design of a consistent, reliable, robust overall system. So, the network protocols are also different for wireless.

1.9 Communication Protocols and Routing

Some basics that are useful for understanding sensor nets are presented here.

Headers

Each message generally has a header identifying its source node, destination node, length of the data field, and other information. This is used by the nodes in proper routing of the message. In encoded messages, parity bits may be included. In packet routing networks, each message is broken into packets of fixed length. The packets are transmitted separately through the network and then reassembled at the destination. The fixed packet length makes for easier routing and satisfaction of QoS. Generally, voice communications use circuit switching, while data transmissions use packet routing.

In addition to the information content messages, in some protocols (e.g., FDDI) the nodes transmit special frames to report and identify fault conditions. This can allow network reconfiguration for fault recovery. Other special frames might include route discovery packets or ferrets that flow through the network, for example, to identify shortest paths, failed links, or transmission cost information. In some schemes, the ferret returns to the source and reports the best path for message transmission.

When a node desires to transmit a message, handshaking protocols with the destination node are used to improve reliability. The source and destination might transmit alternately as follows: request to send, ready to receive, send message, message received. Handshaking is used to guarantee QoS and to retransmit messages that were not properly received.

Switching

Most computer networks use a store-and-forward switching technique to control the flow of information. Then, each time a packet reaches a node, it is completely buffered in local memory, and transmitted as a whole. More sophisticated switching techniques include wormhole, which splits the message into smaller units known as flow control units or flits. The header flit determines the route. As the header is routed, the remaining flits follow it in pipeline fashion. This technique currently achieves the lowest message latency. Another popular switching scheme is virtual-cut-through. Here, when the header arrives at a node, it is routed without waiting for the rest of the packet. Packets are buffered either in software buffers in memory or in hardware buffers, and various sorts of buffers are used including edge buffers, central buffers, and so forth.

Multiple Access Protocols

When multiple nodes desire to transmit, protocols are needed to avoid collisions and lost data. In the ALOHA scheme, first used in the 1970s at the University of Hawaii, a node simply transmits a message when it desires. If it receives an acknowledgment, all is well. If not, the node waits a random time and re-transmits the message. In Frequency Division Multiple Access (FDMA), different nodes have different carrier frequencies. Since frequency resources are divided, this decreases the bandwidth available for each node. FDMA also requires additional hardware and intelligence at each node. In Code Division Multiple Access (CDMA), a unique code is used by each node to encode its messages. This increases the complexity of the transmitter and the receiver. In Time Division Multiple Access (TDMA), the RF link is divided on a time axis, with each node being given a predetermined time slot it can use for communication. This decreases the sweep rate, but a major advantage is that TDMA can be implemented in software. All nodes require accurate, synchronized clocks for TDMA.

Open Systems Interconnection Reference Model (OSI/RM)

The International Standards Organization (ISO) OSI/RM architecture specifies the relationship between messages transmitted in a communication network and applications. The development of this open standard has encouraged the adoption by different developers of standardized compatible systems interfaces. Figure 1.15 shows the seven layers of OSI/RM. But in wireless not all 7 layers are used. The layers used in wireless systems are discussed in section 1.10.

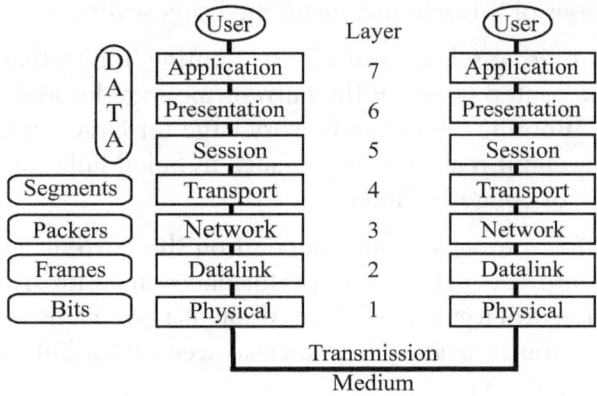

FIGURE 1.15 The OSI Reference Model.

Each layer is self-contained so that it can be modified without unduly affecting other layers. The Transport Layer provides error detection and correction. Routing and flow control are performed in the Network Layer. The Physical Layer represents the actual hardware communication link interconnections. The Applications Layer represents programs run by users.

Routing

Since a distributed network has multiple nodes and services many messages, and each node is a shared resource, many decisions must be made. There may be multiple paths from the source to the destination. Therefore, message routing is an important topic. The main performance measures affected by the routing scheme are throughput (quantity of service) and average packet delay (quality of service). Routing schemes should also avoid both deadlock and livelock. Routing methods can be fixed (i.e., pre-planned), adaptive, centralized, distributed, broadcast, and so on. Perhaps the simplest routing scheme is the token ring. Here, a simple topology and a straightforward fixed protocol result in very good reliability and precomputable QoS. A token passes continuously around a ring topology. When a node desires to transmit, it captures the token and attaches the message. As the token passes, the destination reads the header and captures the message. In some schemes, it attaches a "message received" signal to the token, which is then received by the original source node. Then, the token is released and can accept further messages. The token ring is a completely decentralized scheme that effectively uses TDMA. Though this scheme is very reliable, one can see that it results in a waste of network capacity. The token must pass once around the ring for each message. Therefore, there are various modifications of this scheme, including using several tokens, and so forth.

Fixed routing schemes often use Routing Tables that dictate the next node to be routed to, given the current message location and the destination node. Routing tables can be very large for large networks, and cannot take into account real-time effects such as failed links, nodes with backed-up queues, or congested links.

Adaptive routing schemes depend on the current network status and can take into account various performance measures, including cost of transmission over a given link, congestion of a given link, reliability of a path, and time of transmission. They can also account for link or node failures.

Routing algorithms can be based on various network analysis and graph theoretic concepts in computer science (e.g., A-star tree search) or in operations research, including shortest-route, maximal flow, and minimum-span problems. Routing is closely associated with dynamic programming and the optimal control problem in feedback control theory. Shortest path routing schemes find the shortest path from a given node to the destination node. If the cost, instead of the link length, is associated with each link, these algorithms can also compute minimum cost routes. These algorithms can be centralized (find the shortest path from a given node to all other nodes) or decentralized (find the shortest path from all nodes to a given node). There are certain well-defined algorithms for shortest path routing, including the efficient Dijkstra algorithm, which has polynomial complexity. The Bellman-Ford algorithm finds the path with the least number of hops.

Deadlock and Livelock

Large-scale communication networks contain cycles (circular paths) of nodes. Moreover, each node is a shared resource that can handle multiple messages flowing along different paths. Therefore, communication nets are susceptible to deadlock, wherein all nodes in a specific cycle have full buffers and are waiting for each other. Then, no node can transmit because no node can get free buffer space, so all transmission in that cycle comes to a halt. Livelock, on the other hand, is the condition wherein a message is continually transmitted around the network and never reaches its destination. Livelock is a deficiency of some routing schemes that route the message to alternate links when the desired links are congested, without taking into account that the message should be routed closer to its final destination. Many routing schemes are available for routing with deadlock and livelock avoidance.

Flow Control

In queuing networks, each node has an associated queue or buffer that can stack messages. In such networks, flow control and resource assignment are important. The objectives of flow control are to protect the network from problems related to overload and speed mismatches, and to maintain QoS, efficiency, fairness, and freedom from deadlock. If a given node A has high priority, its messages might be preferentially routed in every case, so that competing nodes are choked off as the traffic of A increases. Fair routing schemes avoid this. There are several techniques for flow control: in buffer management, certain portions of the buffer space are assigned for certain purposes. In choke packet schemes, any node sensing congestion sends choke packets to other nodes telling them to reduce their transmissions.

Isarithmic schemes have a fixed number of "permits" for the network. A message can be sent only if a permit is available. In window or kanban schemes, the receiver grants "credits" to the sender only if it has free buffer space. Upon receiving a credit, the sender can transmit a message. In Transmission Control Protocol (TCP) schemes, a source linearly increases its transmission rate as long as all its sent messages are acknowledged for. When it detects a lost packet, it exponentially decreases its transmission rate. Since lost packets depend on congestion, TCP automatically decreases transmissions when congestion is detected.

1.10 Wireless Network Architecture

The general functions of networks are bit pipes of data, MAC for sharing of a common medium, routing, synchronization, and error control. Figure 1.16 shows the wireless layers needed.

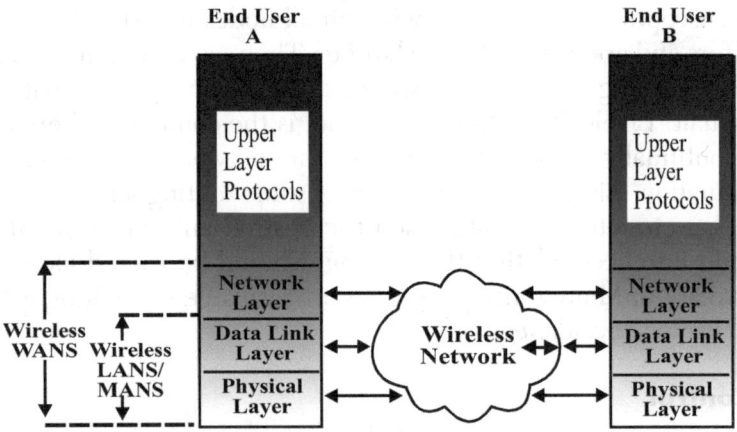

FIGURE 1.16 Wireless layers for LAN/MAN/WAN.

Network Models

A wireless sensor network consists of hundreds or thousands of low cost nodes which could either have a fixed location or be randomly deployed to monitor the environment. Due to their small size, they have a number of limitations. Sensors usually communicate with each other using a multi hop approach. The flowing of data ends at special nodes called base stations (sometimes they are also referred to as sinks). A base station links the sensor network to another network (like a gateway) to disseminate the data sensed for further processing. Base stations have enhanced capabilities over simple sensor nodes since they must do complex data processing; this justifies the fact that base stations have workstation/laptop class processors, and of course enough memory, energy, storage, and computational power to perform their tasks well. Usually, the communication between base stations is initiated over high bandwidth links.

One of the biggest problems of sensor networks is power consumption, which is greatly affected by the communication between nodes. To solve this issue, aggregation points are introduced to the network. This reduces the total number of messages exchanged between nodes and saves some energy. Usually, aggregation points are regular nodes that receive data from neighboring nodes, perform some kind of processing, and then forward the filtered data to the next hop. Similar to aggregation points is clustering. Sensor nodes are organized into clusters, each cluster having a "cluster head" as the leader. The communication within a cluster must travel through the cluster head, which then is forwarded to a neighboring cluster head until it reaches its destination, the base station. Another method for saving energy is setting the nodes to go idle (into sleep mode) if they are not needed and wake up when required. Of course, the challenge is to find a pattern at which energy consumption is made evenly for all the nodes in the network.

Due to sensors' limited capabilities, there are a lot of design issues that must be addressed to achieve an effective and efficient operation of wireless sensor networks.

Energy Saving Algorithms

Since sensor nodes use batteries for power that are difficult to replace when consumed (often sensor nodes are deployed in remote and hostile environments), it is critical to design algorithms and protocols in such a way to utilize minimal energy. To do so, implementers must reduce communication between sensor nodes, simplify computations, and apply lightweight security solutions.

Location Discovery

For many applications tracking an object requires knowing the exact or approximate physical location of a sensor node in order to link sensed data with the object under investigation. Furthermore, many geographical routing protocols need the location of sensor nodes to forward data among the network. Location discovery protocols must be designed in such a way that minimum information is needed to be exchanged among nodes to discover their location. Since sensor nodes are energy constrained, solutions like GPS are not recommended. After all, cost is another factor that influences design; try to keep the cost at minimum levels since most sensor nodes are usually needed for many applications. If the cost is high, the adoption and spread of sensor technology will be prohibited.

Security

Is it possible to introduce a new technology without addressing security? Of course not! However, as with all other technologies, security is not the top priority when designing something new. This approach is acknowledged by almost everyone, and it is erroneous, but they keep doing it anyway. Security solutions are constrained when applying them to sensor networks. For example, cryptography requires complex processing to provide encryption to the transmitted data. Secure routing, secure discovery, and verification of location, key establishment and trust setup, and attacks against sensor nodes, secure group management, and secure data aggregation are some of the many issues that need to be addressed in a security context.

1.11 WSN Sensors Introduction

Sensor networks are the key to gathering the information needed by smart environments, whether in buildings, utilities, industries, homes, shipboards, transportation systems, automation, or elsewhere. Recent terrorist and guerilla warfare countermeasures require distributed networks of sensors that can be deployed using, for example, aircraft, and have self-organizing capabilities. In such applications, running wires or cabling is usually impractical. A sensor network is required that is fast and easy to install and maintain.

IEEE 1451 Smart Sensors

Wireless sensor networks satisfy these requirements. Desirable functions for sensor nodes include: ease of installation, self-identification, self-diagnosis, reliability, time awareness for coordination with other nodes, some software functions and Digital Signal Processing (DSP), and standard control protocols and network interfaces. There are many sensor manufacturers and many networks on the market today. It is too costly for manufacturers to make special transducers for every network on the market. Different components made by different manufacturers should be compatible. Therefore, in 1993 the IEEE and the National Institute of Standards and Technology (NIST) began work on a standard for Smart Sensor Networks. IEEE 1451, the Standard for Smart Sensor Networks, was the result. The objective of this standard is to make it easier for different manufacturers to develop smart sensors and to interface those devices to networks.

Smart Sensor, Virtual Sensor

Figure 1.17 shows the basic architecture of IEEE 1451. Major components include STIM, TEDS, TII, and NCAP, as detailed in the figure. A major outcome of IEEE 1451 studies is the formalized concept of a smart sensor. A smart sensor is a sensor that provides extra functions beyond those necessary for generating a correct representation of the sensed quantity. Included might be signal conditioning, signal processing, and decisionmaking/alarm functions. A general model of a smart sensor is shown in Figure 1.18. Objectives for smart sensors include moving the intelligence closer to the point of measurement; making it cost-effective to integrate and maintain distributed sensor systems; creating a confluence of transducers, control,

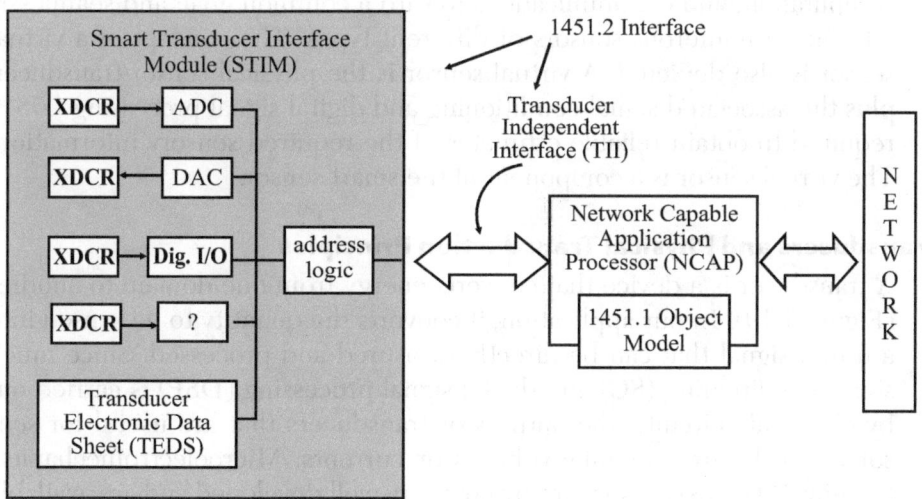

FIGURE 1.17 The IEEE 1451 standard for Smart Sensor Networks.

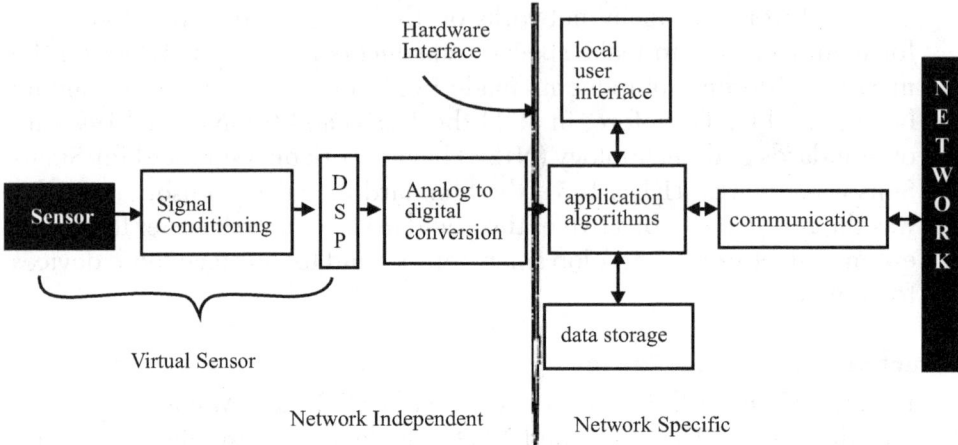

FIGURE 1.18 A general model of a smart sensor [IEEE 1451].

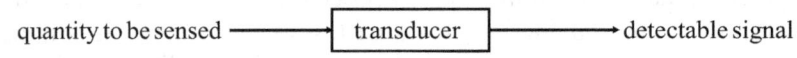

FIGURE 1.19 Sensory transducer.

computation, and communications toward a common goal; and seamlessly interfacing numerous sensors of different types. The concept of a virtual sensor is also depicted. A virtual sensor is the physical sensor/transducer, plus the associated signal conditioning and digital signal processing (DSP) required to obtain reliable estimates of the required sensory information. The virtual sensor is a component of the smart sensor.

Transducers and Physical Transduction Principles

A transducer is a device that converts energy from one domain to another (Figure 1.19). In our application, it converts the quantity to be sensed into a useful signal that can be directly measured and processed. Since much signal conditioning (SC) and digital signal processing (DSP) is carried out by electronic circuits, the outputs of transducers that are useful for sensor networks are generally voltages or currents. Microelectromechanical Systems (MEMS) sensors are by now very well developed and are available for most sensing applications in wireless networks. **Mechanical Sensors** include those that rely on direct physical contact.

The Piezoresistive Effect

The Piezoresistive effect converts an applied strain to a change in resistance that can be sensed using electronic circuits such as the Wheatstone bridge. The relationship is

$$\triangle R/R = S \in$$

with R the resistance, $\in$ the strain, and S the gauge factor which depends on quantities such as the resistivity and the Poisson's ratio of the material. Metals and semiconductors exhibit piezoresistivity. The piezoresistive effect in silicon is enhanced by doping with boron (p-type silicon can have a gauge factor up to 200). With semiconductor strain gauges, temperature compensation is important.

The Piezoelectric Effect

The Piezoelectric effect converts an applied stress (force) to a charge separation or potential difference. Piezoelectric materials include barium titanate, PZT (Lead Zirconate Titanate - piezoelectric ceramic material), and single-crystal quartz. The relation between the change in force F and the change in voltage V is given by

$$\triangle V = k \triangle F$$

where k is proportional to the material charge sensitivity coefficients and the crystal thickness, and inversely proportional to the crystal area and the material relative permittivity. The piezoelectric effect is reversible, so that a change in voltage also generates a force and a corresponding change in thickness. Thus, the same device can be both a sensor and an actuator.

Tunneling Sensing

Tunneling sensing depends on the exponential relationship between the tunneling current I and the tip/surface separation z given by

$$I = I_0\, e^{-kz}$$

where k depends on the tunnel barrier height in ev. Tunneling is an extremely accurate method of sensing nanometer-scale displacements, but its highly nonlinear nature requires the use of feedback control to make it useful.

Capacitive Sensors

Capacitive sensors typically have one fixed plate and one movable plate. When a force is applied to the movable plate, the change in capacitance C is given as

$$\triangle C = \in A / \triangle d$$

with the resulting displacement, A the area, and $\in$ the dielectric constant. Changes in capacitance can be detected using a variety of electric circuits and converted to a voltage or current change for further processing. **Inductive sensors**, which convert displacement to a change in inductance, are also often useful. **Magnetic and Electromagnetic Sensors** do not require direct physical contact and are useful for detecting proximity effects.

The Hall Effect

The Hall Effect relies on the fact that the Lorentz Force deflects flowing charge carriers in a direction perpendicular to both their direction of flow and an applied magnetic field (i.e., vector cross product). The Hall voltage induced in a plate of thickness T is given by

$$V_H = R I_x B_z / T$$

with R the Hall coefficient, I_x the current flow in direction x, and B_z the magnetic flux density in the z direction as in Figure 1.20. R is 4-5 times larger in semiconductors than in most metals. **The Magnetoresistive effect** is a related phenomenon depending on the fact that the conductivity varies as the square of the applied flux density.

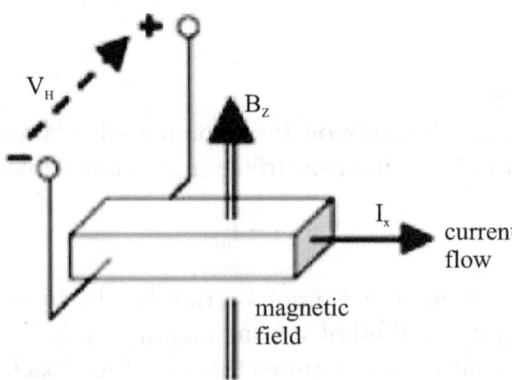

FIGURE 1.20 The Hall Effect.

Magnetic Field Sensors can be used to detect the remote presence of metallic objects. **Eddy-Current Sensors** use magnetic probe coils to detect defects in metallic structures such as pipes. **Thermal Sensors** are a family of sensors used to measure temperature or heat flux. Most biological organisms have developed sophisticated temperature sensing systems.

Thermo-Mechanical Transduction

Thermo-Mechanical Transduction is used for temperature sensing and regulation in homes and automobiles. On changes in temperature T, all materials exhibit (linear) thermal expansion of the form,

$$\triangle L/L = \mu \, \triangle T$$

with L the length and μ the coefficient of linear expansion. One can fabricate a strip of two joined materials with different thermal expansions. Then, the radius of curvature of this thermal bimorph depends on the temperature change.

Thermoresistive Effects

Thermoresistive effects are based on the fact that the resistance R changes with temperature T. For moderate changes, the relation is approximately given by many metals,

$$\triangle R/R = \mu_R \, \triangle T$$

with μ_R the temperature coefficient of resistance. Hence, silicon is useful for detecting temperature changes.

Thermocouples

Thermocouples are based on the thermoelectric Seebeck effect, whereby if a circuit consists of two different materials joined together at each end, with one junction hotter than the other, current flows in the circuit. This generates a Seebeck voltage given approximately by,

$$V \sim = \mu \, (T1 - T2) + (T1^2 - T2^2)$$

with T_1, T_2 the temperatures at the two junctions. The coefficients depend on the properties of the two materials. Semiconductor thermocouples generally have higher sensitivities than do metal thermocouples. Thermocouples are inexpensive and reliable, and so are much used. Typical thermocouples have outputs on the order of $50 \, \mu V/ \, °C$ and some are effective for temperature ranges of $-270 \, °C$ to $2700 \, °C$.

Resonant Temperature Sensors

Resonant temperature sensors rely on the fact that single-crystal SiO_2 exhibits a change in resonant frequency depending on temperature change. Since this is a frequency effect, it is more accurate than amplitude-change effects and has extreme sensitivity and accuracy for small temperature changes.

Optical Transducers

Optical transducers convert light to various quantities that can be detected. In the photoelectric effect one electron is emitted at the negative end of a pair of charged plates for each light photon of sufficient energy. This causes a current to flow. In photoconductive sensors, photons generate carriers that lower the resistance of the material. In junction-based photo sensors, photons generate electron-hole pairs in a semiconductor junction that cause current flow. This is often misnamed the photovoltaic effect. These devices include photodiodes and phototransistors. Thermopiles use a thermocouple with one junction coated in a gold or bismuth black absorber, which generates heat on illumination.

Solar cells are large photodiodes that generate voltage from light. Bolometers consist of two thermally sensitive resistors in a Wheatstone bridge configuration, with one of them shielded from the incident light. Optical transducers can be optimized for different frequencies of light, resulting in infrared detectors, ultraviolet detectors, and so on. Various devices, including accelerometers, are based on optical fiber technology, often using timeof-flight information.

Chemical and Biological Transducers

Chemical and biological transducers cover a very wide range of devices that interact with solids, liquids, and gases of all types. Potential applications include environmental monitoring, biochemical warfare monitoring, security area surveillance, medical diagnostics, implantable biosensors, and food monitoring. Effective use has been shown for NOx (from pollution), organophosphorus pesticides, nerve gases (Sarin, etc.), hydrogen cyanide, smallpox, anthrax, CO_X, SO_X, and others.

Chemiresistors

Chemiresistors have two interdigitated finger electrodes coated with specialized chemical coatings that change their resistance when exposed to certain chemical challenge agents. The electrodes may be connected directly to an FET (Field Effect Transistor), which amplifies the resulting signals for good noise rejection. This device is known as an interdigitated-gate electrode FET (IGEFET), as shown in Figure 1.21. Arrays of chemiresistors, each device with a different chemically active coating, can be used to increase specificity for specific challenge agents. Digital signal processing, including neural network classification techniques, is important in correct identification of the agent.

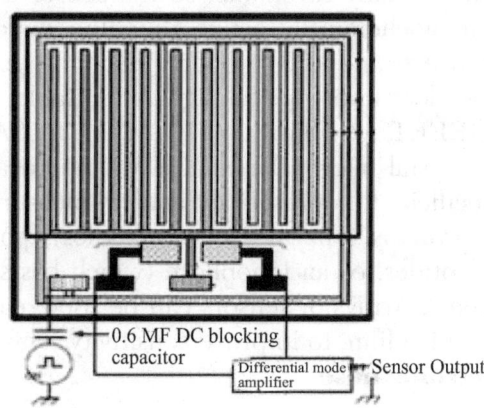

FIGURE 1.21 IGEFET structure.

Metal-Oxide Gas Sensors

Metal-oxide gas sensors rely on the fact that adsorption of gases onto certain semiconductors greatly changes their resistivities. In thin-film detectors, a catalyst such as platinum is deposited on the surface to speed the reactions and enhance the response. Useful as sensors are the oxides of tin, zinc, iron, zirconium, and so forth. Gases that can be detected include CO_2, CO, H_sS, NH_3, and ozone. Reactions are of the form

$$O_2 + 2e^- \rightarrow 2O^-$$

so that adsorption effectively produces an electron trap site, effectively depleting the surface of mobile carriers and increasing resistance.

Electrochemical Transducers

Electrochemical transducers rely on currents induced by oxidation or reduction of a chemical species at an electrode surface. These are among the simplest and most useful of chemical sensors. An electron transfer reaction occurs that is described by O, with O the oxidized species, R the reduced species, and z the charge on the ion involved. The resulting current density is given in terms of z.

Biosensors

Biosensors of a wide variety of types depend on the high selectivity of many biomolecular reactions; that is, molecular binding sites of the detector may only admit certain species of analyte molecules. Unfortunately, such reactions are not usually reversible, so the sensor is not reusable. These devices have a biochemically active thin film deposited on a platform device that converts induced property changes (e.g., mass, resistance) into detectable electric or optical signals. Suitable conversion platforms include the IGEFET, ion-sensitive FET (ISFET), SAW (Surface Acoustic Wave), quartz crystal microbalance (QCM), microcantilevers, and so on. To provide specificity to a prescribed analyte measurand, for the thin film one may use proteins (enzymes or antibodies), polysaccharide, nucleic acid, oligonucleotides, or an ionophore (which has selective responses to specific ion types). Arrays of sensors can be used, each having a different biochemically active film, to improve sensitivity. This has been used in the so-called "**electronic nose**."

The Electromagnetic Spectrum can be used to fabricate remote sensors of a wide variety of types. Generally the wavelength suitable for a particular application is selected based on the propagation distance, the level of detail and resolution required, the ability to penetrate solid materials or certain mediums, and the signal processing difficulty. Doppler techniques allow the measurement of velocities. Millimeter waves have been used for satellite remote monitoring. Infrared is used for night vision and sensing heat. IR motion detectors are inexpensive and reliable. Electromagnetic waves can be used to determine distance using time-of-flight information. Radar uses RF waves and Lidar uses light (laser). The velocity of light is $c = 299.8 \times 10$ m/s. GPS uses RF for absolute position localization. Visible light imaging using cameras is used in a broad range of applications but generally requires the use of sophisticated and computationally expensive DSP techniques including edge detection, thresholding, segmentation, pattern recognition, motion analysis, and so forth.

Acoustic Sensors

Acoustic sensors include those that use sound as a sensing medium. Doppler techniques allow the measurement of velocities. Ultrasound often provides more information about mechanical machinery vibrations, fluid leakage, and impending equipment faults than do other techniques. Sonar uses sound to determine distance using time-of-flight information. It is effective in media other than air, including underwater. Caution should be used in that the propagation speed of acoustic signals depends on the medium. The speed of sound at sea level in a standard atmosphere is $c_s = 340.294$ m/s. Subterranean echoes from earthquakes and tremors can be used to glean information about the earth's core as well as about the tremor event, but deconvolution techniques must be used to remove echo phenomena and to compensate for uncertain propagation speeds. The acoustic spectrum is shown in Figure 1.22.

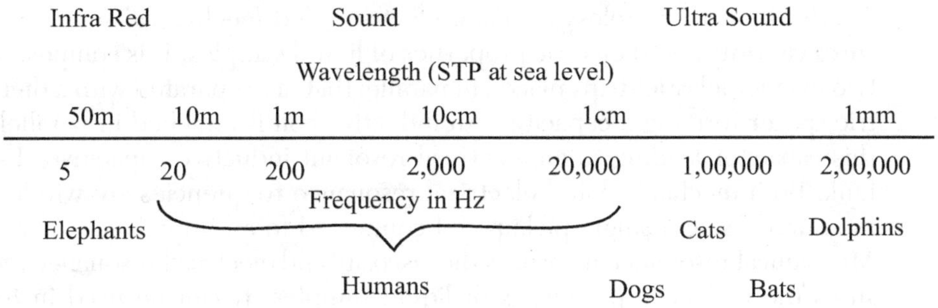

Infra Red		Sound			Ultra Sound	
		Wavelength (STP at sea level)				
50m	10m	1m	10cm	1cm		1mm
5	20	200	2,000	20,000	1,00,000	2,00,000
			Frequency in Hz			
Elephants					Cats	Dolphins
			Humans		Dogs	Bats

FIGURE 1.22 The acoustic spectrum.

Acoustic Wave Sensors

Acoustic wave sensors are useful for a broad range of sensing devices. These transducers can be classified as surface acoustic wave (SAW), thickness-shear mode (TSM), flexural plate wave (FPW), or acoustic plate mode (APM). The SAW is shown in Figure 1.23 and consists of two sets of interdigitated fingers at each end of a membrane, one set for generating the SAW and one for detecting it. Like the IGEFET, these are useful platforms to convert property changes such as mass into detectable electrical signals. For instance, the surface of the device can be coated with a chemically or biologically active thin film. On presentation of the measurand to be sensed, adsorption might cause the mass m to change, resulting in a frequency shift given by the Sauerbrey equation

$$\triangle f = k\, f_0^2\, \triangle m/A$$

with f_0 the membrane resonant frequency, constant k depending on the device, and A the membrane area.

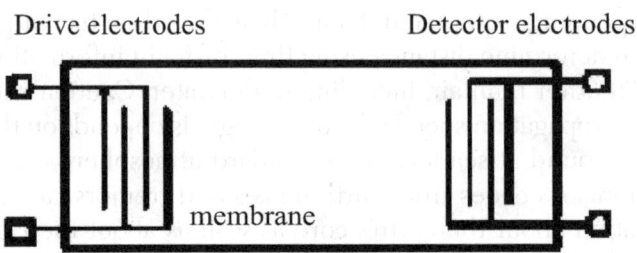

FIGURE 1.23 SAW Sensor.

Hybrid passive wireless resonant electrical and mechanical sensors

A hybrid passive wireless resonant electrical and mechanical sensor measures viscosity and dielectric properties of liquid samples. It is composed of two magnetoelastic strips placed in parallel that are separated with a dielectric spacer forming a capacitor. An inductive coil is attached in parallel to this capacitor leading to an electrical resonant inductive capacitive (L-C) tank. Both mechanical and electrical resonance frequencies are wirelessly measured using a single pickup coil connected to an impedance analyzer. Mechanical resonance measures the viscosity and electrical resonance measures the dielectric properties of liquid samples. It can be used in food quality monitoring and control such as adulteration detection of oils, milk adulterated with water, and glucose in water.

Sensors for Smart Environments

Table 1.3 shows which physical principles may be used to measure various quantities. MEMS sensors are by now available for most of these measurands.

Table 1.3 Measurement of Various Quantities for WSNs

Measurements for wireless sensor networks		
Measurand		**Transduction Principle**
Physical properties	Pressure	Piezoresistive, capacitive
	Temperature	Thermistor, thermo-mechanical, thermocouple
	Humidity	Resistive, capacitive
	Flow	Pressure change, thermistor
Motion properties	Position	E-mag, GPS, contact sensor
	Velocity	Doppler, Hall effect, optoelectronic

Measurements for wireless sensor networks		
Measurand		**Transduction Principle**
Contact Properties	Angular velocity	Optical encoder
	Acceleration	Piezoresistive, piezoelectric, optical fiber
	Strain	Piezoresistive
	Force	Piezoelectric, piezoresistive
	Torque	Piezoresistive, optoelectronic
	Slip	Dual torque
	Vibration	Piezoresistive, piezoelectric, optical fiber, sound, ultrasound
Presence	Tactile/contact	Contact switch, capacitive
	Proximity	Hall effect, capacitive, magnetic, seismic, acoustic, RF
	Distance/range	E-mag (sonar, radar, lidar), magnetic, tunneling
	Motion	E-mag, IR, acoustic, seismic (vibration)
Biochemical	Biochemical agents	Biochemical transduction
Identification	Personal features	Vision
	Personal ID	Fingerprint, retinal scan, voice, heat plume, vision motion analysis

Comparing Sensor Nodes to Ad Hoc Wireless Networks

An ad hoc network is a local area network (LAN) that is built spontaneously as devices connect. Instead of relying on a base station to coordinate the flow of messages to each node in the network, the individual network nodes forward packets to and from each other. In Latin, ad hoc literally means "for this," meaning "for this special purpose," and also, by extension, improvised. In the Windows operating system, ad hoc is a communication mode (setting) that allows computers to directly communicate with each other without a router. Wireless sensor networks share similarities (and differences) with ad hoc wireless networks. The main similarity is the multi-hop communication method.

Both consist of wireless nodes, but they are different in the factors such as failure, energy drain, unique global IDs, Data-centric, query-based addressing vs. address-centric, and resource limitations like memory, power, and processing.

The differences among the two types of networks are listed as follows:

- More nodes are deployed in a sensor network, up to a hundred or a thousand nodes, than in an ad hoc network that usually involves far fewer nodes.

- Sensor nodes are more constrained in computational, energy, and storage resources than ad hoc.

- Sensor nodes can be deployed in environments without the need of human intervention and can remain unattended for a long time after deployment.

- Neighboring sensor nodes often sense the same events from their environment, thus forwarding the same data to the base station, resulting in redundant information.

Aggregation and in-network processing often require trust relationships between sensor nodes that are not typically assumed in ad hoc networks.

Sensor Technologies and the Selection of Sensors

A sensor should be selected depending on the application in mind. Sensors are of two types: limit detectors and qualitative and quantitative analysis measuring elements. The limit detection type of sensors gives out logic 1 or 0. This operates as a watchdog in the system and annunciates in fast response at the occurrence of a monitored parameter. Specially this could be an active element used for a particular purpose as a feedback circuit in a power supply, or a Hall effect switch detecting the presence of a magnetic field of rated a magnetic flux and direction. The other example is a PIR detector detecting presence of an object and movement of the same. A qualitative and quantitative analysis type of sensor could do the job rather slowly and deliver its response in a required electrical form. Table 1.4 indicates different sensor-based technologies.

There are certain things which the design has to ensure so that product does not fail due to faulty selection of the sensor involved.

1. Identifying the sensing parameter

The parameter could be in any form such as voltage, current, frequency, temperature, pressure, light, touch, presence, sound, and chemical reaction to name a few. Any of these needs to be converted to electrical form for electronic analysis. "A thermistor varies its resistance depending on the thermal stress over it. A thermistor as an element in a potential divider circuit results out a temperature corresponding to the potential across it. An engineer working on this needs to analyze the range of operation of the thermistor, and needs a way to get the safe operating range of the sensor, to observe the performance of the thermistor to see what time it takes to convert the parameter, and the linearity and response of the sensor for the physical change occurring across it. While selecting an IR sensor for sensing the proximity of objects, which emits an electromagnetic field or a beam of electromagnetic radiation, and looks for changes in the field or return signal, one essentially needs to carefully consider the "sensing frequency" at which sensing data is provided, the "range" that gets covered in the form of angle/distance with "resolution," "size" depending on the device upon which such a proximity sensor will be mounted, and operating environment/ conditions under which such a sensor will function." Such a sensor can have high reliability and long functional life due to the absence of mechanical parts and the lack of physical contact between the sensor and the object.

2. Reliability

Ensuring a robust design with maintenance-free packaging is very important. This is because sensors are that part of the device which mostly comes into contact with the world, and hence has to have the ability to withstand harsh environments without losing performance.

Table 1.4 Different Sensor-Based Technologies

Sensors	Technology	Usage
Highly sensitive six-axis sensor	Augmented Realty technology and Remote play technology	Provides completely new ways to play and interact with games
Proximity sensor	Capacitive multipoint touch technology	Orientation, capacitive touch
Sudden motion sensor	Sudden motion sensor technology	To help protect the hard disk from damage against sudden vibration or accelerated movement
Motion sensors (gyro-scopes, accelerometers and magnetic sensors)	9-axis sensor fusion technology	Less power consumption, increase in battery life, fusion of motion sensor by manufacturer without any restriction
Motion sensor	DSP technology	Remote playback control

Sensors	Technology	Usage
Accelerometer, IR sensor	Motion sensing technology, RF technology	For private listening with built-in headphone jack and compatibility of the device with a remote from any manufacturer
Color sensor	Triluminos display technology	For auto color calibration
Capacity sensor	Capacitive input technology	Easy one-touch scrolling and quick system control through a customizable gadget bar
Eco sensor	Motion sensing technology, smart control and smart recognition	Allows easy and smooth working of the set top box and remote control with the following smart interactions: face recognition, hand gesture recognition and voice recognition
Exmor R CMOS sensor		For image processing
Fingerprint sensor	Fingerprint technology	To unlock the phone and use iTunes
Motion sensor	Motion sensor technology	To save power when the phone is not being carried around by checking for updates less often (motion sensor)
Gyroscope, accelerometer, light sensor, magnetometer, hall sensor, proximity sensor (with cellular model)	TruVivid technology	Fuses sensor glass with the sensor attached as a film
Ambient light sensor, gyroscope, magnetometer, proximity sensor, tilt sensor	In cell touch technology	No discrete touch panel, technology is integrated in the display
Kinect sensor	Kinect real motion, voice, vision technology	Track up to six skeletons at once, perform heart rate tracking, track gestures performed with an Xbox One controller, and scan Quick Response (QR) codes to redeem Xbox Live gift cards

3. Easy integration

From the developer's perspective, using ADC, employing filtering equations, extracting data, and then calibrating the firmware is troublesome. This is why, to reduce the design effort, pre- calibrated sensors which support common interfaces like IIC or SPI, and so on, are preferable. A common example is the use of NTC or PTC thermistors versus a DHT11 sensor to read temperature and humidity data. DHT11 temperature and humidity sensors feature a temperature and humidity sensor complex with a calibrated digital signal output. By using the exclusive digital signal acquisition technique and temperature and humidity sensing technology, it ensures high reliability and excellent long-term stability. This sensor includes a resistive type humidity measurement component and an NTC temperature measurement component, and connects to a high performance 8-bit microcontroller, offering excellent quality, fast response, anti-interference ability, and cost effectiveness. The SP1202S01RB sensor by National Semiconductor is used for measuring the liquid level utilizing a pressure sensor. Most importantly, this sensor is being used for sensing the quantity of water in draught affected areas.

4. Software

There are also some sensors whose role mainly depends on the software. These analyze the environment by sensing some parameters and make use of a software to come to a decision. Fuzzy logic plays a vital role in such sensors. The place where the sensor plays a small role but circuitry and software play a vital role is the load cell for weighing applications. The load cell is a wheat stone bridge to develop a differential potential at the stress applied on one of the wings of the bridge. The developed potential puts in few microvolts prone to get affected by noise due to external disturbances, and varies due to vibrations created over the platform. ADCs employed to read these parameters need to be properly selected. Software to reject the noise as well as average the weight data to produce a calibrated weight accurately is as important as a physical sensor.

Summary

- Sensor networks are highly distributed networks of wireless sensor nodes, deployed in large numbers to monitor the environment or system.

- A wireless sensor network (WSN) is a collection of nodes organized into a cooperative network. Each node consists of processing capability (one or more microcontrollers, CPUs, or DSP chips), may contain multiple types of memory (program, data, and flash memories), have a RF transceiver (usually with a single omni-directional antenna), have a power source (e.g., batteries and solar cells), and accommodate various sensors and actuators.

- The wireless standards used by WSN are 802.15.4 and Zigbee.

- A spectrum is a range of electromagnetic radiation. There are three approaches: ISM band, narrow band, and spread spectrum.

- The wireless sensor networks consist of a data acquisition network and a data distribution network, monitored and controlled by a management center.

- The general functions of networks are bit pipes of data, Medium Access Control for sharing of a common medium, routing, synchronization, and error control.

- The wireless layers needed are physical, data link, network, and upper layers.

- Sensor networks are the key to gathering the information needed by smart environments, whether in buildings, utilities, industries, homes, shipboards, transportation systems, automation, or elsewhere.

- A transducer is a device that converts energy from one domain to another.

Questions

1. What is a wireless sensor network?

2. Explain different network topologies.

3. What are the advantages and uses of WSNs?

4. Write about the historical development of wireless standards.

5. What is the sensor?

6. What is the sensor network?

7. What is the sensor node?

8. What is the definition of a wireless sensor network?

9. List some of applications of WSNs.

10. List some of difficulties in WSN research.

11. What are the basic requirements of WSNs?

12. Define spectrum.

13. Explain about the wireless network architecture.

14. Write about smart sensors.

15. Explain in detail about different sensors.

16. Differentiate between wireless sensor networks and ad hoc wireless networks.

17. What are the factors influencing the selection of sensors?

Further Reading

1. *Fundamentals of Wireless Sensor Networks Theory and Practice* by Waltenegus Dargie

2. *A Guide to Wireless Sensor Networks* by S. Swapna Kumar

3. *Wireless Sensor and Actuator Networks* by Nayak

References

1. *http://www.ni.com/white-paper/7142/en/*

2. *http://wireless.ictp.it/wp-content/uploads/2012/02/Zennaro.pdf*

3. Saranraj, Karuppuswami, Harikrishnan Arangali, and Premjeet Chahal. "A Hybrid Electrical-Mechanical Wireless Magnetoelastic Sensor for Liquid Sample Measurements." Paper presented at the Electronic Components and Technology Conference (ECTC), 2016 IEEE 66th, IEEE, 2016.

2

NODE HARDWARE ARCHITECTURE

This chapter discusses different node hardware architectures. At the end of the chapter, one would be able to select the node architecture suitable for their project.

2.1 Architecture of Wireless Sensor Nodes

Wireless sensor nodes are the essential building blocks in a wireless sensor network for sensing, processing, and communication. The node stores and executes the communication protocols as well as data processing algorithms. It consists of sensing, processing, communication, and power subsystems as shown in Figure 2.1. It is a trade-off between flexibility and efficiency both in terms of energy and performance.

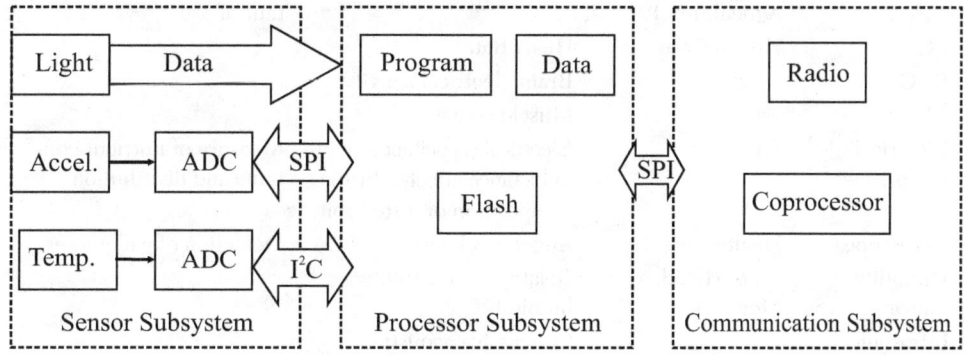

FIGURE 2.1 Architecture of a wireless sensor node.

Sensing Subsystem

The sensing subsystem integrates all the kinds of sensors, needed to measure the parameters. Table 2.1 shows the sensors and their application area.

Table 2.1 Sensors and Their Application Area

Sensor	Application Area	Sensed Event	Explanation
Accelerometer		2D and 3D acceleration of movements of people and objects	Volcano activities
	Structural Health Monitoring (SHM)		Stiffness of a structure
	Health care		Stiffness of bones, limbs, joints; Motorfluctuation in Parkinson's disease
	Transportation		Irregularities in rail, axle box, or wheels of a train system
	Supply Chain Management (SCM)		Detect ion of fragile objects during transportation
Acoustic emission sensor	SHM	Elastic waves generated by the energy released during crack propagation	Measures micro-structural changes or displacements
Acoustic sensor	Transportation & Pipelines	Acoustic pressure vibration	Vehicle detection; Measures structural irregularities; Gas contamination
Capacitance sensor	Precision Agriculture (PA)	Solute concentration	Measures the water content of soil
ECG	Health Care	Heart Rate	
EEG		Brain electrical activity	
EMG		Muscle activity	
Electrical sensors	PA	Electrical capacitance or inductance affected by the composition of tested soil	Measure of nutrient contents and distribution
Gyroscope	Health care	Angular velocity	Detection of gall phases
Humidity sensor	PA & Health Monitoring (HM)	Relative and absolute humidity	
Infrasonic sensor		Concussive acoustic waves—earthquake or volcanic eruption	

Sensor	Application Area	Sensed Event	Explanation
Magnetic sensor	Transportation	Presence, intensity, direction, rotation, and variation of a magnetic field	Presence, speed, and density of a vehicle on a street; congestion
Oximeter	Health care	Blood oxygenation of a patient's hemoglobin	Cardiovascular exertion and trending of exertion relative to activity
pH sensor	Pipeline (water)	Concentration of hydrogen ions	Indicate the acid and alkaline content of water to measure of cleanliness
Photo acoustic spectroscopy	Pipeline	Gas sensing	Detects gas leak in a pipeline
Piezoelectric cyclinder	Pipeline	Gas velocity	A leak produces a high frequency noise that produces a vibration
Soil moisture sensor	PA	Soil moisture	Fertilizer and water management
Temperature sensor	PA&HM	Pressure exerted on a fluid	
Passive Infrared Sensor	Health care & HM	Infrared radiation from objects	Motion detection
Seismic sensor		Measure primary and secondary seismic waves (Body wave, ambient vibration)	Detection of earth quake
Oxygen sensor	Health care	Amount and proportion of oxygen in the blood	
Blood flow sensor	Health care	The Doppler shift of a reflected ultrasonic wave in the blood	

Analog to Digital Converter (ADC)

An ADC converts the output of a sensor, which is a continuous, analog signal, into a digital signal. It requires two steps:

1. The analog signal has to be quantized. Allowable discrete values are influenced. It depends on the frequency and magnitude of the signal and by the available processing and storage resources.

2. The sampling frequency is needed to convert to an equivalent digital signal. The Nyquist sampling rate does not suffice because of noise and transmission error.

The resolution of ADC is an expression of the number of bits that can be used to encode the digital output.

$$Q = \frac{E_{pp}}{2^M}$$

where Q is the resolution in volts per step (volts per output code); Epp is the peak-to-peak analog voltage; M is the ADC's resolution in bits. Its main purpose is to execute instructions pertaining to sensing, communication, and self-organization. It consists of a processor chip in which nonvolatile memory stores the program instructions and active memory temporarily stores the sensed data.

The Processor Subsystem

The processor subsystem can be designed by employing one of the three basic computer architectures.

1. Von Neumann architecture

2. Harvard architecture

3. Super-Harvard (SHARC) architecture

Von Neumann Architecture

It provides a single memory space for storing program instructions and data. It provides a single bus to transfer data between the processor and the memory. It has slow processing speed, as each data transfer requires a separate clock. This architecture is shown in Figure 2.2.

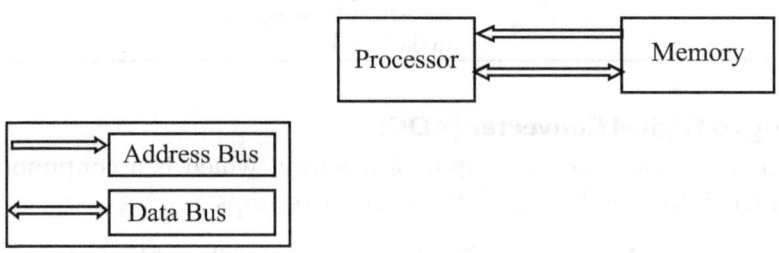

FIGURE 2.2 Von Neumann architecture.

Harvard Architecture

It provides separate memory spaces for storing program instructions and data. Each memory space is interfaced with the processor with a separate data bus, and program instructions and data can be accessed at the same time. It has a special single instruction, multiple data (SIMD) operation, a special arithmetic operation, and a bit reverse.

It supports multi-tasking operating systems but does not provide virtual memory protection. This architecture is shown in Figure 2.3.

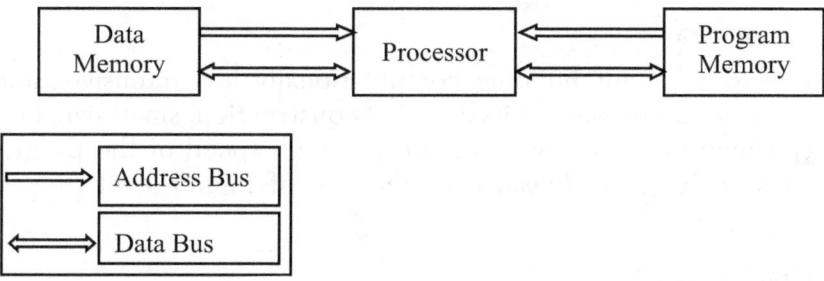

FIGURE 2.3 Harvard architecture.

Super-Harvard Architecture (SHARC)

It is an extension of the Harvard architecture. It is shown in Figure 2.4. It adds two components to the Harvard architecture:

1. An internal instruction cache temporarily stores frequently used instructions, which enhances performance, and an underutilized program memory can be used as a temporary relocation place for data.

2. With Direct Memory Access (DMA), costly CPU cycles can be invested in a different task. The program memory bus and data memory bus are accessible from outside the chip.

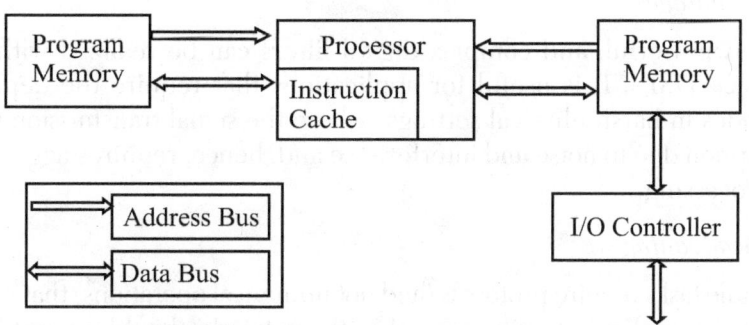

FIGURE 2.4 SHARC architecture.

Structure of the Microcontroller

It integrates the components such as the CPU core, volatile memory (RAM) for data storage, ROM, EPROM, EEPROM or Flash memory, parallel I/O interfaces, discrete input and output bits, a clock generator, one or more internal analog-to-digital converters, and serial communications interfaces.

Advantages

It is suitable for building computationally less intensive, standalone applications, because of its compact construction, small size, low-power consumption, and low cost, and the high speed of the programming eases debugging, because of the use of higher-level programming languages.

Disadvantages

It is not as powerful and as efficient as some custom-made processors (such as DSPs and FPGAs). In some applications (simple sensing tasks but with large scale deployments), one may prefer to use architecturally simple but energy-and cost-efficient processors.

Digital Signal Processor

The main function is to process discrete signals with digital filters. Its filters mimmize the effect of noise on a signal or enhance or modify the spectral characteristics of a signal. While analog signal processing requires complex hardware components, digital signal processors (DSP) require simple adders, multipliers, and delay circuits. DSPs are highly efficient. Most DSPs are designed with Harvard Architecture.

Advantages

It is powerful, and complex digital filters can be realized with commonplace DSPs. It is useful for applications that require the deployment of nodes in harsh physical settings (where the signal transmission suffers corruption due to noise and interference and, hence, requires aggressive signal processing).

Disadvantages

Some tasks require protocols (and not numerical operations) that require periodic upgrades or modifications (i.e., the networks should support flexibility in network reprogramming).

Application-Specific Integrated Circuit (ASIC)

ASIC is an IC that can be customized for a specific application. There are two types of design approaches: full-customized and half-customized.

Full-customized ICs are built with some logic cells, circuits, or layouts which are custom made in order to optimize cell performance and include features which are not defined by the standard cell library. They are expensive and require a long design time. Half-customized ASICs are built with logic cells that are available in the standard library. In both cases, the final logic structure is configured by the end user. An ASIC is a cost-efficient solution, flexible and reusable.

Advantages

It is a relatively simple design; it can be optimized to meet a specific customer demand, and multiple microprocessor cores and embedded software can be designed in a single cell.

Disadvantages

It requires high development costs and a lack of re-configurability

Applications

ASICs are not meant to replace microcontrollers or DSPs but to complement them. They are able to handle rudimentary and low-level tasks and decouple these tasks from the main processing subsystem.

Field Programmable Gate Array (FPGA)

The distinction between ASICs and FPGAs is not always clear. FPGAs are more complex in design and more flexible to program. FPGAs are programmed electrically, by modifying a packaged part. Programming is done with the support of circuit diagrams and hardware description languages, such as VHDL and Verilog.

Advantages

- It requires higher bandwidth compared to DSPs.
- It is flexible in its application.
- It supports parallel processing.
- It works with floating point representation.
- It has greater flexibility of control.

Disadvantages

It is complex and the design and realization process is costly.

Comparison

Working with a microcontroller is preferred if the design goal is to achieve flexibility. Working with the other mentioned options is preferred if power consumption and computational efficiency is desired. DSPs are expensive, large in size and less flexible; they are best for signal processing, with specific algorithms. FPGAs are faster than both microcontrollers and digital signal processors and support parallel computing, but their production cost and programming difficulty make them less suitable. ASICs have higher bandwidths; they are the smallest in size, perform much better, and consume less power than any of the other processing types, but have a high cost of production owing to the complex design process.

Communication Interfaces

The choice is often between serial interfaces: Serial Peripheral Interface (SPI), General Purpose Input/Output (GPIO), Secure Data Input/Output (SDIO), and Inter-Integrated Circuit (I2C). Among these, the most commonly used buses are SPI and I2C.

Serial Peripheral Interface

SPI (Motorola, in the mid-80s) is a high-speed, full-duplex, synchronous serial bus. It does not have an official standard, but use of the SPI interface should conform to the implementation specification of other interfaces. The SPI bus defines four pins:

1. MOSI (MasterOut/SlaveIn): It is used to transmit data from the master to the slave when a device is configured as a master.

2. MISO (MasterIn/SlaveOut): The slave generates this signal, and the recipient is the master.

3. SCLK (Serial Clock) is used by the master to send the clock signal that is needed to synchronize transmission. It is used by the slave to read this signal to synchronize transmission.

4. CS (Chip Select) is used to communicate via the CS port.

Both master and slave devices hold a shift register. Every device in every transmission must read and send data. SPI supports a synchronous communication protocol. The master and the slave must agree on the timing. Master and slave should agree on two additional parameters:

1. Clock polarity (CPOL) defines whether a clock is used as high- or low-active.

2. Clock phase (CPHA) determines the times when the data in the registers is allowed to change and when the written data can be read. Table 2.2 shows the different modes of SPI.

Table 2.2 Different Modes of Serial Peripheral Interface

SPI Mode	CPOL	CPHA	Description
0	0	0	SCLK is low-active. Sampling is allowed on odd clock edges. Data changes on even clock edges.
1	0	1	SCLK is low-active. Sampling is allowed on even clock edges. Data changes on odd clock edges.
2	1	0	SCLK is high-active. Sampling is allowed on odd clock edges. Data changes on even clock edges.
3	1	1	SCLK is high-active. Sampling is allowed on even clock edges. Data changes on odd clock edges.

Inter-Integrated Circuit (I²C)

Every device type that uses I²C must have a unique address that will be used to communicate with a device. In earlier versions, a 7 *bit address* was used, allowing 112 devices to be uniquely addressed. Due to an increasing number of devices, it *is insufficient.* Currently I²C uses *10 bit addressing.* I²C is a *multi-master half-duplex synchronous serial* bus with only two bidirectional lines (unlike SPI, which uses four): Serial Clock (SCL) and Serial Data (SDA). Since each master generates its own clock signal, communicating devices must *synchronize their clock speeds.* A slower slave device could wrongly detect its address on the SDA line while a faster master device is sending data to a third device. I²C requires arbitration between master devices *wanting* to send or receive data at the same time and *no* fair arbitration *algorithm. It* is rather that the master that holds the SDA line low for *the longest time wins* the medium. I²C enables a device to read data *at a byte level* for fast communication. The device can hold the SCL low until it completes reading or sending the next byte, which is called *handshaking*. The *aim* of I²C is *to minimize costs* for connecting devices and accommodate lower transmission speeds. I²C defines two speed modes:

1. *A fast-mode:* A bit rate of up to 400Kbps.

2. *High-speed mode:* A transmission rate of up to 3.4 Mbps.

They are downward compatible to ensure communication with older components. Table 2.3 gives the comparison between SPI and I²C.

Table 2.3 Comparison between SPI and I²C

SPI	I²C
4 lines enable full-duplex transmission.	2 lines reduce space and simplify circuit layout; lower costs.
No addressing is required due to CS.	Addressing enables multi-master mode; Arbitration is required.
Allowing only one master avoids conflicts	Multi-master mode is prone to conflicts.
Hardware requirement support increases with an increasing number of connected devices, hence, it is costly.	Hardware requirement is independent of the number of devices using the bus.
The master's clock is configured according to the slave's speed but speed adaptation slows down the master.	Slower devices may stretch the clock latency but keep other devices waiting.
Speed depends on the maximum speed of the slowest device.	Speed is limited to 5.4MHz.
Heterogeneous register size allows flexibility in the devices that are supported.	Homogeneous register size reduces overhead.
Combined registers imply every transmission should be read AND written.	Devices that do not read or provide data are not forced to provide potentially useless bytes.
The absence of an official standard leads to application specific implementations.	Official standard eases integration of devices since developers can rely on a certain implementation.

Buses are essential highways to transfer data, and due to the concern for size, only serial buses can be used. Serial buses demand high clock speeds to gain the same throughput as parallel buses. Serial buses can also be bottlenecks (e.g., Von Neumann architecture) or may not scale well with the processor speed (e.g., I²C). Delays due to contention for bus access become critical, for example, if some of the devices act unfairly and keep the bus occupied.

2.2 Components of Wireless Sensor Node Architecture

Wireless sensor networks comprise a number of spatially distributed sensor nodes which cooperate to monitor the physical qualities of a given environment. Commercial wireless sensor node products are typically composed of a single microcontroller and a number of other components. The wireless sensor node architecture divides processing power among multiple microcontrollers, with the intention of increasing flexibility, reducing cost, providing fault tolerance, improving development processes, and conserving energy. A wireless sensor node is composed of four basic components [Figure 2.5]: a sensing unit, a processing unit (microcontroller), a transceiver unit, and a power unit.

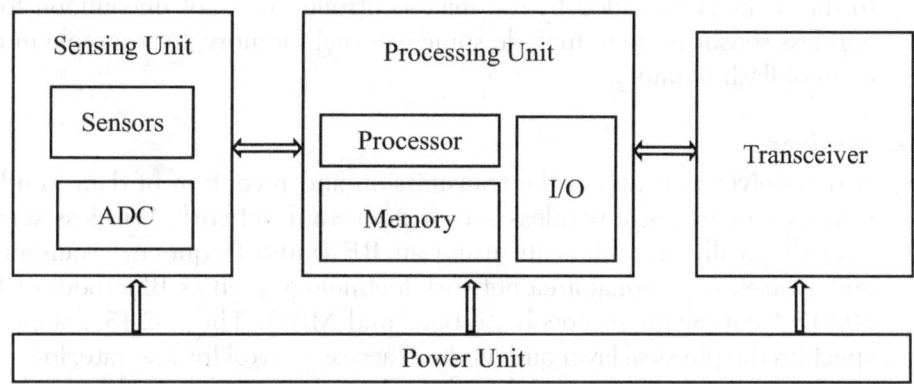

FIGURE 2.5 The components of a wireless sensor node.

In addition to the previous units, a wireless sensor node may include a number of application-specific components, for example, a location detection system or mobilizer; for this reason, many commercial sensor nodes include expansion slots and support serial wired communication.

Sensing Unit

A sensor is a device that measures some physical quantity and converts it into a signal to be processed by the microcontroller. A wide range of sensor types exist including seismic, thermal, acoustic, visual, infrared, and magnetic. Some of the sensors are discussed in the previous chapter. Sensors may be passive (sensing without active manipulation of the environment) or active (using active manipulation/probing of the environment to sense data, e.g., radar) and may be directional or omni-directional. A wireless sensor node may include multiple sensors providing complimentary data. The sensing of a physical quantity such as those described typically results in the production of a continuous analog signal, and for this reason, a sensing unit is typically composed of a number of sensors and an analog to digital convertor (ADC) which digitizes the signal.

Microcontroller

A microcontroller provides the processing power for, and coordinates the activity of, a wireless sensor node. Unlike the processing units associated with larger computers, a microcontroller integrates processing with some memory provision and I/O peripherals; such integration reduces the need for additional hardware, wiring, energy, and circuit board space. In addition to the memory provided by the microcontroller, it is not uncommon for a wireless sensor node to include some external memory, for example in the form of flash memory.

Transceiver

A transceiver unit allows the transmission and reception of data to other devices connecting a wireless sensor node to a network. Wireless sensor nodes typically communicate using an RF (radio frequency) transceiver and a wireless personal area network technology such as Bluetooth or the 802.15.4 compliant protocols ZigBee and MiWi. The 802.15.4 standard specifies the physical layer and medium access control for low-rate, low-cost wireless communications while protocols such as ZigBee and MiWi build upon this by developing the upper layers of the OSI Reference Model. The Bluetooth specification crosses all layers of the OSI Reference Model and is also designed for low-rate, low-cost wireless networking. Wireless sensor communications tend to operate in the RF industrial, scientific, and medical (ISM) bands, which are designed for unlicensed operation.

Power Source

Wireless sensor nodes must be supported by a power unit which is typically some form of storage (that is, a battery) but may be supported by power scavenging components (for example, solar cells). Energy from power scavenging techniques may only be stored in rechargeable (secondary) batteries, and this can be a useful combination in wireless sensor node environments where maintenance operations like battery changing are impractical. To conserve energy a power unit may additionally support power conservation techniques such as dynamic voltage scaling.

2.3 Common Wireless Sensor Node Architecture

The following is a discussion about the two commonly used sensor nodes. Sharing a number of common features, the two devices are built around different microcontrollers and are arranged in considerably different configurations. Like many wireless sensor platforms, the sensor in Figure 2.6 (a) uses 802.15.4 wireless sensor communications in the 2.4GHz ISM band, allowing the devices to be used unlicensed regardless of their global location. Support is provided for a high data rate (250kbps) and secure (AES-128) radio communications, and every node has the capacity to operate as a router to forward data received from other nodes. The Figure 2.6 (a) sensor node does not directly include any sensor functionality but instead supplies an expansion connector compatible with a range of sensor boards including light, temperature, acoustic, and magnetic sensing. It is powered by an ATMEGA 128 microcontroller, has 4kB of RAM and 128kB of flash memory, and supports a range of operating systems for wireless sensor nodes including TinyOS, SOS, Mantis OS, and Nano-RK.

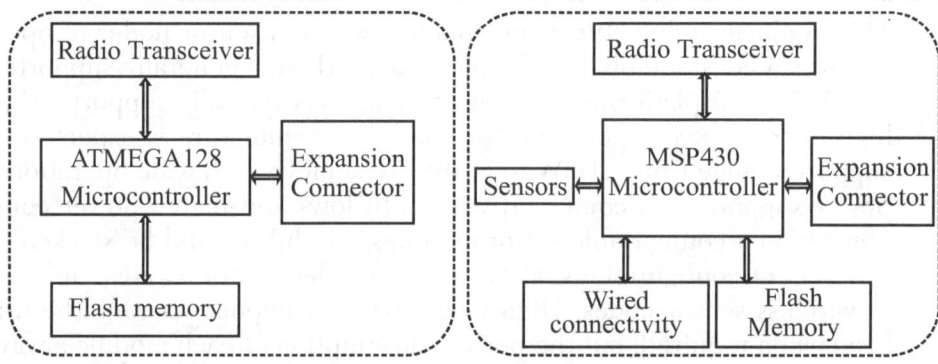

FIGURE 2.6 (a) WSN usingAtmega 128 MC.   **FIGURE 2.6** (b) WSN using MSP430MC.

Like Figure 2.6(a), the Figure 2.6(b) sensor node uses 802.15.4 wireless sensor communications in the 2.4GHz ISM band and provides support for secure radio communications at 250kbps. This board provides onboard sensing for humidity, temperature, and light as well as an expansion connector for other sensing devices, displays, and digital peripherals. The Figure 2.6(b) node is powered by an 8MHz Texas Instruments MSP 430 microcontroller with 10kB of RAM and has 48kB of flash memory. It supports a range of operating systems for wireless sensor nodes including TinyOS, SOS, Mantis OS, and Contiki. The two node architectures both use a single, central processor to which a number of additional components are connected. Each node contains the identified transceiver and sensor units (either directly or through use of expansion cards) and is powered by some external energy source.

2.4 Modular Sensor Node Architectures

While the previous two commercial wireless sensor node architectures generally follow a similar design involving a single microcontroller and a number of connected components, recent research has begun to consider alternative architectures for wireless sensor nodes. Such alternative architectures are typically described as modular or layered architectures and offer a range of benefits including energy conservation, knowledge reuse, and real-time performance. The most prominent of these alternative architectures are presented in the following sections.

Modular Architectures for Improved Power Management

This modular architecture aims to allow wireless sensor nodes to operate over a considerably larger power range than is generally supported in commercial platforms. Traditional platforms typically support either high or low power operation; the modular architecture is expected to support a range from <1mW to >10W. To achieve such wide operational power support, it is combined with both low- and higher-power components in a configurable set of modules. Modules could be stacked in a variety of configurations, allowing the production of an adaptable set of wireless sensor nodes. High performance components could be utilized when required, but the power consumption of each module always reflects their role within the overall system and each consumes minimal power when unused. Every module within a stack contains a low-power, always-on, slowly clocked microcontroller directed over an I²C bus for the purposes of power management, module discovery, and channel allocation.

The modular architecture failed to meet the originally intended low power requirements (due to a combination of factors including discrepancies between actual power consumption and datasheet-based predictions, power drainage during wake-up, and ADC performance), but the system did demonstrate that a modular design can reduce power consumption by removing the burden of power management from the central processor.

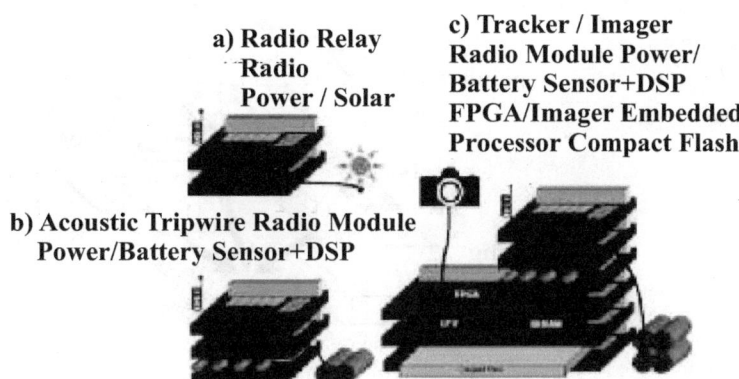

a) Radio Relay
    Radio
    Power / Solar

c) Tracker / Imager
Radio Module Power/
Battery Sensor+DSP
FPGA/Imager Embedded
Processor Compact Flash

b) Acoustic Tripwire Radio Module
Power/Battery Sensor+DSP

FIGURE 2.7 Modular architecture.

Modular Architectures for Improved Development

A modular sensor platform in Figure 2.8 is used to reduce the repetition involved in developing knowledge and infrastructures for sensor networks. The stack enables creation of sensor nodes through the combination of a master board and a series of other boards with specific sensing modalities communicating through a time division multiplexing (TDMA) scheme.

A number of applications have been developed using "The Stack" for their underlying node architecture. When compared against existing technologies for this purpose, nodes developed with "The Stack" produced an almost identical result set at a considerably lower cost and in real time. The advantages of modular sensor architectures are for improving flexibility, adaptability, and the redesign process. These divide sensor node functionality into four layers: communication, processing, power supply, and sensing/actuating. Each layer of the architecture may have any number of implementations which may be reused or interchanged when developing new applications. Furthermore, within a single WSN application, multiple heterogeneous nodes may interact despite their differing hardware composition.

Interchangeability of layers was facilitated through use of standard physical and electrical interfaces creating vertical connections. Each implementation of any layer of the architecture provides the required connectors in a standard location, allowing layers to be quickly and easily slotted together to create connections.

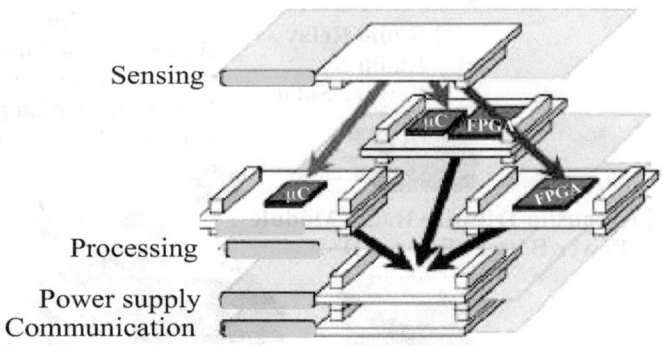

FIGURE 2.8 Interchangeability of modules using the four-layer node architecture.

Modular Architectures for Improved Performance

A more complex modular sensor node architecture provides a reconfigurable node platform with additional provision for real-time processing. It uses a combination of serial and parallel commumcations to provide a scalable high-performance bus, controlled using TDMA by complex programmable logic devices (CPLDs). The use of both serial and parallel commumcations provides support for high-performance real-time point-to-point and multipoint-to-point exchanges, while provision of a dedicated CPLD on each module for communications allows the module's central processor to engage in asynchronous communication allowing processing to continue on real-time tasks.

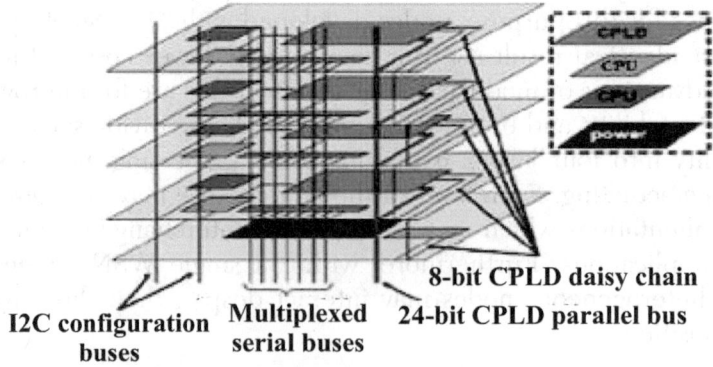

FIGURE 2.9 More complex modular sensor node architecture.

In particular, the communication channel architecture used by the Figure 2.9 node is particularly efficient, improving throughput and reducing end-to-end communication delays when compared against more traditional solutions.

2.5 Pic Node Architecture

The previously described architectures all utilize the concept of modularity to improve upon existing wireless sensor node platforms. Each of the architectures divides a wireless sensor node into multiple modules or layers to provide some benefit (for example, reduced power consumption, knowledge reuse, or real-time capabilities). While a small number of the architectures were developed with multiple goals in mind, each had a primary area of focus to which their architecture was tailored. The PIC farm architecture aims to develop a modular architecture that goes some way to meet each of the goals of the previous platforms.

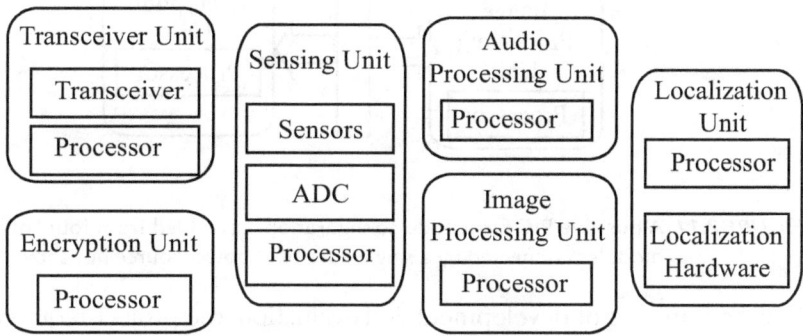

FIGURE 2.10 A selection of potential PIC farm components.

The PIC Farm project envisages the development of a pool of readily available off-the-shelf modules [Figure 2.10] which can be assembled in a "plug and play" or "Lego style" manner [Figure 2.11]. Each module would encapsulate some aspect of wireless sensor node functionality, providing both the required hardware and software. For example, a radio transceiver module might include a transceiver device alongside a small microcontroller programmed to control the operation of the radio, including hardware features such as power management and application-specific features such as the encapsulation of data into a given format.

While many of the existing research platforms combine multiple networking interfaces (for example, use of SPI and I²C technologies), the PIC farm project uses a more minimal approach suggesting use of a single data bus. For the purposes of this research the selected network technology is I²C, but it is assumed that any technology can be used as long as it meets the requirements of all intended exchanges within a node. Restriction of the networking capabilities in this way reduces the cost and complexity of node hardware with the intention of providing a more lightweight architecture than some of those described thus far.

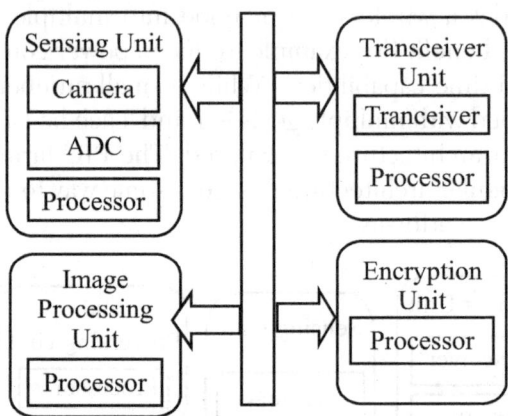

FIGURE 2.11 A possible PIC farm node configuration assembled from four "off the shelf" modules connected by a single data bus (power source not shown).

For the purposes of development and evaluation, the project required assembly of a specific PIC farm architecture. A dummy PIC farm node was assembled using three PIC 16F88 chips. Although not connected to sensor node components such as a radio and transceiver, each chip was intended to function as if part of a PIC farm unit, simulating functionality such as sensing, encryption, and transmission.

Component Selection

The PIC Farm board was assembled from the following components:

PIC 16F88

The PIC Farm aims to support distributed processing within a wireless sensor node. For this reason, development of the dummy node utilized multiple low capability PIC microcontrollers. While the PIC Farm project should be extendable to configurations composed of multiple different chips, the dummy node was assembled from three identical 16F88 processors. The PIC Farm dummy node is intended to simulate a simple wireless sensor node in which three core functionalities are required:

Sensing

An abstract sensing functionality is required. Such functionality should potentially represent any sensing activity: from the very simple (e.g., temperature sensing) to the more complex (e.g., visual, camera-based sensing).

Radio operations

It is assumed that the wider sensor network within which the dummy node is considered to operate utilizes some form of radio transmissions. For this reason, the dummy node should simulate radio operations.

Processing

Wireless sensor nodes vary in the processing provision required: a node within a security conscious network may require encryption while others with complex sensors may require specialized data analysis (for example, face detection, audio localization). The dummy node aims to simulate an abstract processing operation which could represent a variety of tasks.

The identification of the three core functionalities resulted in the decision to include three PICs within the dummy node: each chip can potentially represent any one of the required functionalities. The PIC devices are cheap, low capability, low power microcontrollers (compared to popular alternatives for wireless sensor node architectures, for example, the MSP 430 consumes 150-300 μA at 1 MHz while a low end PIC consumes around 76 μA at the same frequency), making them an ideal selection for a distributed sensor node architecture. The 16F88 [Figure 2.12] is a low to mid range PIC device providing the smallest possible memory and processing requirements for the intended implementation at a low cost while ensuring minimal resource wastage. In addition, the selected devices provide support for a number of networking technologies including hardware support for I^{2}C.

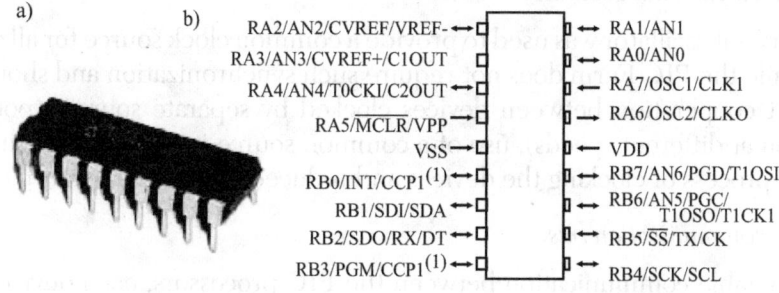

FIGURE 2.12 The PIC 16F88: a) Photograph of the device b) Pin allocations for the device.

LED output

For each PIC on the board, an array of three LEDs was supplied for debug/output purposes.

In addition to the three LED arrays associated with the processors on the board, an additional LED array was connected to the MAC bus (see as follows) to allow easy debugging. Although not representative of any sensor node component, the presence of the LEDs facilitated the PIC Farm development process. Each LED on the board was connected to a resistor to limit current consumption. While the use of LEDs provides very limited output, they are considerably simpler than some alternative output mechanisms (for example, LCD displays or external memory). The simplicity of LEDs as a method of output allowed easy output even in the early stages of development.

Push button input

In addition to the LED outputs, the dummy node was equipped with a push button input component purely for development purposes. While a typical wireless sensor node would be expected to operate independently of any user and would therefore not be equipped with such input, the dummy node was required to demonstrate developed functionality in response to user input in order to confirm that the expected behavior occurred. The button was connected to a single device and was generally used for the purpose of generating new tasks. For example, while developing the MAC mechanism, the appropriate node was programmed to lock the bus for a period of time in response to a button press. Use of a push button for this purpose was ideal, as it was considerably simpler than many other forms of input device (e.g., microphone, light sensor).

Shared external oscillator

A crystal oscillator was used to provide a common clock source for all devices. While the PIC Farm does not require such synchronization and should support cooperation between devices clocked by separate sources (potentially even at different speeds), use of a common source in this manner simplified the process of clocking the devices and reduced hardware requirements.

I^2C communication bus

To enable communication between the PIC processors, each device is connected to an I^2C bus on pins B1 and B4. The I^2C bus is intended for data exchange using the RPC mechanism.

Medium access control (MAC) bus

As the I²C bus depends on the sharing of a communication medium between multiple devices, some form of access control is required. The PIC farm node uses an out-band medium access control (MAC) bus, separate to the I²C medium, for the purpose of ensuring effective use of the shared medium.

Traditional wireless sensor nodes follow a centralized architecture in which components are managed by a single microcontroller. Recent research has considered a number of modular architectures for sensor nodes; such architectures are typically designed with a single specific improvement in mind, for example knowledge reuse or energy efficiency. The PIC Farm project aims to build upon existing work by developing a simple, flexible architecture in which self-contained functional blocks can be assembled in a "Lego style" manner to quickly create a variety of wireless sensor nodes appropriately tailored to their intended application. To consider the viability of such an architecture, a dummy node has been developed using the modular architecture proposed by the PIC farm using three low-power, low -capability PIC microcontrollers.

2.6 IMote Node Architecture

The IMote sensor node architecture is a *multi-purpose* architecture (Figure 2.13) consisting of

1. A power management subsystem,

2. A processor subsystem,

3. A sensing subsystem,

4. A communication subsystem, and

5. An interfacing subsystem.

Amultiple-sensor board of the IMote Node architecture contains (Figure 2.14):

A12-bit, four channel ADC

- A high-resolution temperature/humidity sensor
- A low-resolution digital temperature sensor
- A light sensor
- The I²C bus is used to connect *low* data rate sources
- The SPI bus is used to interface *high* data rate sources

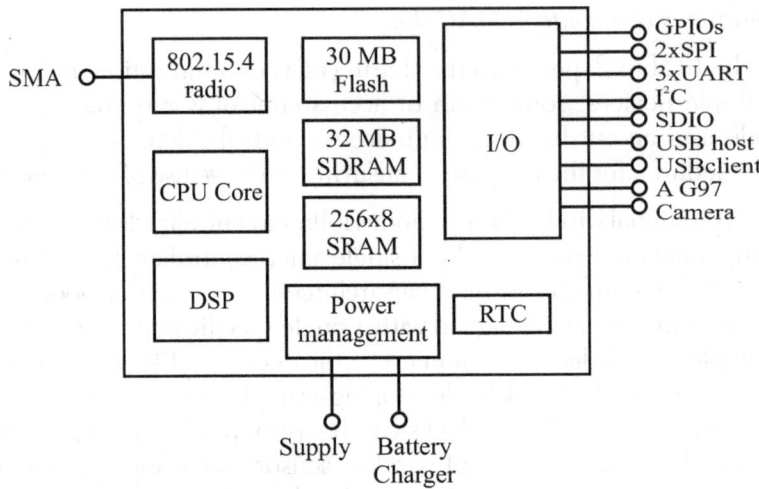

FIGURE 2.13 The IMote node architecture.

The processing subsystem provides

- *Main processor* (microprocessor)
- Operates in low voltage (0.85v) and low frequency (13MHz) mode
- Dynamic voltage scaling (104MHz–416MHz)
- Sleep and deep sleep modes
- Thus enabling low power operation
- Coprocessor (a DSP)
- Accelerates multimedia operations—computation intensive

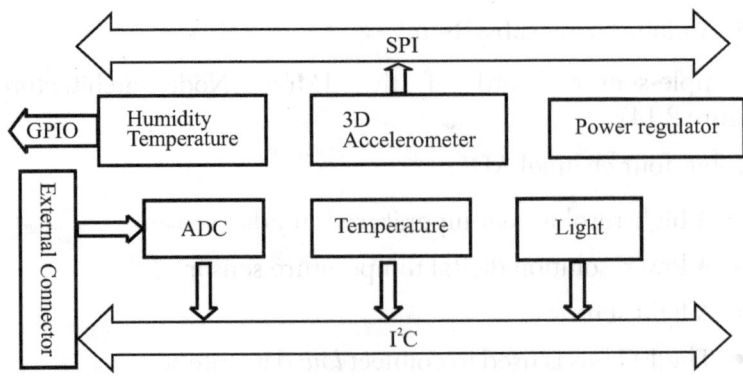

FIGURE 2.14 The IMote node architecture.

2.7 XYZ Node Architecture

It consists of the four subsystems:

1. Power subsystem

2. Communication subsystem

3. Mobility subsystem

4. Sensor subsystem

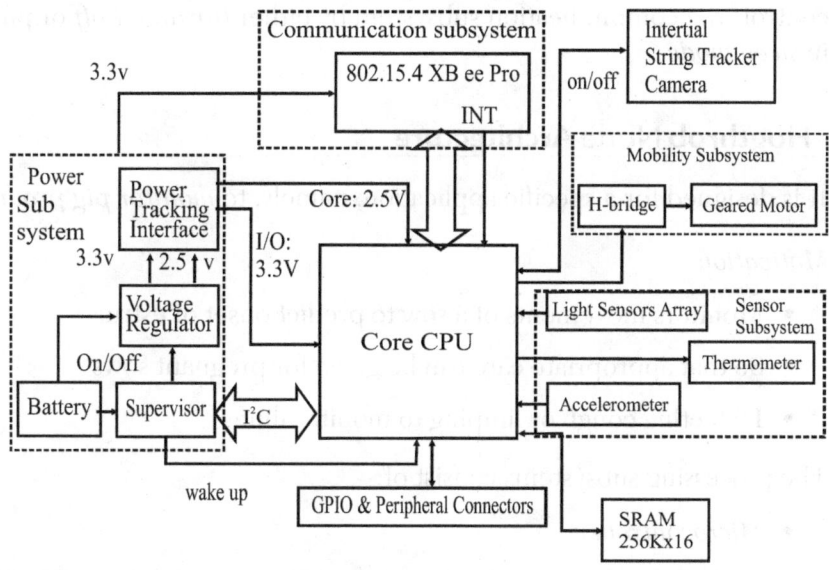

FIGURE 2.15 The XYZ node architecture.

The processor subsystem is based on the ARM7TDMI core microcontroller

- F_{max} = 58MHz
- Two different modes (32bits and 16bits)
- Provides an on-chip memory of 4KB boot ROM and a 32KB RAM
- Can be extended by up to 512KB of flash memory

Peripheral components:

- DMA controller
- Four 10-bit ADC inputs

- Serial ports (RS232, SPI, I²C, SIO)

- 42 multiplexed general purpose I/O pins

The communication subsystem is connected to the processing subsystem through an SPI interface.

RF transceiver

When an RF message has been successfully received, the *SPI interface* enables the radio to wake up a sleeping processor. The processor subsystem controls the commumcation subsystem by either *turning it off* or putting it in *sleep mode.*

2.8 Hogthrob Node Architecture

It is designed for a specific application, namely, to *monitor pig production.*

Motivation

- Monitors movements of a sow to predict onset of estrus

- So that appropriate care can be given for pregnant sows

- Detecting cough or limping to monitor illness

The processing subsystems consist of:

- *Microcontroller*

    ○ Performs less complex, less energy-intensive tasks

    ○ Initializes the FPGA and functions as an external timer and an ADC converter

- *Field Programmable Gate Array*

    ○ Executes the sow monitoring application

    ○ Coordinates the functions of the sensor node

There are a number of interfaces supported by the processing subsystem, including:

- The I²C interface for the sensing subsystem

- The SPI interface for the communication subsystem

- The JTAG interface for in-system programmability and debugging
- The serial (RS232) interface for interaction with a PC

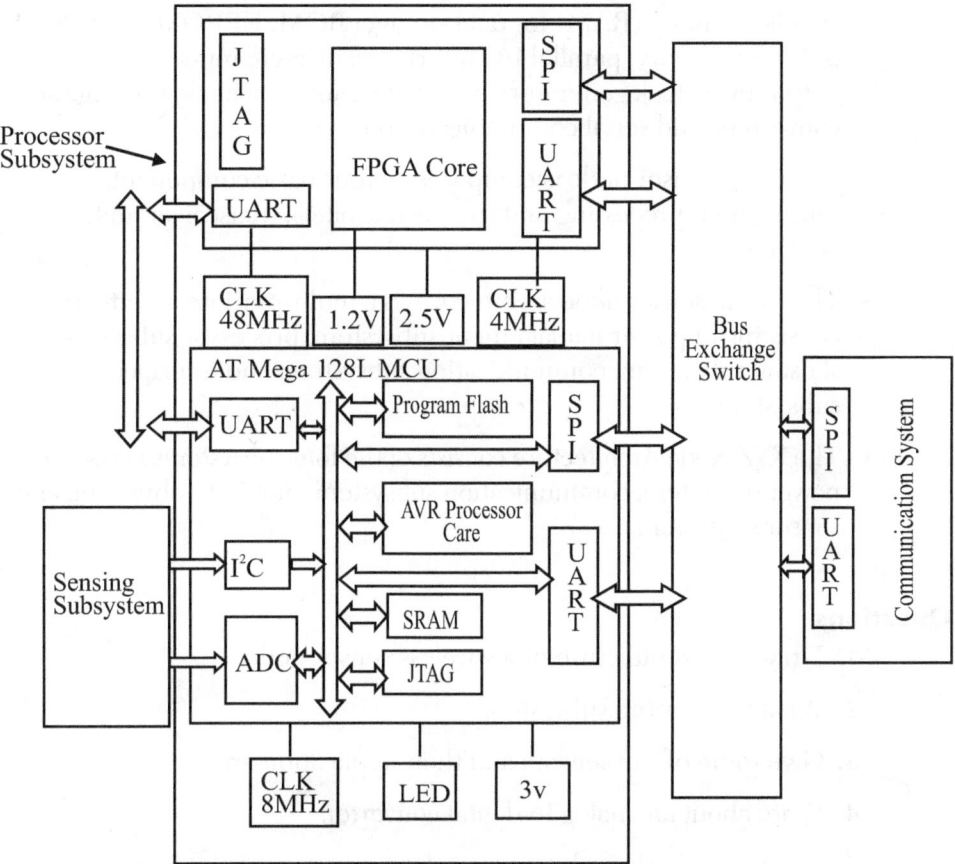

FIGURE 2.16 The Hogthrob node architecture.

Summary

- Wireless sensor nodes are the essential building blocks in a wireless sensor network for sensing, processing, and communication.

- The sensing subsystem integrates all kinds of sensors needed to measure the parameters.

- Von Neumann architecture provides a single memory space for storing program instructions and data.

- Harvard architecture provides separate memory spaces for storing program instructions and data.

- Microcontrollers integrate the components such as the CPU core; volatile memory (RAM) for data storage; ROM, EPROM, EEPROM, or Flash memory; parallel I/O interfaces; discrete input and output bits; clock generator; one or more internal analog-to-digital converters; and serial communications interfaces.

- A wireless sensor node is composed of four basic components: sensing unit, processing unit (microcontroller), transceiver unit, and power unit.

- The IMote sensor node architecture is a multi-purpose architecture consisting of power management subsystem, processor subsystem, sensing subsystem, communication subsystem, and interfacing subsystem.

- The XYZ Node Architecture consists of the four subsystems known as power subsystem, communication subsystem, mobility subsystem, and sensor subsystem.

Questions

1. Draw the architecture of a wireless sensor node.

2. What is a sensing subsystem?

3. Give some of the sensors and their application areas.

4. Write about an analog to digital converter.

5. What are 3 basic architectures of a processor subsystem?

6. What is Von Neumann architecture? Draw its blocks.

7. What is Harvard architecture? Draw its blocks.

8. What is Super-Harvard architecture? Draw its blocks.

9. List advantages and disadvantages of microcontrollers.

10. List advantages and disadvantages of Digital signal Processors.

11. List advantages and disadvantages of ASIC.

12. List advantages and disadvantages of FPGA.

13. What are the communication interfaces used in nodes?

14. Write different modes of serial peripheral interface.

15. Write a short note on IIC Inter-Integrated circuits.

16. Compare SPI and IIC.

17. Explain with the help of a diagram the components of wireless sensor node architecture.

18. Draw the diagram of two different common sensor node architectures.

19. With a diagram explain about the modular sensor node architectures.

20. Draw PIC node architecture and explain each block.

21. Explain about IMote Node architecture with the help of a diagram.

22. Explain about XYZ Node architecture with the help of a diagram.

23. Explain about Hogthrob Node architecture with the help of a diagram.

Further Reading

1. *Wireless Sensor Networks: Architectures and Protocol* by Edgar H. Callaway Jr.

2. *Fundamentals of Wireless Sensor Networks Theory and Practice* by Waltenegus Dargie and Christian Poellabauer

3. *Protocols and Architecture for Wireless Sensor Networks* by Holger Karl and Andreas Willig.

References

1. *http://www3.nd.edu/~cpoellab/teaching/cse40815/Chapter3.pdf*

2. *http://comp.ist.utl.pt/ece-wsn/doc/slides/sensys-ch2-single-node.pdf*

CHAPTER 3

SOFTWARE ARCHITECTURE

This chapter mainly discusses the software architectures used in wireless sensor networks.

3.1 Introduction to Software Architecture

Wireless sensor networks are aggregates of numerous small sensor nodes. Each node can send messages through the network to the information sink or ultimate controlling device. The nodes can also forward messages from other nodes, perform network organization tasks, and complete a variety of other functions.

The applications of WSNs vary widely. WSNs could be used in industrial settings for machine control and environment monitoring. Other applications could be medical monitoring of a patient's health from a variety of perspectives. The military is highly interested in sensor networks for intelligence gathering, while WSNs have possible applications in aerospace for the structural integrity of planes.

The sensor units contain several types of sensors for measuring temperature, light, acceleration, angular velocity, geo-orientation, and so on. A microcontroller or programmable logic device, combined with A/D converters, collects and processes the sensor signals and assembles sensor data and control frames. A serial interface connects the sensor unit to a radio module, which converts the sensor data frames to radio mes sages and sends them to the gateway.

All radio modules within a given range establish a wireless sensor network (WSN). The IEEE 802.15.4 / Zigbee standard is one of the most promising candidates for designing WSNs which need to be self-organized and self-healing, that is, nodes automatically establish and maintain connectivity among themselves. Mesh networking protocols provide new capabilities where each node operates not only as a direct source or sink, but also as a message forwarder for other nodes that do not have direct connectivity with their communication peers. Providing reliable wireless connectivity, stability, and scalability, while at the same time coping with the limitations imposed by low cost, battery-powered sensor nodes presents a multitude of challenging research problems.

3.2 Operating System (OS) Requirements

The following are the requirements of an operating system used for a wireless sensor network.

- Small physical size and low power consumption
  Devices have limited memory and power resources

- Concurrency intensive operation
  Need to be able to service packets on the fly in real time

- Limited hardware parallelism and controller hierarchy
  Limited number and capability of controllers
  Unsophisticated processor memory switch level interconnect

- Diversity in design and usage
  Provide a high degree of software modularity for application specific sensors

- Robust Operation
  OS should be reliable, and assist applications in surviving individual device failures

3.3 Wireless Sensor Network Characteristics

The concept of wireless sensor networks implies a number of WSN characteristics which heavily influence the software architecture. Specifically, WSNs must be self organizing, perform cooperative processing, energy optimized, and modular. These four requirements in particular impact heavily on the form of the software architecture.

1. *Self organization*

The large number of nodes in a WSN renders direct manipulation by a user for network organization impractical. A user could not go through thousands of nodes directing the network configuration and clustering. Subsequently, the nodes must be capable of organizing the network and partitioning it for efficient operation given the environment and network attributes.

Additionally, the nodes of a sensor network must be robust. The aggregate formed by the nodes must have a high up time. The large number of nodes in a network along with unattended operation complicates any attempt at a fault tolerant design. Sensor networks with wired connections do not necessarily rely on other nodes to transmit data. This reduces the need for redundancy and the robustness of individual nodes.

In contrast, wireless sensor network nodes transmit information from node to node with a small amount of processing in between. Consequently, individual nodes must be highly robust, while the organization of the network must tolerate individual device failure. Variations in the network topology can affect the degree of network vulnerability to failures, necessitating complex routines to implement fault tolerance.

2. *Concurrency, Cooperative Processing*

The nodes in a network primarily direct information flow through the network to various data sinks, the points to which data from the network is fed. Each sensor node may possess a limited amount of memory, so the buffering of data is impractical. Additionally, the node performs a number of simultaneous operations: capturing, processing, and transmitting sensor data, while simultaneously forwarding data from other nodes in multi-hop or bridging situations.

WSNs also provide a unique opportunity for cooperative processing. Cooperative processing can reduce network traffic through data aggregation and preprocessing. For example, the establishment of a wireless network might involve the triangulation of a new node when it joins a network to establish the node's position.

3. *Energy Efficiency*

Wired sensor networks have the luxury of external power sources, such as power over the Ethernet. The nodes of wireless networks have no practical way of utilizing an external energy source, which would in any case be contrary to the point of a WSN. A sensor network may also be distributed in hostile or remote environments. Energy efficiency dictates the minimization of communication between nodes. Therefore, the choice of protocols and network configuration are key in terms of network life span.

Protocol-related energy savings are directly related to the physical, link, and network layers.

Additional power savings come from an operating system (OS) for the nodes which supports advanced power management and lower power task scheduling. Power sensitive task scheduling can minimize power use through nonlinear battery effects. Advanced power management would put any hardware not in use to sleep, minimizing power consumption.

4. *Modularity*

Sensor nodes in a network tend to be specific, and therefore contain only the hardware needed for the application. The range of possible applications dictates a large variance in the hardware required for sensor nodes. Accordingly, the software for the nodes must exhibit a high degree of modularity.

3.4 Software Architecture Components

The nature of a sensor network lends itself to a service-oriented, component-based framework. Applications split into sensor, node, and network applications, providing the basis for fundamental application layers in a sensor network.

Sensor applications interface with the sensors, local data, and hardware on a node, along with the operating system. Sensor applications form the base layer and provide the basic functions of a sensor node.

Node applications use the basic functions provided by sensor applications to perform middleware tasks for network buildup, maintenance, and localization. Network applications deal with the services and tasks of the network as a whole. Network applications thereby act as an interface to the layer administrating to the network.

Middleware

Middleware refers "to the software layer between operating system and sensor application on the one hand and the distributed application which interacts over the network on the other hand." The design of middleware aims to be scalable, adaptive, generic, and reflective.

Scalable middleware performs optimization based on resource constraints at runtime. The nature of wireless sensor networks calls for lightweight middleware, or middleware which has low communication and computation requirements. By performing optimizations at runtime, the interfaces of middleware are customized.

The sensor network changes as nodes move, necessitating runtime adaptations of the middleware to exchange and run components as needed by the application. Localized algorithms can be used to enhance system scalability and robustness in the face of interactions between sensor nodes. These algorithms can also provide reflective middleware, which changes the behavior of layers on the fly instead of exchanging them.

Generic middleware attempts to reduce overhead imposed by using generic interfaces for middleware components. This implies the customization of the application interfaces and features, allowing for interpretation by middleware and compile-time optimization. Generic interfaces also allow for the standardization of system services to diverse applications. Conversely, while middleware interfaces may be generic, the interfaces of the application component on a specific sensor node are anything but. In short, middleware acts as an abstraction layer to help hide software specifics from the application layer.

A simple use case can help determine the software architecture. The client application requests data from the network about surface conditions in a certain area. The client first sends a request to a surrogate proxy for the desired information. The proxy communicates with the appropriate nodes, which in turn then determine the surface conditions in the area using cooperative algorithms. The proxy takes the information returned from the nodes, translates it, and sends it back to the client.

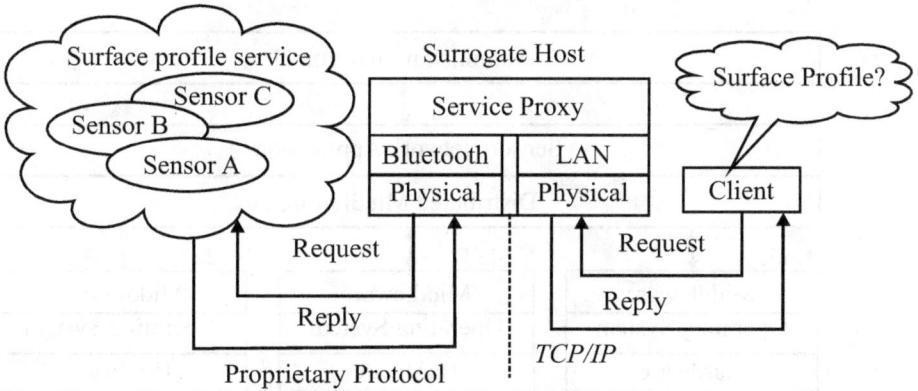

FIGURE 3.1 Surrogate architecture in sensor networks.

Figure 3.1 illustrates the use case. Such architecture can be realized using the node application structure shown in Figure 3.2, along with the sensor network architecture shown in Figure 3.3.

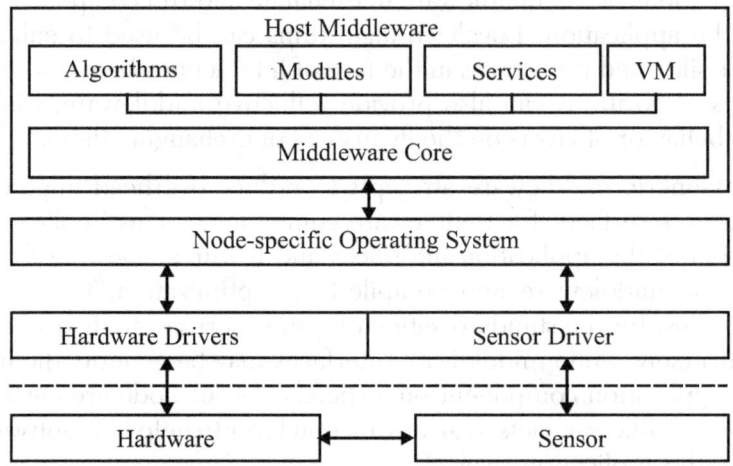

FIGURE 3.2 Node application structures.

Figure 3.2 illustrates the division of node applications into three layers. The lowest layer handles hardware specifics, such as hardware and sensor drivers. The node operating system acts as a buffer layer between the hardware specifics and the host middleware application layer. The operating system layer handles the processes which relate strictly to the node operation, while the host middleware handles processes concerning the services offered by the node to the network.

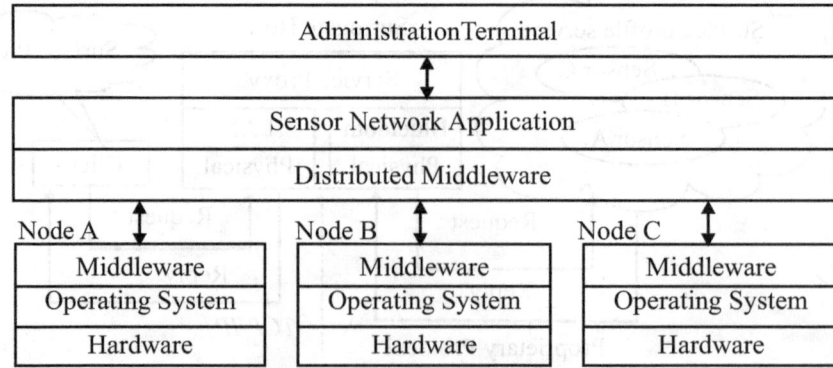

FIGURE 3.3 Overall software architecture of a sensor net.

The middleware is comprised of four different components which are called as needed, with the option to add additional modules for security or routing. The VM or Virtual machine component enables platform independent program execution, while algorithms define the behavior of modules.

The general overall software architecture of the sensor net is shown in Figure 3.3. The individual nodes interact with the distributed middleware layer to perform the functions dictated by the sensor network application. The administration terminal is a connection-point-independent external actor which evaluates results from the sensor network application. The diagram specifically illustrates the behavior of the sensor network application, which cannot assign tasks to individual nodes. Instead the layer abstract shown indicates that the distributed middleware handles tasks for the entire network and acts as network service coordinator.

Architectural Issues

While the architecture presented certainly presents a solid and basic design for a sensor network, it does not reflect some of the requirements such as energy efficiency, which can be a significant effect of software architecture. Network topology control in particular proves effective in extending network life and increasing network capacity.

Additionally, the amount of energy available to each sensor cannot support long range communication, necessitating a tiered network structure. The need for such a network structure then influences the software architecture design.

Finally, the use of a single tier architecture increases network load on nodes surrounding the command node. The increased traffic on these key nodes decreases their life span, and in turn shortens the life span of the entire network. The adoption of a multi-tiered network structure can reduce the energy consumption imbalance if the network supports sufficient fault-tolerance.

3.5 A Cluster-Based, Service-Oriented Architecture

The factors mentioned above can influence the software architecture of a wireless sensor network when taken into account. A second architecture uses clustering to handle the factors mentioned and provide for application of Quality of Service (QoS) management.

Architectural Description

One of the more favored network architectures for wireless sensor networks involves clustering. A cluster is a set of adjacent sensors which are grouped together and interface with the rest of the network through a gateway, or cluster head. Gateways are higher energy nodes which maintain the network in the cluster, perform data aggregation, and organize sensors into subsets.

Clusters exhibit dynamic behavior. Clusters form and are modified on the fly depending on conditions and node availability. During cluster formation, one node is elected as the gateway. It is important to note that while clusters can overlap spatially, one node cannot belong to multiple clusters.

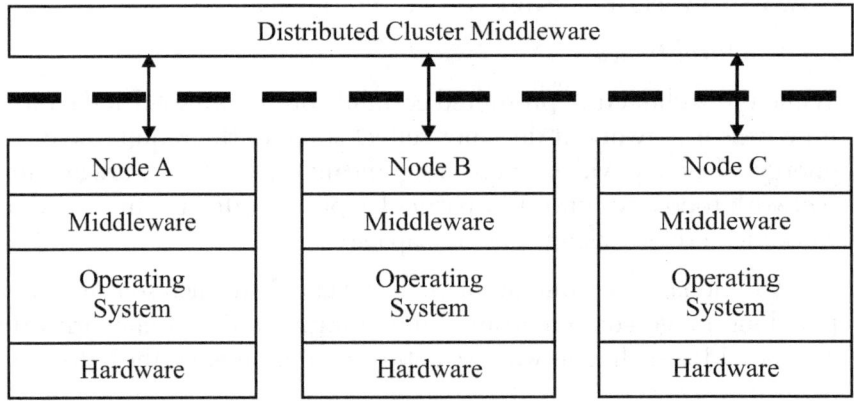

FIGURE 3.4 Cluster-software-based architecture.

Given the presence of clustering, a cluster can be regarded as the base unit for the software architecture. In this case, data collection would be performed in a distributed manner. However, in order to dynamically manage these clusters, the architecture of the middleware needs be fairly complex. Figure 3.4 displays the software architecture at the cluster level. Figure 3.5 shows a cluster-based middleware architecture.

As shown in Figure 3.5, the architecture proposed contains an abstraction modeled as a Virtual Machine. The Virtual Machine provides the same service as in the first architecture, that of hardware-independent program execution. However, this Virtual Machine splits down into two additional layers: the resource management layer and the cluster layer.

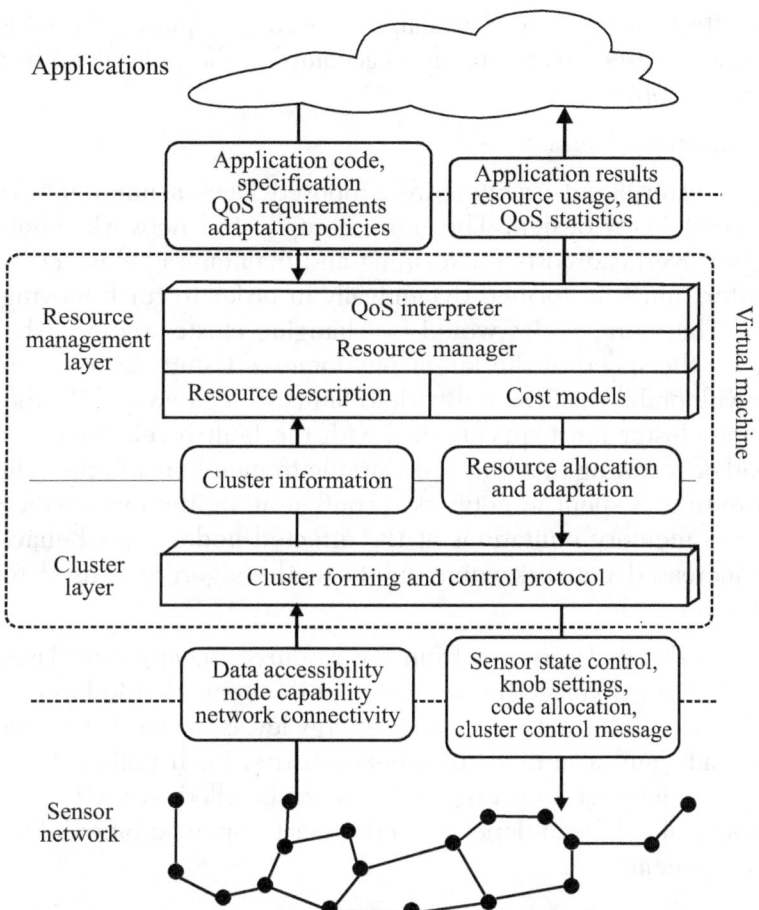

FIGURE 3.5 Middleware architecture for cluster-based WSNs.

The cluster layer encompasses the distributed cluster middleware illustrated in Figure 3.4. This software layer forms clusters from the collection of sensor nodes surrounding the target area. The exact factors controlling the initial formation of clusters can vary depending on the application of the sensor network, and do not have significant impact on the software architecture.

The resource management layer controls resource allocation and adaptation to meet QoS requirements for the sensor network application. Resource management is an important part of QoS in distributed applications such as a wireless sensor network. Environmental and system changes can affect the amount of available resources, requiring the middleware to reallocate resources on the fly to accomplish the tasks given by the sensor network application.

Architectural Issues

The cluster-based architecture proposed faces a number of challenges inherent in its design. The more complicated network topology incurs higher overhead costs for forming and maintaining clusters. Specifically, clusters must be formed dynamically in order to track moving phenomena. Therefore, nodes would be changing cluster membership depending on the speed of the target phenomena. Using clusters also increases the vulnerability of the network to faults. If a gateway fails, the members of the cluster must quickly deal with the fault by electing a new cluster head or resorting to ad hoc networking to members of other clusters. The more time is spent in network reconfiguration, the more data can be lost due to memory limitations at the effected nodes. This behavior implies an increased network vulnerability to the algorithms used for network formation.

Overhead also comes from the resource management layer. In order to effectively manage the resources of the network, the layer must gather and update information on node energy levels, network connectivity, cluster loads, and a number of other statistics. Such polling can result in a dramatic increase to overhead if it is not handled correctly. Obviously the amount of overhead depends on the exact implementation of the resource management.

3.6 Software Development For Sensor Nodes

Sensor node software in particular lends itself well to an iterative form of development. As with any real-time embedded system, the code must be optimized to perform within certain parameters.

Due to the sensitivity of the performance, efficiency, and lifetime of a WSN to the algorithms controlling network configuration, an iterative design, implementation, and test phase is required. Since the applications of WSNs vary so greatly, the algorithms involved in a specific type of WSN must be optimized through a large amount of calculation and design.

The development pattern for node applications favors the highly iterative design pattern shown in Figure 3.6. The design method closely follows standard software engineering practices. Of note, component interface optimization is performed during the design stage. Evaluation through the monitoring of results leads to additional iterations. The development pattern results in a specific application for the node comprised of specially tailored parts.

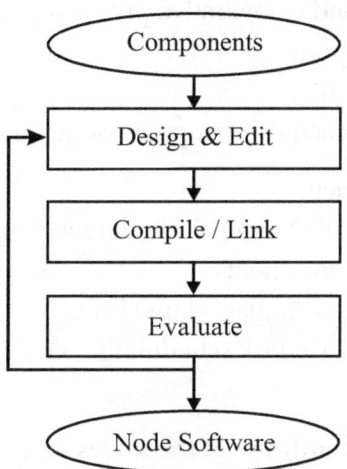

FIGURE 3.6 Proposed node software development process.

Numerous software difficulties must be solved before wireless sensor networks may be considered a mature technology. Chief among these problems stands the formation, creation, and testing of a robust, efficient software architecture that can fulfill all of the goals and requirements needed. Tiny OS and ZigBee are discussed in detail now.

3.7 Tiny OS

Tiny OS is a microthreaded OS that draws on previous work done for lightweight thread support and efficient network interfaces. It is a two-level scheduling structure, that is, with long-running *tasks* that can be interrupted by hardware *events* and a small, tightly integrated design that allows the crossover of software components into hardware. TinyOS was designed by Berkeley. The following are the characteristics of TinyOS.

    a. Provides a good framework
        i. Concurrency intensive
        ii. Efficient modularity

b. Self organizing
   i. Large numbers of nodes
      1. Robust operation—no redundancy
   ii. Fault tolerance—If a node fails, will the network recover?

c. Concurrency and Cooperative processing
   i. Limited memory
   ii. Triangulation
   iii. Sensor data acquisition, processing, targeting—data aggregation

d. Energy efficiency
   i. Nodes in a WSN have limited power supplies
   ii. Hostile Environments
   iii. Architecture can play a large role
   iv. Power sensitive task scheduling

e. Modularity
   i. The actual hardware of the nodes may vary, for example, temp sensors, hall effect, proximity, and so on.
   ii. While the drivers and software on each node may be specific, the interface presented to the distributed software should be generic.

f. Client
   i. Connects at any place on the network

g. Sensor Application
   i. Requests certain information from the network

h. Network software
   i. Handles request, maintenance, and so forth.

i. Node-specific software and hardware

j. Middleware
   i. "The software layer between operating system and sensor application on the one hand and the distributed application which interacts over the network on the other hand."
   ii. In essence, the complexity and layer architecture of the middleware defines the software architecture of the entire system.

k. Scalable

  i. Should perform optimization based on resource constraints at run-time

l. Adaptive

  i. Changes in the network and event under observation call for change.

  ii. Network restructuring

    a. Distributed application task reallocation

    b. Supports scalability and robustness

m. Reflective

  i. Changes the actual behavior of layers on the fly

    a. Example: The modification of the routing strategy depending on mobility

n. Generic

  i. Standard or generic interfaces between middleware components

  ii. Customizes application interfaces

It covers the basic TinyOS abstractions, such as hardware abstractions, communication, timers, the scheduler, booting, and initialization.

Platforms/Hardware Abstraction

Hardware abstractions in TinyOS generally follow a three-level abstraction hierarchy, called the HAA (Hardware Abstraction Architecture).

At the bottom of the HAA is the HPL (Hardware Presentation Layer). The HPL is a thin software layer on top of the raw hardware, presenting hardware such as IO pins or registers as nesC interfaces. The middle of the HAA is the HAL (Hardware Abstraction Layer). The HAL builds on top of the HPL and provides higher-level abstractions that are easier to use than the HPL but still provide the full functionality of the underlying hardware. The top of the HAA is the HIL (Hardware Independent Layer). The HIL builds on top of the HAL and provides abstractions that are hardware independent. This generalization means that the HIL usually does not provide all of the functionality that the HAL can. TinyOS supports platforms such as eyesIFXv2, intelmote2, mica2, mica2dot, micaZ, telosb, tinynode, btnode3, and so on.

Scheduler

The TinyOS scheduler has a non-preemptive FIFO policy. In TinyOS, every task has its own reserved slot in the task queue, and a task can only be posted once. A post fails if and only if the task has already been posted. If a component needs to post a task multiple times, it can set an internal state variable so that when the task executes, it reposts itself.

This slight change in semantics greatly simplifies a lot of component code. Rather than test to see if a task is posted already before posting it, a component can just post the task. Components do not have to try to recover from failed posts and retry. The cost is one byte of state per task. Applications can also replace the scheduler, if they wish. This allows programmers to try new scheduling policies, such as priority- or deadline-based. It is important to maintain non-preemptiveness, however, or the scheduler will break all nesC's static concurrency analysis.

Booting/Initialization

TinyOS has interface StdControl split into two interfaces: Init and StdControl. The latter only has two commands: start and stop. In TinyOS, wiring components to the boot sequence would cause them to be powered up and started at boot. That is no longer the case: the boot sequence only initializes components. When it has completed initializing the scheduler, hardware, and software, the boot sequence signals the Boot.booted event. The top-level application component handles this event and start services accordingly.

Virtualization

TinyOS is written with nesC, which introduces the concept of a "generic" or instantiable component. Generic modules allow TinyOS to have reusable data structures, such as bit vectors and queues, which simplify development. More importantly, generic configurations allow services to encapsulate complex wiring relationships for clients that need them.

Timers

TinyOS provides a much richer set of timer interfaces. Timers are one of the most critical abstractions a mote OS can provide, and so expands the fidelity and form that timers take. Depending on the hardware resources of a platform, a component can use 32KHz as well as millisecond granularity timers, and the timer system may provide one or two high-precision timers that fire asynchronously (they have the async keyword). Components can query their timers for how much time remaining before they fire, and can start timers in the future (e.g., "start firing a timer at 1Hz starting 31ms from now"). Timers present a good example of virtualization.

Communication

In TinyOS, the message buffer type is message_t, and it is a buffer that is large enough to hold a packet from any of a node's communication interfaces. The structure itself is completely opaque: a component cannot reference its fields. Instead, all buffer accesses go through interfaces. Send interfaces distinguish the addressing mode of communication abstractions. Active messages are the network HIL. A platform's ActiveMessageC component defines which network interface is the standard communication medium. For example, a mica2 defines the CC1000 active message layer as ActiveMessageC, while the TMote defines the CC2420 active message layer as ActiveMessageC.

There is no longer a TOS_UART_ADDRESS for active message communication. Instead, a component should wire to SerialActiveMessageC, which provides active message communication over the serial port.

Active message communication is virtualized through four generic components, which take the AM type as a parameter: AMSenderC, AMReceiverC, AMSnooperC, and AMSnooping ReceiverC. AMSenderC is virtualized in that the call to send() does not fail if some other component is sending. Instead, it fails only if that particular AMSenderC already has a packet outstanding or if the radio is not in a sending state. Underneath, the active message system queues and sends these outstanding packets.

Sensors

In TinyOS, named sensor components comprise the HIL of a platform's sensors. If a component needs high-frequency or very accurate sampling, it must use the HAL, which gives it the full power of the underlying platform.

Error Codes

In TinyOS the return code is error_t, whose values include SUCCESS, FAIL, EBUSY, and ECANCEL. Interface commands and events define which error codes they may return and why.

Arbitration

Basic abstractions, such as packet communication and timers, can be virtualized. The most pressing example of this is a shared bus on a microcontroller. Many different systems like sensors, storage, and the radio might need to use the bus at the same time, so some way of arbitrating access is needed.

To support these kinds of abstractions, TinyOS introduces the resource interface, which components use to request and acquire shared resources, and arbiters, which provide a policy for arbitrating access between multiple clients. For some abstractions, the arbiter also provides a power management policy, as it can tell when the system is no longer needed and can be safely turned off.

Power Management

Power management is divided into two parts: the power state of the microcontroller and the power state of devices. Microcontroller Power Management is computed in a chip-specific manner by examining which devices and interrupt sources are active. TinyOS provides low-power stacks for the CC1000 (mica2) and CC2420 (micaz, telosb, imote2) radios. Both use a low-power listening approach, where transmitters send long preambles or repeatedly send packets and receivers wake up periodically to sense the channel to hear if there is a packet being transmitted.

Network Protocols

TinyOS provides simple reference implementations of two of the most basic protocols used in mote networks: dissemination and collection. Dissemination reliably delivers small (fewer than 20 byte) data items to every node in a network, while collection builds a routing tree rooted at a sink node. Together, these two protocols enable a wide range of data collection applications. Collection has advanced significantly since the most recent beta release; experimental tests in multiple network conditions have seen very high (>98%) delivery rates as long as the network is not saturated. Software development platforms need to be power-consumption aware, and miserly on memory usage and processing power.

TinyOS, the operating system that runs on motes, is component-based in that you 'assemble' components that you need into the deployed program. These components handle radio communication, sending messages, taking of measurements, timing, and LEDs. Figure 3.7 shows the internal component graphware and Figure 3.8 shows an sample application for temperature and light measurement.

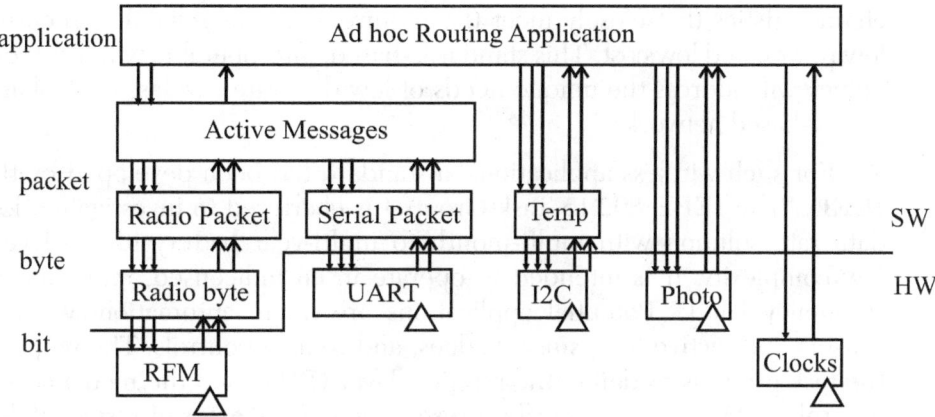

FIGURE 3.7 Internal component graphware.

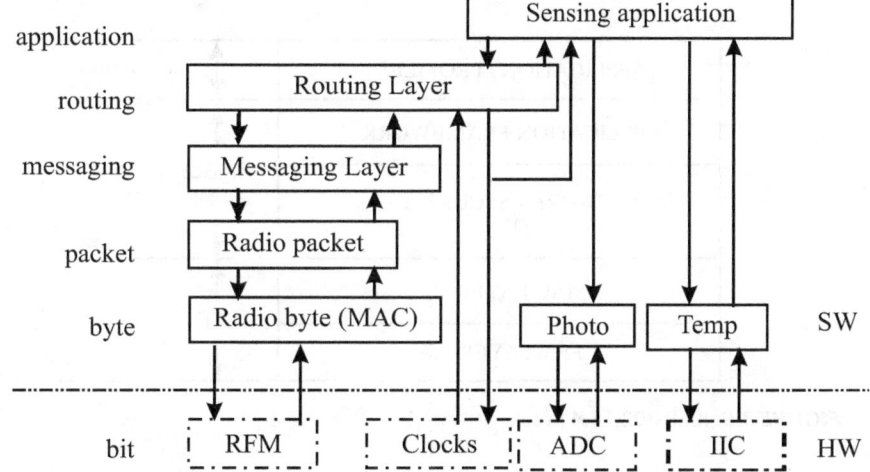

FIGURE 3.8 An example application.

3.8 ZigBee

There are many wireless monitoring and control applications for industrial and home markets which require longer battery life, lower data rates, and less complexity than is available from existing wireless standards. These standards provide higher data rates at the expense of power consumption, application complexity, and cost. What these markets need, in many cases, is a standards-based wireless technology having the performance characteristics that closely meet the requirements for reliability, security, low power, and low cost. This standards-based, interoperable wireless technology will address the unique needs of low data rate wireless control and sensor-based networks.

For such wireless applications, a standard has been developed by the IEEE: "The IEEE 802.15 Task Group 4 is chartered to investigate a low data rate solution with multi-month to multi-year battery life and very low complexity. It is intended to operate in an unlicensed, international frequency band." Potential applications are home automation, wireless sensors, interactive toys, smart badges, and remote controls. The scope of the task group is to define the physical layer (PHY) and the media access controller (MAC). A graphical representation of the areas of responsibility between the IEEE standard, ZigBee Alliance, and User is presented in Figure 3.9.

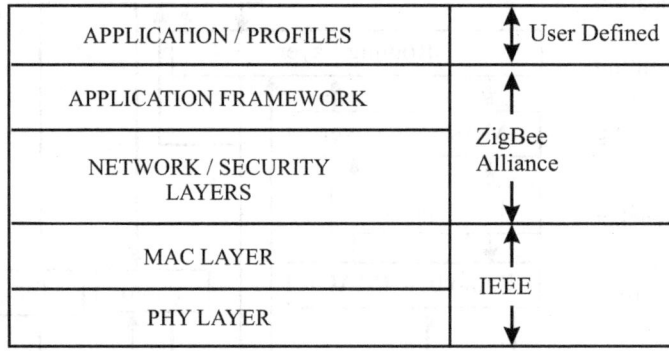

FIGURE 3.9 IEEE 802.15.4 Stack.

Since total system cost is a key factor for industrial and home wireless applications, a highly integrated single-chip approach is the preferred solution of semiconductor manufacturers developing IEEE 802.15.4 compliant transceivers. The IEEE standard at the PHY is the significant factor in determining the RF architecture and topology of ZigBee-enabled transceivers currently sampling. Generally, CMOS is the desired technology to integrate both analog circuitry and high gate count digital circuitry for lower cost with the challenge being RF performance.

For these optimized short-range wireless solutions, the other key elements above the Physical and MAC Layer are the Network/Security Layers for sensor and control integration. The ZigBee Alliance is in the process of defining the characteristics of these layers for star, mesh, and cluster tree topologies. The performance of these networks will complement the IEEE standard while meeting the requirements for low complexity and low power.

IEEE 802.15.4 Overview

The IEEE 802.15.4 standard defines two PHYs representing three license-free frequency bands that include sixteen channels at 2.4 GHz, ten channels at 902 to 928 MHz, and one channel at 868 to 870 MHz. The maximum data rates for each band are 250 kbps, 40 kbps, and 20 kbps respectively. The 2.4 GHz band operates worldwide, while the sub-1 GHz band operates in North America, Europe, and Australia/New Zealand, as in Table 3.1. The IEEE standard is intended to conform to established regulations in Europe, Japan, Canada, and the United States.

Table 3.1 Frequency Bands and Data Rates

PHY	Frequency Band	Channel Numbering	Spreading Parameters		Data Parameters		
			Chip Rate	Modulation	Bit Rate	Symbol Rate	Modulation
868/915 MHz	868-870 MHz	0	300 kchip/s	BPSK	20 kb/s	20 kbaud	BPSK
	902-928 MHz	1 to 10	600 kchip/s	BPSK	40kb/s	40 kbaud	BPSK
2.4 GHz	24-2.4835 GHz	Ilto26	20M chip/s	O-QPSK	250 kb/s	62.5 kbaud	16-ary Orthogonal

Both PHYs use the Direct Sequence Spread Spectrum (DSSS). The modulation type in the 2.4 GHz band is O-QPSK with a 32 PN-code length and an RF bandwidth of 2 MHz. In the sub-1 GHz bands, BPSK modulation is used with a 15 PN-code length and operates in an RF bandwidth of 600 kHz in Europe and 1200 kHz in North America.

RF Design Considerations

A representative sub-1 GHz transceiver is shown in Figure 3.10. The IC contains a 900 MHz physical layer (PHY) and portion of the media access controller (hardware-MAC). The remaining MAC functions (software-MAC) and the application layer are executed on an external microcontroller.

All PHY functions are integrated on the chip with minimal external components required for a complete radio. A low-cost crystal is used as a reference for the PLL and to clock the digital circuitry. To optimize energy consumption in sleep mode while still keeping an accurate time base, a Real Time Clock reference can be used.

ZMD44101

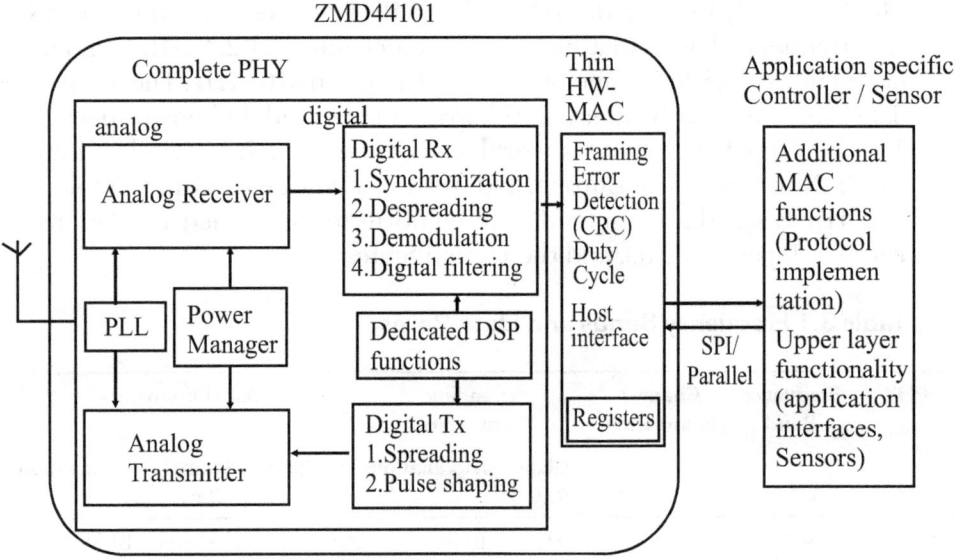

FIGURE 3.10 Sub-IGHZ transceiver block diagram.

The analog portion of the receiver converts the desired signal from RF to the digital baseband. Synchronization, dispreading, and demodulation are done in the digital portion of the receiver. The digital part of the transmitter does the spreading and baseband filtering, whereas the analog part of the transmitter does the modulation and conversion to RE The three main analog blocks — the direct-conversion receiver, direct-conversion transmitter, and fractional-N PLL—are discussed as follows.

The choice of the receiver architecture is mainly a compromise between performance, cost (considering both silicon area and external components), and power consumption. A direct-conversion receiver (DCR) architecture (or Zero-IF architecture) was selected as there is no image frequency and IF filtering required. Further advantages are that the channel select filters are low-pass filters, instead of band-pass filters, and the baseband frequency is the lowest possible. The DCR architecture provides the additional benefits of lower cost, complexity, and power consumption.

The transmitter architecture is also direct-conversion. Since BPSK modulation is used, only one baseband path is required. A differential architecture was used to minimize common mode noise. The output can be single-ended or differential. The single-ended output was selected for the advantages of lower cost, an on-chip TR switch, and the elimination of the requirement for an external balun.

Table 3.1 shows the channel allocation in the sub-1 GHz bands of the IEEE standard, which sets the required bandwidth and frequency resolution. This had major impact on the PLL topology. The goal was one PLL circuit for the 868/915 MHz bands using a fixed crystal frequency. To meet these requirements, a fractional-N PLL architecture was chosen. An additional benefit is that the software-controlled fractional-N PLL provides the adaptability to meet future worldwide spectrum expansion in the range of 860 to 930 MHz.

Zigbee Network Considerations

The features of the PHY include receiver energy detection, link quality indication, and clear channel assessment. Both contention-based and contention-free channel access methods are supported with a maximum packet size of 128 bytes, which includes a variable payload up to 104 bytes. Also employed are 64-bit IEEE and 16-bit short addressing, supporting over 65,000 nodes per network. The MAC provides network association and disassociation, has an optional superframe structure with beacons for time synchronization, and a guaranteed time slot (GTS) mechanism for high priority communications. The channel access method is carrier sense multiple access with collision avoidance (CSMA-CA).

ZigBee defines the network, security, and application framework profile layers for an IEEE 802.15.4-based system. ZigBee's network layer supports three networking topologies; star, mesh, and cluster tree, as shown in Figure 3.11. Star networks are common and provide for very long battery life operation. Mesh, or peer-to-peer, networks enable high levels of reliability and scalability by providing more than one path through the network. Cluster-tree networks utilize a hybrid star/mesh topology that combines the benefits of both for high levels of reliability and support for battery-powered nodes.

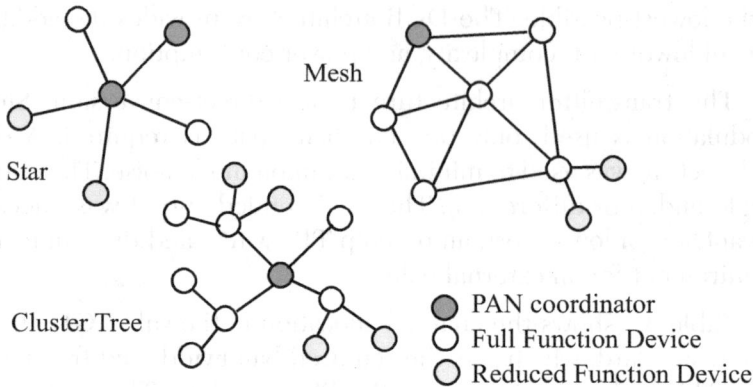

FIGURE 3.11 ZigBee network topologies.

To provide for low-cost implementation options, the ZigBee Physical Device type distinguishes the type of hardware based on the IEEE 802.15.4 definition of reduced function device (RFD) and full function device (FFD). An IEEE 802.15.4 network requires at least one FFD to act as a network coordinator. The description of each Physical Device type is found in Table 3.2.

Table 3.2 ZigBee Physical Device Types

Reduced Function Device	Full Function Device
Limited to star topology	Can function in any topology
Cannot become network coordinator	Capable of being network coordinator
Talks only to network coordinator (FFD)	Capable of being a coordinator
Simple implementation—min RAM and ROM	Can talk to any other device (FFD/RFD)
Generally battery powered	Generally line powered

An RFD is implemented with mimmum RAM and ROM resources and designed to be a simple send and/or receive node in a larger network. With a reduced stack size, less memory is required, and thus a less expensive IC. ZigBee RFDs are generally battery powered. RFDs can search for available networks, transfer data from its application as necessary, determine whether data is pending, request data from the network coordinator, and sleep for extended periods of time to reduce battery consumption. RFDs can only talk to an FFD, a device with sufficient system resources for network routing. The FFD can serve as a network coordinator, a link coordinator, or as just another communications device. Any FFD can talk to other

FFDs and RFDs. FFDs discover other FFDs and RFDs to establish communications, and are typically line powered.

The ZigBee Logical Device type distinguishes the Physical Device types (RFD or FFD) deployed in a specific ZigBee network. The Logical Device types are ZigBee Coordinators, ZigBee Routers, and ZigBee End Devices. The ZigBee Coordinator initializes a network, manages network nodes, and stores network node information. The ZigBee Router participates in the network by routing messages between paired nodes. The ZigBee End Device acts as a leaf node in the network and can be an RFD or FFD. ZigBee application device types distinguish the type of device from an end-user perspective as specified by the Application Profiles.

ZigBee's self-forming and self-healing mesh network architecture permits data and control messages to be passed from one node to another node via multiple paths. This feature extends the range of the network and improves data reliability. This peer-to-peer capability may be used to build large, geographically dispersed networks where smaller networks are linked together to form a 'cluster tree' network. ZigBee provides a security toolbox to ensure reliable and secure networks. Access control lists, packet freshness timers, and 128-bit encryption protect data transmission and ZigBee wireless networks.

Zigbee Applications

ZigBee networks consist of multiple traffic types with their own unique characteristics, including periodic data, intermittent data, and repetitive low latency data. The characteristics of each are as follows:

- Periodic data—usually defined by the application such as a wireless sensor or meter. Data typically is handled using a beaconing system whereby the sensor wakes up at a set time and checks for the beacon, exchanges data, and goes to sleep.

- Intermittent data—either an application or external stimulus such as a wireless light switch. Data can be handled in a beaconless system or disconnected. In a disconnected operation, the device will only attach to the network when communications is required, saving significant energy.

- Repetitive low latency data—uses time slot allocations such as a security system. These applications may use the guaranteed time slot (GTS) capability. GTS is a method of QoS that allows each device a specific duration of time as defined by the PAN coordinator in the superframe to do whatever it requires without contention or latency.

For example, an automatic meter reading application represents a periodic data traffic type with data from water or gas meters being transmitted to a line-powered electric meter and passed over a powerline to a central location. Using the beaconing feature of the IEEE standard, the respective RFD meter wakes up and listens for the beacon from the PAN coordinator, and if received, the RFD requests to join the network. The PAN coordinator accepts the request. Once connected, the device passes the meter information and goes to sleep. This capability provides for very low duty cycles and enables multi-year battery life. Intermittent traffic types, such as wireless light switches, connect to the network when needed to communicate (i.e., turn on a light). For repetitive low latency applications, a guaranteed time slot option provides for Quality of Service with a contention-free, dedicated time slot in each superframe that reduces contention and latency. Applications requiring timeliness and critical data pas sage may include medical alerts and security systems. In all applications, the smaller packet sizes of ZigBee devices results in higher effective throughput values compared to other standards.

ZigBee networks are primarily intended for low-duty cycle sensor networks (<1%). A new network node may be recognized and associated in about 30 ms. Waking up a sleeping node takes about 15 ms, as does accessing a channel and transmitting data. ZigBee applications benefit from the ability to quickly attach information, detach, and go to deep sleep, which results in low power consumption and extended battery life.

This combined the characteristics of the IEEE 802.15.4 standard with the maturing ZigBee specification in defining the wireless profiles for low-data rate monitoring and control applications. The capabilities of both will result in the availability of a technology tailored specifically for the low-power, low-cost, and low-complexity applications in industries and the home today and in the future. Zigbee is built upon the foundations provided by the IEEE 802.15.4 standard.

Full Function Device (FFD)

- Capable of being the PAN Coordinator
- Implements processing of "Association Request"
- Implements processing of "Orphan Notification"
- Implements processing of "Start Request"
- Implements processing of "Disassociation Notification"

Reduced Function Device (RFD)

- Can only associate and communicate with an FFD
- Reduced stack removes optional components

Zigbee Network Layer

The application layer consists of three parts: the Application Sublayer (APS), the Application Framework (AF), and the endpoints. The Application Sublayer interfaces the Zigbee application layer to the Zigbee networking layer and it provides a common set of data transport services to all the endpoints. There are also a couple of other services that the APS provides. Zigbee stack is shown in Figure 3.12 and Zigbee network is shown in Figure 3.13.

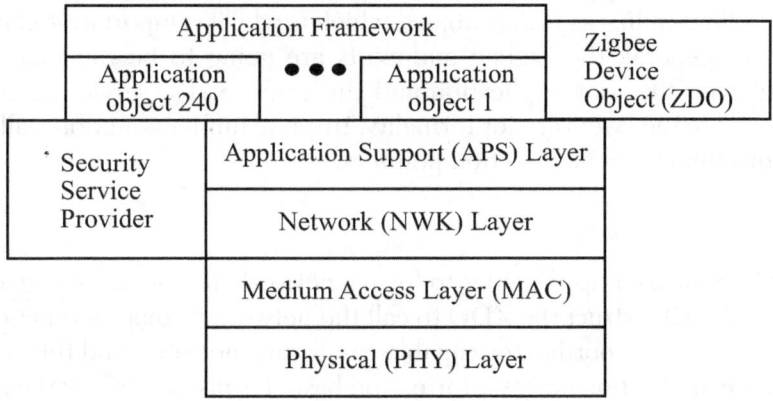

FIGURE 3.12 Zigbee stack.

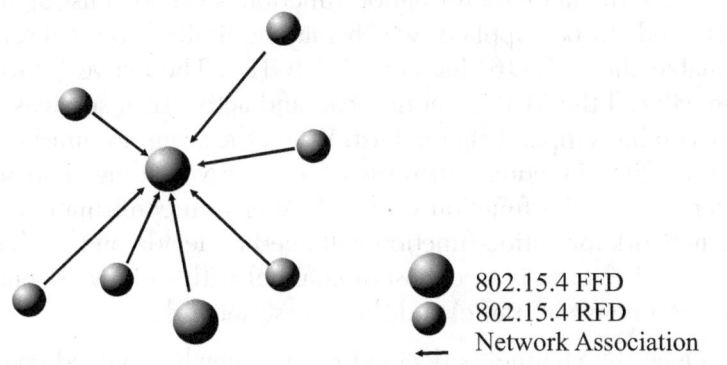

FIGURE 3.13 Zigbee network.

The Application Framework is a glorified multiplexer and container for all of the endpoints. All of the endpoints register themselves with the AF, and when a data frame comes into the application layer, the AF will check its destination endpoint and forward it there.

The endpoints are what most people associate with Zigbee. Each endpoint houses what's called an application object, which is basically a device profile with whatever extra functionality you decide to add. When the device is started, all the endpoints will register themselves with the application framework and provide descriptions of their device profile and their capabilities. Endpoint 0 is a special endpoint and always contains the Zigbee Device Object (ZDO). This object implements the Zigbee Device Profile, which has multiple functions, one of them being the network manager.

The user application can manage the network by making requests and handling callbacks to this object, which is why it's important to know about it. In general, the Zigbee endpoints are going to be the main interface between the user application and the stack. SAPs (service access points) between the layers are for formality. In a real implementation, call the functions directly or via function pointers.

Network Formation

When the user app decides to form a network instead of joining an existing one, it will instruct the ZDO to call the network formation function. Only a router that is coordinator-capable can form a network, and this is indicated in the application layer's information base. It's just a term for the app layer's configuration table.

When the network formation function is called, a list of allowed channels needs to be supplied, which may be limited to a subset of the total available channels (16 channels @ 2.4GHz). The network formation function will call the MAC's energy scan and active scan services and perform scans on the supplied channel list. When the scans are finished, the MAC's scan confirm function will return the energy readings and network scan descriptors to the function via the MAC's scan confirmation. From there, the network formation function will need to decide on the channel to join. The usual criteria is to choose a channel with the lowest energy reading (lowest amount of traffic) and the fewest networks.

Once the channel is decided on, the newly crowned coordinator will decide on a PAN ID and set the channel in the radio. The final step is for the NWK layer to call the MAC start service which configures the MAC layer. After that, confirmations go back all the way up to the user app. It is shown in Figure 3.14.

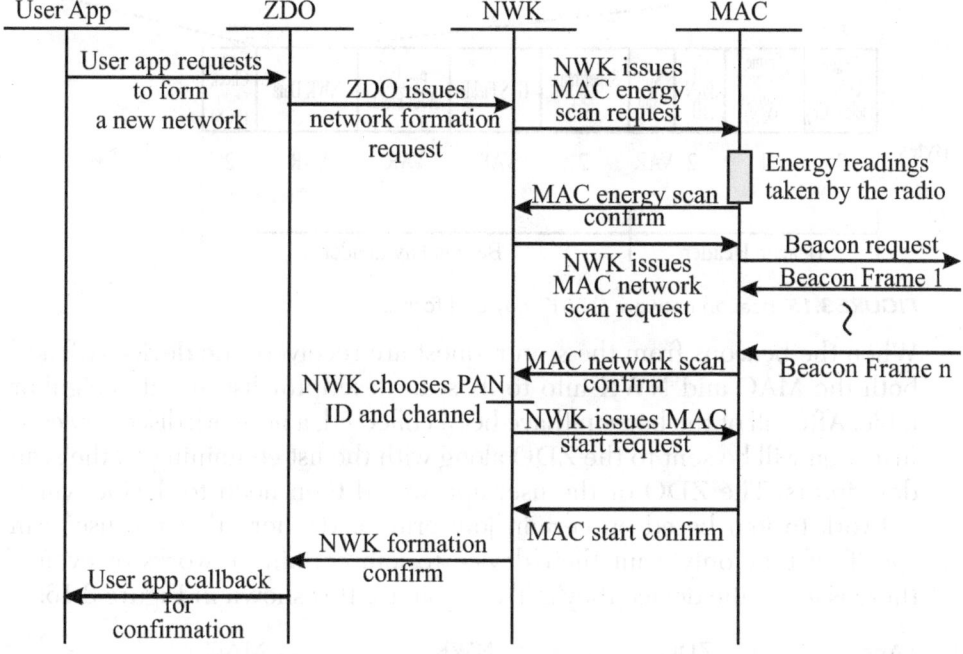

FIGURE 3.14 Network formation sequence diagram.

Network Discovery

As the name implies, the Zigbee network discovery service is used to discover the existing networks on the current channel. It's mostly just used when the device is started to find out if there are any suitable networks to join, although it can also be called at any time via the user app.

When a network discovery is requested by the ZDO (or user app), the discovery function will call the MAC's active scan service which, in turn, will broadcast a beacon request. When other devices see the beacon request, they will respond with an 802.15.4 beacon frame. The beacon frame contains MAC information about the responding device as well as a beacon payload for generic data. Within that payload, the responding device will include Zigbee network information such as the protocol ID and version, amount of routers and end devices allowed to join, the device profile that is being used, and other somewhat useful information. It is shown in figure 3.15.

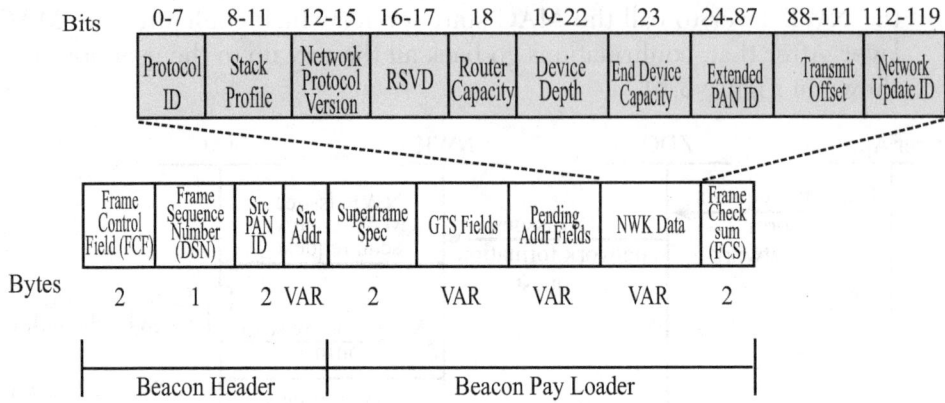

FIGURE 3.15 Beacon network (NWK) payload format.

When the beacons from the scan request are received, the device will add both the MAC and NWK info to its scan descriptor list and its neighbor table. After all of the beacons have been collected, a network discovery confirmation will be sent to the ZDO along with the list containing all the scan descriptors. The ZDO or the user app would then need to decide which network to join based on certain join critera. It's here that the user can specify if they only want their device to join certain networks or even if there is a specific device they'd like to join to. It is shown in Figure 3.16.

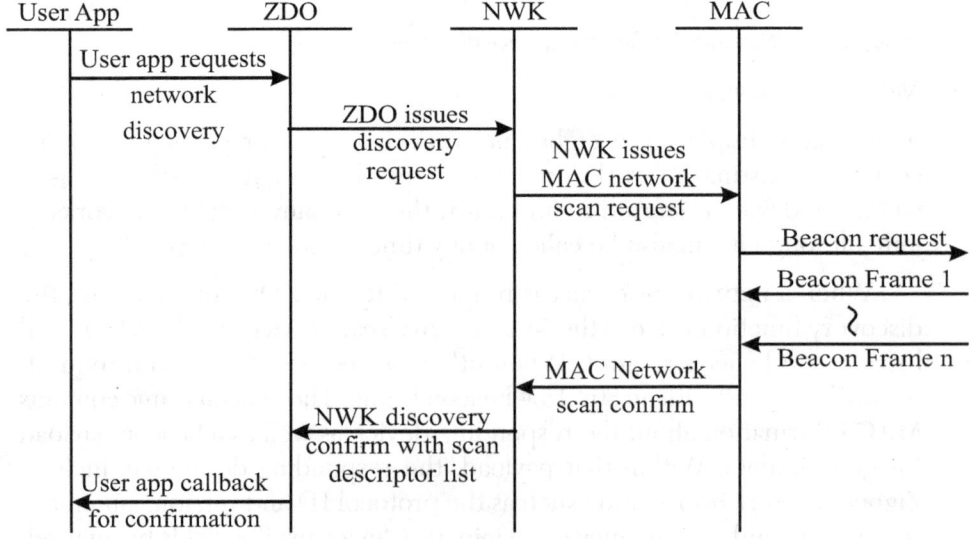

FIGURE 3.16 Network discovery sequence diagram.

Network Join

Joining a device or allowing a device to join is probably one of the most complicated processes in Zigbee. There are actually two sides to the networkjoin function: the child side which sends the request and the parent side which processes the request and sends the response.

Network join Child

The first part of the join process for the child is to do a network discovery. This is usually done when the device is first started and is not associated with any network as mentioned previously. Once the network discovery is finished and the potential parent has been decided on according to the join criteria, then it's time for the network join process to start.

When the potential parent has been chosen, a network join request is called by the ZDO. The network join request will call the MAC's association service and issue an association request to the potential parent. From there, the procedure follows the MAC's association sequence until the association response is received from the potential parent.

When this response is received, it will get passed up to the network layer via the MAC's association response. If the join was successful, the device will update its NWK and MAC information tables to include the new network address, PAN ID, and also update the neighbor table to specify its parent. Once the administrative work is taken care of, the network join confirmation is sent up to the ZDO where it can inform the application about the join status. If the join status was unsuccessful, then the ZDO/user app will choose another potential parent from the neighbor table and retry the join procedure until it eventually joins a network or runs out of potential parents.

One of the last things that occurs after a successful join is that the device will broadcast a device announcement informing everyone on the network that it has joined the network as well as it's 16-bit network address and 64-bit IEEE address. This is important because if the device was previously joined to the network with a different network address, the other devices will be able to find out from its IEEE address and can clear all references to the old network address. Also, the address info will be added to everyone's address map, which tracks all the devices on the network. It is explained in Figure 3.17.

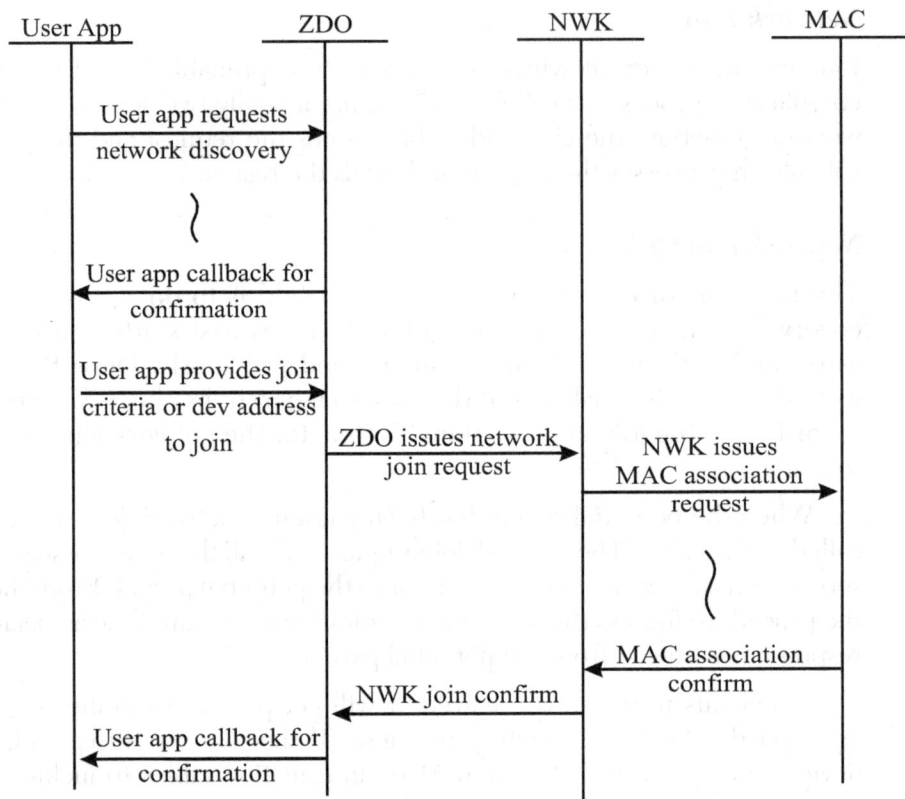

FIGURE 3.17 Network join sequence diagram—child side.

Network Join Parent

The parent side of the join process is slightly easier. When a MAC association request arrives at the potential parent, it sends an indication to the network layer that a device is trying to join. The potential parent will then search its neighbor table to see if the 64-bit IEEE address already exists. If it does, then that means that the device was already previously joined and the parent will just issue the same network address to it. If not, and the parent is allowing devices to join it, then it will simply add the device to its neighbor table specifying that it's a child device and generate a new network address for it. This all gets packaged up and sent out as a MAC association response. Again, the rest goes according to the MAC's association service. It is shown in Figure 3.18.

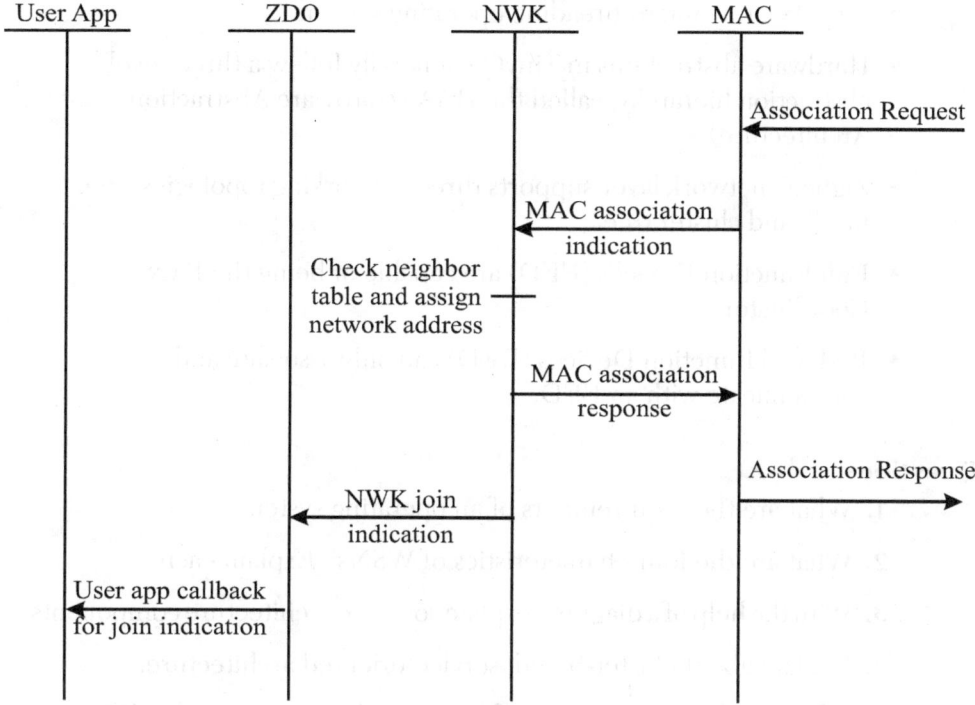

FIGURE 3.18 Network join sequence diagram—parent side.

Summary

- Wireless sensor networks are aggregates of numerous small sensor nodes.

- The IEEE 802.15.4 / Zigbee standard is one of the most promising candidates for designing WSNs, which need to be self-organized and self-healing; that is, nodes automatically establish and maintain connectivity among themselves.

- WSNs must be self organizing, perform cooperative processing, and be energy optimized and modular.

- Middleware refers to the software layer between the operating system and sensor application on the one hand and the distributed application which interacts over the network on the other hand.

- A cluster is a set of adjacent sensors which are grouped together and interface with the rest of the network through a gateway or cluster head.

- TinyOS a tiny microthreading operating system.

- Hardware abstractions in TinyOS generally follow a three-level abstraction hierarchy, called the HAA (Hardware Abstraction Architecture).

- Zigbee's network layer supports three networking topologies: star, mesh, and cluster tree.

- Full Function Devices (FFD) are capable of being the PAN Coordinator.

- Reduced Function Devices (RFD) can only associate and communicate with an FFD.

Questions

1. What are the requirements of an operating system?

2. What are the four characteristics of WSNs? Explain each.

3. With the help of a diagram, explain software architecture components.

4. Explain about cluster-based, service-oriented architecture.

5. What are the sensor node software development processes?

6. List the characteristics of TinyOS.

7. Explain the TinyOS abstractions.

8. Give frequency bands and data rate of IEEE 802.15.4.

9. What are design factors to consider in RF design?

10. Write about Zigbee network considerations.

11. Compare RFD and FED.

12. Explain Zigbee stacks.

13. In detail write about the Zigbee network layer.

14. Give the network beacon payload format.

Further Reading

1. *Wireless Sensor Networks: Technology, Protocols, and Applications* by Kazem Sohraby, Daniel Minoli, and Taieb Znati

2. *Wireless Sensor Networks: Architecture and Applications* by Dr. Anis Koubaa

3. *Wireless Sensor Network Designs* by Anna Hac

References

1. *http://anp.tu-sofia.bg/djiev/Networks_Wireless.htm*

2. *http://www.freaklabs.org/index.php/blog/zigbee/*

CHAPTER 4

WIRELESS BODY SENSOR NETWORKS

This chapter discusses the wireless body sensor network for medical field application.

4.1 Introduction to Wireless Body Sensor Networks

Improving the quality of healthcare and the prospects of "aging in place" using wireless sensor technology requires solving difficult problems in scale, energy management, data access, security, and privacy. An aging baby boom generation is stressing the healthcare system, causing hospitals and other medical caregivers to look for ways to reduce costs while maintaining quality of care. It is economically and socially beneficial to reduce the burden of disease treatment by enhancing prevention and early detection. This requires a long-term shift from a centralized, expert-driven, crisis-care model to one that permeates personal living spaces and involves informal caregivers, such as family, friends, and community. Systems for enhancing medical diagnosis and information technology often focus on the clinical environment, and depend on the extensive infrastructure present in traditional healthcare settings. The expense of high fidelity sensors limits the number that are available for outpatient deployment, and some require specialized training to operate.

For healthcare applications, wireless sensor networks can be deployed inexpensively in existing structures without IT infrastructure. Data are collected automatically, enabling daily care and longitudinal medical monitoring and diagnosis. The wireless devices can integrate with a wide variety of environmental and medical sensors.

While addressing some of the needs of distributed healthcare, WSNs also present their own challenges to being practical, robust platforms for pervasive deployment. Privacy and security of collected medical data may be jeopardized by careless use of a wireless medium. Without smart power management, battery-powered sensors have short lifetimes of a few days or require continual maintenance.

A smart medical home is a system of room labs outfitted with infrared sensors, computers, biosensors, and video cameras. The goal of the system is to develop an integrated personal health system that collects data 24 hours a day and presents it to health professionals. The sensors such as a portable 2-lead ECG, pulse oximeter, wearable Pluto mote with built-in accelerometer, and a module with an accelerometer, a gyroscope, and an electromyogram sensor for stroke patient monitoring are included in the medical node.

One of the most interesting areas for the implementation of the WSN is in the medical field, because there are different challenges which are associated with monitoring the human body. The human body responds to its environment as well as external conditions in its life every day. Thus, in order to monitor all these features, one must apply the monitoring and sensor networks in order to really diagnose what gets the sensors on the body surface, depending on the frequency of monitoring. The name associated with this implementation is Body Sensor Networks (BSN). This BSN technology may offer the possibility of developing a detailed diagnosis of the patient, because the network would be able to monitor all vital signs and synthesize all relevant information for more effective patient care.

4.2 Architecture of Body Sensor Networks

A key requirement for healthcare systems is the ability to operate continuously over long time periods and still integrate new technologies as they become available. Mobile Body Networks are wireless sensor devices worn by a resident which provide activity classification or physiological sensing, such as an ECG, pulse oximeter, or accelerometers. The network is tailored to the patient's own medical needs, and can provide notifications (for example, alerts to take medicine) using network wearable interface with a color LCD. Figure 4.1 shows wearable body networks, emplaced wireless sensors, user interfaces, and back end processing elements for body sensor network architecture.

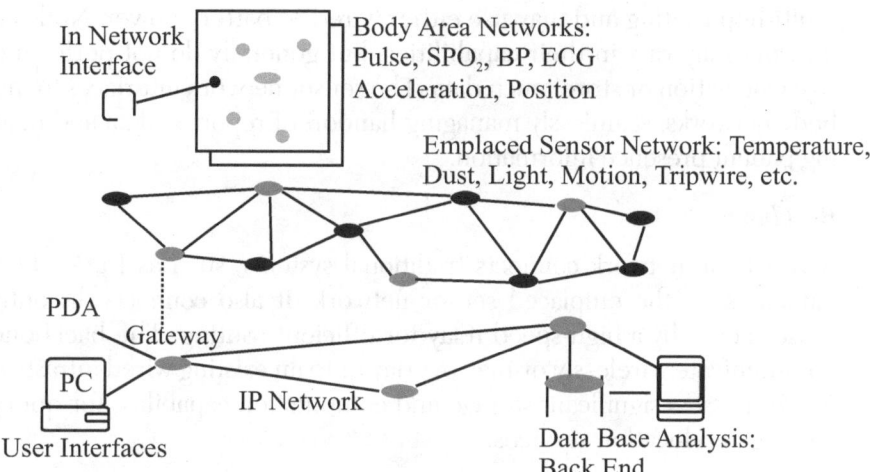

FIGURE 4.1 Body sensor network architecture.

Body Network and Subsystems

This network comprises tiny portable devices equipped with a variety of sensors (such as heart rate, heart rhythm, temperature, oximeter, accelerometer), and performs biophysical monitoring, patient identification, location detection, and other desired tasks. These devices are small enough to be worn comfortably for a long time. Their energy consumption should also be optimized so that the battery is not required to be changed regularly. They may use "kinetic" recharging. Actuators notify the wearer of important messages from an external entity. For example, an actuator can remind an early Alzheimer's patient to check the oven because sensors detect an abnormally high temperature. Or, a tone may indicate that it is time to take medication. The sensors and actuators in the body network are able to communicate among themselves. A node in the body network is designated as the gateway to the emplaced sensor network. Due to size and energy constraints, nodes in this network have little processing and storage capabilities.

Emplaced Sensor Network

This network includes sensor devices deployed in the environment (rooms, hallways, furniture) to support sensing and monitoring, including temperature, humidity, motion, acoustics, camera, and so forth. It also provides a spatial context for data association and analysis. All devices are connected to a more resourceful backbone. Sensors communicate wirelessly using multi-hop routing and may use either wired or battery power. Nodes in this network may vary in their capabilities, but generally do not perform extensive calculation or store much data. The sensor network interfaces to multiple body networks, seamlessly managing handoff of reported data and maintaining patient presence information.

Backbone

A backbone network connects traditional systems, such as PDAs, PCs, and databases, to the emplaced sensor network. It also connects discontinuous sensor nodes by a high-speed relay for efficient routing. The backbone may communicate wirelessly or may overlay onto an existing wired infrastructure. Nodes possess significant storage and computation capability, for query processing and location services.

Back End Databases

One or more nodes connected to the backbone are dedicated databases for long-term archiving and data mining. If unavailable, nodes on the backbone may serve as network databases.

Human Interfaces

Patients and caregivers interface with the network using PDAs, PCs, or wearable devices. These are used for data management, querying, object location, memory aids, and configuration, depending on who is accessing the system and for what purpose. Limited interactions are supported with the on-body sensors and control aids. These may provide memory aids, alerts, and an emergency commumcation channel. PDAs and PCs provide richer interfaces to real time and historical data. Caregivers use these to specify medical sensing tasks and to view important data.

Body networks contain a designated gateway device that mediates interaction with the surrounding WSN. This modularizes the system's interaction with the body network to ease its integration. Data are streamed directly or multi-hop through the emplaced network to the gateways for storage, analysis, or distribution to user interfaces. Emplaced Sensors are deployed in living spaces to sense environmental quality, such as temperature, dust, and light, or resident activities. Motion and tripwire sensors, in particular, provide a spatial context for activities and enable location tracking.

Due to their low cost, small form factor, and limited power budget, the devices answer queries for local data and perform limited processing and caching. Though some deployment environments may enable the use of mains power, we do not require it so as to support ad hoc retrofitting of existing structures. Figure 4.2(a) shows the lightweight stack resident on sensor devices.

Body sensor networks need to support dynamically adding new devices to the network, which register their capabilities and are initialized. This flexibility allows the system to change over time as sensors are developed or new pathologies require monitoring. Gateway software stacks serve as a commumcation backbone and application level gateway between the wireless sensor and IP networks. Owing to their greater resources, these devices perform major aspects of system operation related to dynamic privacy, power management, query management, and security. The gateway software stack is shown in Figure 4.2(b).

Back end programs perform online analysis of sensor data, feeding back behavior profiles to aid context-aware power management and privacy. A database provides long-term storage of system configuration, user information, privacy policies, and audit records.

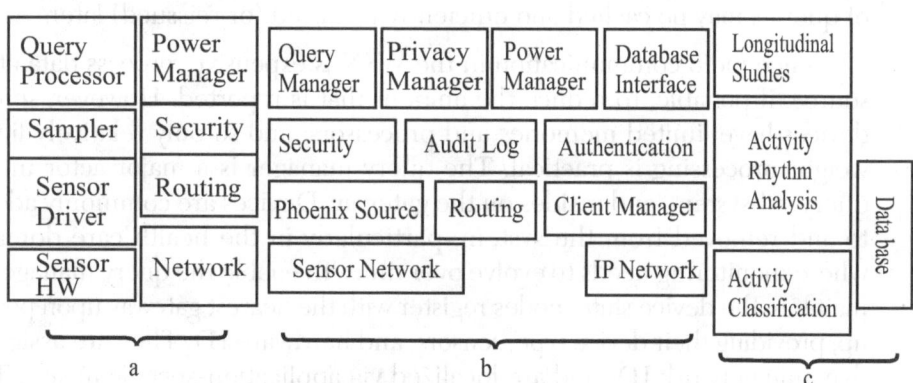

FIGURE 4.2 a) Sensor device software stack   b) Gateway software stack
                c) Back end analysis and storage.

Figure 4.2 shows the body sensor network software architecture for sensor devices, gateways, and back end servers.

Activity rhythm analysis processes sensor data stored in the database and learns behavior patterns of residents. These are used to detect deviations from personal norms that may signal a short- or long-term decline in resident health. The back end is extensible to new analyses using a modular framework, wherein programs consume input sensor streams, filter and process them, and produce output streams in the database for other modules to use. These are composed hierarchically from low-level sensor streams to high-level inference of symptoms and diseases. User interfaces allow doctors, nurses, residents, family, and others to query sensor data, subject to enforced privacy policies. A patient-tracking GUI for a nurse's station, and a query issuer for a PDA that graphs sensor data in real time. These programs are not trusted components; they must connect through the gateway and do not have direct access to the database. This makes it easier to develop and deploy new interfaces customized to the application's needs.

A query system satisfies the requirements of the application domain, and is reconfigurable in network sensing and processing, dynamic query origination by embedded devices, and high level abstractions for expressing queries. The back end system, user interfaces, and embedded devices all issue queries using a common network protocol, in which queries are uniquely identified by <source ID, query ID> tuples. Originators may request a snapshot of the current value or a periodic stream of a sensing modality. To reduce repetitive query parsing overhead on resource constrained motes, both types of queries may be cached and efficiently restarted (or reissued) later.

Since radio communication in the WSN is expensive, process data at its source, if possible, to reduce the amount that is reported. However, sensor devices have limited memories and processors, and so only relatively lightweight processing is practical. The query manager is a major actor in the query subsystem, and resides on the gateway. Devices are commonly added to and removed from the system, particularly in the health care domain, where monitoring needs to evolve over time. To enable the query manager to maintain the device state, nodes register with the nearest gateway upon power up, providing their device type, sensors, and hardware ID. They are assigned dynamic network IDs and are localized via application-specific means. The query manager issues background queries to devices as they are added to the network to satisfy the system's core management and tracking functionalities. Examples of background queries in the body sensor network are:

- All devices sample and report their battery supply voltage every four hours, but only if it is below 2.8V (indicating imminent failure);

- Motion, tripwire, and contact switch sensors report activations on demand, but no more often than every 100ms to debounce or dampen spurious bursts;

- Pulse oximetry devices, which are intermittently switched on, collect heart rate and SpO_2 samples every 250ms, but report them every 750ms, each an average of three samples until the device is switched off; and,

- ECG sensors immediately begin reporting a stream of raw samples every 20ms, using full buffering to reduce network load and energy usage.

The query manager is the main point of access for user interfaces, translating between higher level query abstractions. Connected users receive a list of active devices that is updated in real time as registrations are received. A request for sensor information about person P must be mapped to a device (or group of devices) D for execution. Some have static associations, such as a wearable device owned or assigned to a user. Likewise for locations L in which fixed sensor nodes are placed. But since networks for assisted living are more human oriented and heterogeneous than most other types of WSNs, many sensor types require dynamic binding based on a person's context (location, activity, etc).

Motes that are part of body networks necessarily use batteries (or scavenge energy from motion). Consequently, energy efficiency is an important design issue and an application demands some particular requirements on power management. First, sensors are used to detect and collect information on residents, so they should adapt their operational states according to changes in the resident's behavior. Second, power management should provide openness to system administrators, who should be able to set policies unique for particular applications. Third, individual sensing modalities, as well as radio components, should be controllable. For example, the system may want to set a high rate for temperature sensing, a low rate for light sensing, turn off other sensor types, and set a duty cycle mode for the radio. Last, in a heterogeneous network with diverse sensor nodes, such as ECG, motion, and weight sensors, power management should adaptively control each according to its own characteristics and context, including location.

For battery powered devices, two types of power management operations are designed, that is, those based on administrator directives and those which rely on context awareness. First, administrators can directly control each sensor available on a mote. Sensors can be turned on/off or their rates set for each sensor type, and the radio is similarly controlled. Also, the effective period of each command can be set. A typical command may be, "Mote 1 turns off the light sensor, but senses the temperature every 2 seconds for the following 2 hours." Initially, administrators define some context policies for power management, such as "when the resident is sleeping or outside the apartment, turn off all sensors in the living room and reduce the temperature sensing rate to hourly in all rooms." This context-aware operation provides more efficient power management that adapts to the resident's behavior. Another advantage of context-aware operation is that the power subsystem is open for administrators to define their own context policies, according to application demands.

Overall Wireless Sensor Systems for Patient Monitoring

Intelligent wireless patient monitoring system frameworks include real-time sensing of the patient's vital parameters using the motes, and wireless transmission of such critical information over radio frequencies to the base station. Subsequent data processing on a PC will allow detection of certain medical emergencies, and automatic alerting of medical staff. The overall framework is shown in Figure 4.3.

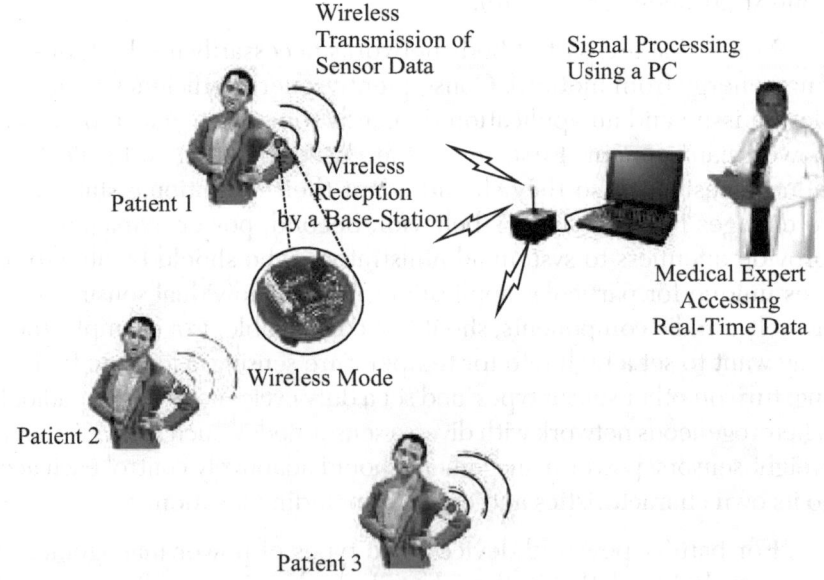

FIGURE 4.3 Overall framework for patient monitoring.

4.3 Bio Signal Monitoring Using Wireless Sensor Networks

A medical node usually consists of four subsystems, as shown in Figure 4.4.

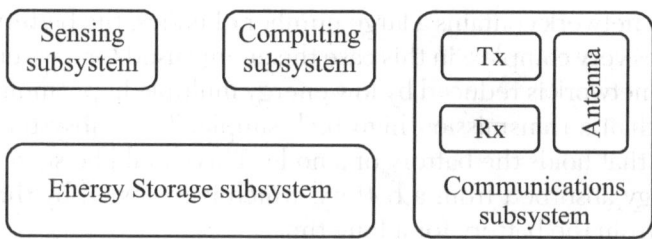

FIGURE 4.4 Wireless body sensor node.

Computing Subsystem

This is a microcontroller unit, which is responsible for the control of sensors and the implementation of commumcation protocols. The microcontroller is usually operated under different operating modes for power management purposes.

Communications Subsystem

Issues relating to standard protocols, which depend on application variables, are obtained as the operating frequency and types of standards to be used (Zigbee, Bluetooth, among others). This subsystem consists of a short range radio which is used to communicate with other neighboring nodes and outside the network. The radio can operate in the modes of transmitter, receiver, standby, and sleep mode.

Sensing Subsystem

This is a group of sensors or actuators and links nodes outside the network. The power consumption can be determined using low energy components.

Energy Storage Subsystem

One of the most important features in a wireless sensor network is related to energy efficiency. Hardware developers in a WSN must provide various techniques to reduce energy consumption. Due to this factor, power consumption of the network must be controlled by two modules:

1. power module (which computes the energy consumption of different components)

2. battery module (which uses this information to compute the discharge of the battery)

When a network contains a large number of nodes, the battery replacement becomes very complex; in this case the energy used for a wireless communications network is reduced by low energy multiple hops (multi-hop routing) rather than a transmission high tech simple. This subsystem consists of a battery that holds the battery of a node. This should be seen as the amount of energy absorbed from a battery, which is reviewed by the high current drawn from the battery for a long time.

WSN Classification and Operation Mode

A wireless sensor network can be classified depending on the application and its programming, as well as its functionality in the field sensing, and so forth. WSNs are classified as follows:

Homogeneous refers to when all nodes have the same hardware; otherwise, it is called heterogeneous.

Autonomous refers to when all nodes are able to perform self-configuration tasks without the intervention of a human.

Hierarchical refers to when nodes are grouped for the purpose of communicating or are otherwise shut down; in this classification it is common to have a base station that works as a bridge to external entities.

Static refers to when nodes are static; otherwise, they are dynamic.

Flexibility. The wireless environment can be totally changed due to interference from other microwaves or forms of materials in the environment, among other conditions, which is why most of the nodes can fail at any time; therefore, the network should seek new paths in real time, must reconfigure the network, and in turn re-calibrate the initial parameters.

Efficiency. This item is very important because the network to be implemented must be efficient to work in real time, and must be reliable and robust to interference from the same nodes or signals from other devices. The network should be tightly integrated with the environment where it will work.

Scalability. When a wireless sensor network is dynamic, due to its topology or application adding nodes is an important factor for the smooth operation of data storage.

WSN Functional Levels

WSN networks are classified into three functional levels: the level of control, the level of the communications network, and the field level, as shown in Figure 4.5.

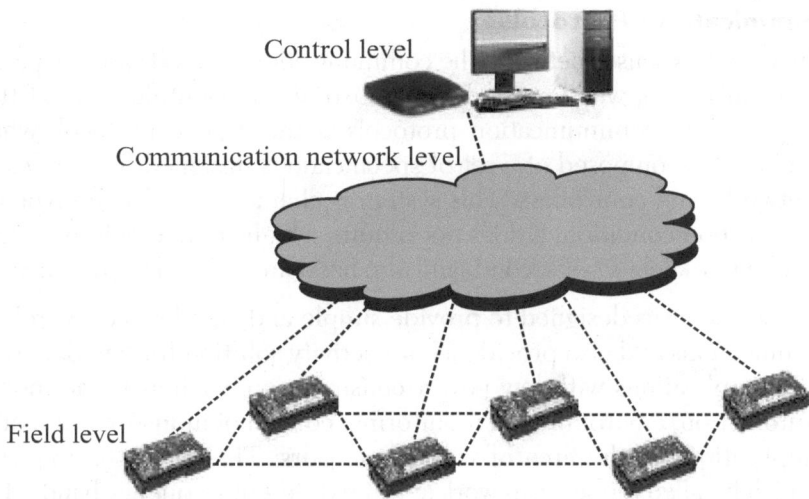

FIGURE 4.5 Architecture of a WSN for medical uses.

The field level consists of a set of sensors and actuators that interact directly with the environment. The sensors are responsible for obtaining types of data, for example, thermal, optical, acoustic, seismic, and so on. The actuators on the other hand receive orders which are the result of processing the information gathered by the sensors so it can be run later. The communication network establish a communication link between the field level and the level of control. Nodes that are part of a WSN communications subsystem are grouped into three categories: Endpoints, Routers, and Gateways. Finally, the level of control consists of one or more control and/or monitoring centers, which uses information collected by the sensors to set tasks that require the performance of the actuators. This control is done through special software to manage network topologies and behavior of the network in diverse environments. One way to consider wireless sensor networks is to organize hierarchically the nodes of the upper level, which are the most complex, and know its location through a transmission technique. The challenges in hierarchically classifying a sensor network are finding relevant quantities to monitor and collect data, access and evaluate information, and so on.

The information needed for intelligent environments or those whose variables are complex to obtain is provided by a distributed network of wireless sensors which are responsible for detecting and for the early stages of the processing hierarchy.

Communications Protocols

In a wireless sensor network, the communication method varies depending on the application, whether medical, industrial, or scientific. One of the most widely used commumcation protocols is the Zigbee protocol, which is a technology composed of a set of specifications designed for wireless sensor networks and controllers. This system is characterized by the type of communication condition; it does not require a high volume of information (just over a few kilobits per second) and also has a limited walking distance.

Zigbee was designed to provide simple and easy low-cost wireless communications and also provide a connectivity solution for low-data transmission applications with low power consumption, such as home monitoring, automation, environmental monitoring, control of industries, and emerging applications in the area of wireless sensors. The IEEE802.15.4 standard, which is called Zigbee, can work at three different frequency bands. This protocol is divided into layers according to the OSI model, where each layer has a specific function depending on the application of the network. The physical layer and the medium access control (MAC) are standardized by the IEEE 802.15 (WPAN), which is a working group under the name of 802.15.4; the higher layers are specified by the Zigbee Alliance.

Some characteristics of the layers are given as follows:

Physical Layer ZigBee / IEEE 802.15.4

The IEEE 802.15.4 physical layer supports unlicensed industrial, scientific, and medical radio frequency bands including 868 MHz, 915 MHz, and 2.4 GHz.

MAC Layer ZigBee / IEEE 802.15.4

At the MAC layer, there are two options to access the medium: beacon-based (based on orientation) and non beacon (based on non-guidance). In the non-oriented option, there is no time for synchronization between Zigbee devices. The devices can access the channel using (CSMA/CA).

Protocol to the network layer /IEEE 802.15.4

Zigbee has multi-hop routing and capabilities designed as an integral part of the system. This function is implemented within the network layer.

The performance of a wireless sensor network is measured depending on the ability to manage energy consumption of all nodes and also the effectiveness in real time transmission of data from the time of sensing to the display of such signs. Depending on the type of environment and resources in a network of wireless sensors, one can define multiple architectures; among the best known are the star, mesh, and cluster-tree networks. The nodes have no knowledge of the topology of the network, which must be "discovered."

A star topology network is characterized by a base station, which can send and receive a message to a number of router nodes. The advantage of this type of network for a WSN is the ease and ability to maintain energy consumption of a router node to a very low level. The disadvantage of this type of topology is the coordinator node (or base station), as it must be within transmission range of all nodes.

Mesh network topology is characterized by allowing any node in the network to transmit to any other node on the network that is within transmission range. This type of topology has an advantage, which is the redundancy and scalability compared to a situation of failure. If the router node gets out of service, other nodes can communicate with each other without depending on the unusable node. The disadvantage of this type of network is that power consumption for nodes that implement multi-hop communication, which generally results in the life of the battery consumption, is too short.

Finally, a cluster-tree network (union of a star and mesh topology, Figure 4.6), is one network that provides versatility to a communications network, while it maintains the ability to have low-power consumption of wireless sensor nodes. This feature allows the power consumption of the entire network to remain.

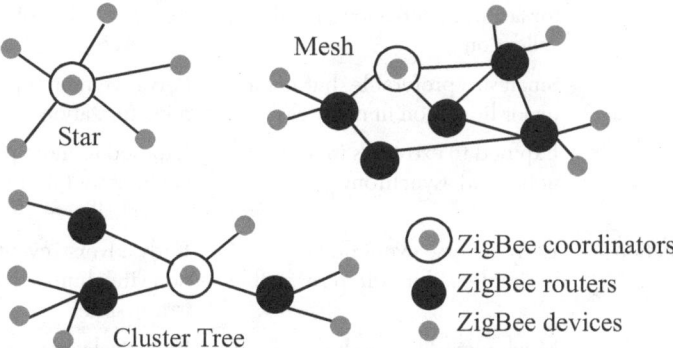

FIGURE 4.6 Network topology.

The position of the sensor nodes in a given area is not predetermined in some situations; this means that the protocols and algorithms used must be capable of self organization (in the case of a changing field). Some designs have protocols for specific design features for main energy saving and management of the interference signal caused by microwaves.

4.4 Differences between Wide Scale WSNs and BSNs

Practically the differences between the BSN and the WSN are very few, but it is very important to note that it is these small differences that allow BSNs to face the challenges posed in the medical field. Table 4.1 presents a summary of the differences between WSNs and BSNs.

Table 4.1 Different Challenges Faced by WSNs and BSNs

Challenges	WSN	BSN
Scale	As large as the environment being monitored (meters / kilometers)	As large as human body parts (millimeters /centimeters)
Node Number	Greater number of nodes required for accurate, wide area coverage	Fewer, more accurate sensor nodes required (limited by space)
Node Function	Multiple sensors, each perform dedicated tasks	Single sensors, each perform multiple tasks
Node Accuracy	Large node number compensates for accuracy and allows result validation	Limited node number with each required to be robust and accurate
Node Size	Small size preferable, but not a major limitation in many cases	Pervasive monitoring and need for miniaturization
Dynamics	Exposed to extremes in weather, noise, and asynchrony	Exposed to more predictable environment, but motion artifacts are a challenge
Event Detection	Early adverse event detection desirable; failure often reversible	Early adverse event detection vital; human tissue failure irreversible
Variability	Much more likely to have a fixed or static structure	Biological variation and complexity means a more variable structure

Challenges	WSN	BSN
Data Protection	Lower level wireless data transfer security required	High-level wireless data transfer security required to protect patient information
Power Supply	Accessible and likely to be changed more easily and frequently	Inaccessible and difficult to replace in implantable setting
Power Demand	Likely to be greater as power is more easily supplied	Likely to be lower as energy is more difficult to supply
Energy Scavenging	Solar and wind power are most likely candidates	Motion (vibration) and thermal (body heat) most likely candidates
Access	Sensors more easily replaceable or even disposable	Implantable sensor replacement difficult and requires biodegradability
Biocompatibility	Not a consideration in most applications	A must for implantable and some external sensors. Likely to increase cost
Context Awareness	Not so important with static sensors where environments are well defined	Very important because body physiology is very sensitive to context change
Wireless Technology	Bluetooth, Zigbee, GPRS, wireless LAN, and RF already offer solutions	Low-power wireless required, with signal detection more challenging
Data Transfer	Loss of data during wireless transfer is likely to be compensated by number of sensors used	Loss of data more significant, and may require additional measures to ensure QoS and Real-time data interrogation capabilities

Topology of a BSN

The application design of a BSN is based regularly in the star topology; this topology has the main advantage of optimizing the energy consumption of the network due to internal nodes called "slaves." Only the coordinator will transmit information received by the sensors, but as a great disadvantage, it has the high possibility of network failure due to the fall of the coordinator node.

4.5 Methodology for Development of Biomedical Signals Acquisition and Monitoring Using WSNs

A three-phase methodology for the development of applications of biomedical signals acquisition is shown in Figure 4.7. The first phase is the acquisition of biomedical signals, whose main objective is to establish a set of features for the proper selection of sensors that will accurately capture the required signal, and at the same time, allow the correct transduction of signals sent. The second stage concerns the correct choice of communication protocol to use and additional features to the network settings such as topology. Finally, one must determine the relevant elements to design the platform for visualization and monitoring of the sensed signals.

FIGURE 4.7 Methodology for development of biomedical signals acquisition and monitoring using WSNs.

Signal Acquisition

The monitoring of biomedical signals requires mechanisms to strengthen, substantiate, and legitimize the information captured by sensors. It should be noted that for the acquisition of biomedical signals, one must meet certain characteristics that do not interfere or alter the information gained. One should take into account sensor components that are responsible for trapping and generating changes in the captured signals.

The concept of biomedical signals focuses on the acquisition of data of common phenomena of the human body, which can reach diagnoses and predict diseases in the short and medium term. A biomedical signal becomes more complex and useful when it captures a common signal. This allows us to argue the importance of establishing and using elements that provide as much information for the analysis of the signal. To define and translate these signals, a set of parameters requires special handling. Because of the complexity and accuracy of bio medical data, signals should have low error rates. The medical sensors should have the ability to capture slight variations in depth to obtain the behavior of the human body.

To acquire biomedical surface signals, such as the humidity or temperature, it should be noted that the structure has characteristics that do not alter sensor data collected by the sensor. It may be the case that if the limit or standard level moisture or temperature are not met, it may yield inaccurate data or oxidation of the sensor to a more advanced level. However, the environments are not extreme in relation to an industrial environment where sensors may be exposed to hostile areas. Only the following types of sensors and their respective form of measurement are known, as seen in Figure 4.8.

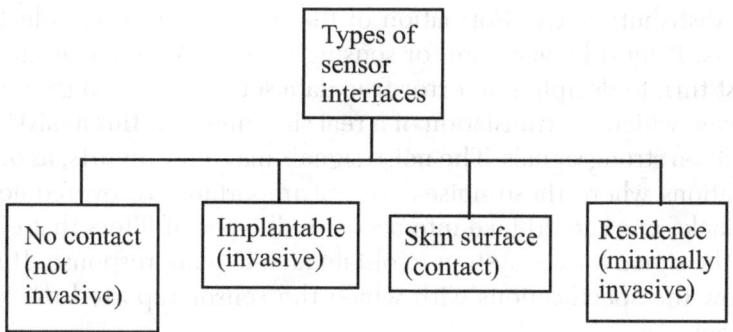

FIGURE 4.8 Types of sensor interfaces.

The sensors can be used in diagnosis of medical diseases or for therapeutic purposes. The sensors respond positively to the demands of an analysis of diagnosis of medical diseases. It must also have high accuracy. In the case of a touch sensor or implantable, it alters the body (negatively affecting the functioning of the body by the presence of an external agent as, in this case, a sensor).

If the sensor must be contact or implantable, this is closely affected by the presence of high humidity or temperature. The design and implementation of protective sensor packages are chosen to protect the sensor from the presence of moisture or temperature at the points where the sensor can be affected, leaving only the part where the sensor makes the sampling. This will protect the information gathered as well as prevent possible damage to the body when placing or deploying foreign agents in the body.

There are some kind of sensors that have direct contact with the body. There may be complications with the replacement of these components. Before deploying sensors, there should be a prior investigation and documentation on the reliability and accuracy of the sensor. It is very complex to make changes or sensor calibrations in real time or rely on technology to which it is connected to maintain proper operation. All this must be properly fulfilling predefined maximum quality standards, taking into account that it is not stopping the functioning of the body.

Processing & Transmission

For optimum performance of a wireless sensor network, it must consider certain variables or characteristics such as: (i) Designing the network topology, (ii) Sensing the environment, (iii) Energy consumption and distribution, (iv) Formation of the network, and (v) Selection of elements. It must be accurate for sensing stability. When analyzing signals, it must turn to decipher an error-free data set. This should give us a straight answer, which is a translation of a real situation. For this analysis, one must count on strong signals. The noise signals may alter reports, as one may find situations where these noises are not important. To overcome this noise obstacle, one should take into account all types of filters that can regenerate the signal for the system to obtain an adequate response. It should also follow the specifications with which the sensor reported the state of the system.

The functionality of a wireless sensor network occurs in large part on the correct and accurate operation of the nodes that comprise it. For the acquisition of signals in a given environment using specific sensors, these sensors are as was seen in the first objective, depending on the application and the environment in which one wants sensing. Based on the basic principles for designing a system for acquiring and processing biomedical signals, the text provides six phases for the data acquisition phase and later emphasizes the hardware design. Figure 4.9 is proposed as follows:

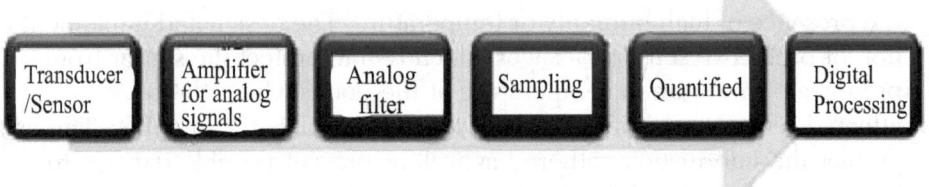

FIGURE 4.9 General block diagram of a procedure analogue to digital.

The function of a node is to sense, process, and commumcate data from the signal for a more detailed study as the network administrator requires. Depending on the topology of the network, each node has a specific function; for example, the case router node can only send or receive a message, but cannot send messages or data to other router nodes. On the other hand there is the coordinator node, which has a dependency on other nodes for the complete management of a network; unlikefor example, router node, this node can send data to different nodes regardless of their classification.

The components that make up a sensor node are mostly very small devices made by MEMS (Micro Electromechanical Systems), in which each plays a vital role in the performance of each node in the network. Some of these components are:

1. Sensing unit and unit performance

2. Processing unit

3. Communications unit

4. Power unit

5. Other

These hardware components should be organized to conduct a proper and effective work without generating any kind of conflict in support of the specific applications for which they were designed. Each sensor node needs an operating system (OS). The operating system operates between the application software and hardware and is regularly designed to be used in workstations and PCs with the following points.

1. The collisions should be avoided whenever possible, since the relay produces unnecessary energy consumption and other potential delays. It is necessary to find an optimal solution to avoid overloading the network and avoid maximum power consumption.

2. The delay of the transmission of sent data packets is very important because it should be broadcasting in continuous time and with the highestpossible quality.

3. The receptor of the network must always be in constant operation (On), for it provides an ideal or hypothetical situation where network only mode when one needs to send or receive packets and minimize the monitoring efforts of spots.

4. There are points in the design of a wireless system such as: efficient use of bandwidth, delay, channel quality, and power consumption.

5. The adaptability and mobility of the network.

Design Coordinators and Router Nodes

Some new technologies in the design and manufacturing of communications devices, such as smaller devices and better yields, have enabled the development of more complete nodes to the field of sensing, transmission, and reception of signals obtained. Currently, there are several devices that meet the requirements demanded for the development of a wireless sensor network. The use of communication modules have helped to design the networks, both in reducing devices included in a node, and the integration of several functions at a level of both hardware and software (i.e., Security Protocols) in a single device.

Acquisition & Visualization

In order to develop a software application that allows the correct visualization of the acquired signals, it must take into account multiple factors to identify the basic features to implement it. One of the first tasks is the selection of the platform for software development, and the parameters to consider are:

- A platform that has the ability to receive a high volume of data
- A platform that allows easy synchronization between hardware and software
- A platform with virtual instrumentation tools

After the selection of the development platform begins the design phase of the application. This stage should establish the visual and written information to be submitted for a proper medical diagnosis. In order to visualize the acquired biomedical signals, the following modules must be designed:

Acquisition Module: This module is responsible for taking the BSN biomedical signals gateway.

Separation Module: This module is responsible for recovering the received frame, and the different signals transmitted (if more than one).

Processing Module: In this module each signal must translate the information received in units of voltage to the unit required by the signal such as temperature and relative humidity, among others.

Display Module: Determines the way in which the signal must be represented.

Graphical User Interface: This module uses integrated display modules to facilitate the analysis of information by the end user.

After finally completing the respective designs, the following steps are implementing for the software and then it is tested to check its proper functioning, as seen in Figure 4.10.

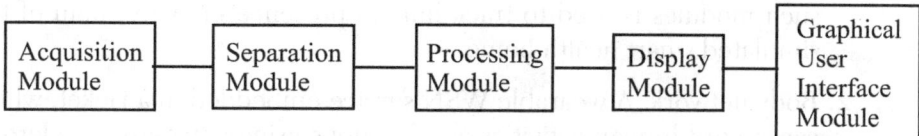

FIGURE 4.10 General block diagram of biomedical signals visualization software.

The impact generated by the use of wireless sensor networks in the quality of patient care is very high. The use of these devices in home care systems can reduce hospitalizations, health professionals' timely interventions can extend patients' lives, and in some cases the use of biofeedback techniques in psychological treatments may overcome difficult phobias. The development of such systems implies challenges to be faced in the area of engineering, such as minimizing energy consumption, since nodes in the network need to survive as long as possible. Another challenge is assuring the reliability of the information transmitted, since any slight variation may generate erroneous diagnoses. Finally, one of the biggest concerns is related to the potential impact of electromagnetic radiation to human bodies subject to the use of such devices.

4.6 Wireless Sensor Networks for Health Monitoring

The medical sensor network system integrates heterogeneous devices, some wearable on the patient and some placed inside the living space. Together they inform the healthcare provider about the health status of the resident. Data is collected, aggregated, pre-processed, stored, and acted upon using a variety of sensors and devices in the architecture (pressure sensor, RFID tags, floor sensor, environmental sensor, dust sensor, etc.). Multiple body networks may be present in a single system. Traditional healthcare provider networks may connect to the system by a gateway, or directly to its database. Some elements of the network are mobile, while others are stationary. Some can use line power, but others depend on batteries. If any fixed computing or communications infrastructure is present it can be used, but the system can be deployed into existing structures without retrofitting.

A. Data acquisition

1. Motion sensor. A low-cost sensor module that is capable of detecting motion and ambient light levels. The module also has a simple one-button and LED user interface for testing and diagnostics. A set of such modules is used to track human presence in every room of the simulated smart health home.

2. **Body network.** A wearable WSN service embedded in a jacket, which can record human activities and locations using a two-axis accelerometer and GPS. The recorded activity data is subsequently uploaded through an access point for archiving, from which past human activities and locations can be reconstructed.

3. **Indoor temperature and luminosity sensor.** These sensors give the environmental conditions of the habitat.

4. **Bed sensor.** The bed sensor is based on an air bladder strip located on the bed, which measures the breathing rate, heart rate, and agitation of a patient.

5. **Pulse-oximeter and EKG.** They are wearable and collect patient vital signs. Heart rate (HR), heartbeat events, oxygen saturation (SpO_2), and electrocardiogram (EKG) are available.

B. Current backbone infrastructure

The current backbone is a gateway between the motes deployed in the home environment and the nurse control station. Motes use a Zigbee compliant (802.15.4) wireless protocol for communication.

C. Database management and data mining

A database (like MySQL/Oracle) serves as a backend data store for the entire system. It is located in a PC connected to the backbone, and stores all the information coming from the infrastructure for longitudinal studies and offline analysis.

D. Graphical user interfaces

A GUI (Figure 4.11), which can run on a PDA, permits a caregiver to request real-time environmental conditions of the living space and the vital signs of the resident. It uses a query management system distributed among the PDA, gateway, and the sensor devices. The interface graphically presents requested data for clear consumption by the user. An LCD interface board was also designed for wearable applications. It presents sensor readings, reminders, and queries, and can accept rudimentary input from the wearer. A final GUI, from a direct medical application based on motion sensors, exists to study the behavioral profile of the user's sleep/wake patterns and life habits, and to detect some pathologies in the early stages.

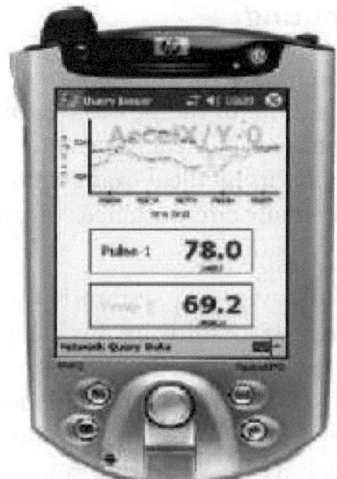

FIGURE 4.11 A GUI displays accelerometer data, patient pulse rate, and environmental temperature.

The system is single hop, as the radio range covers all of the facility. A multi-hop protocol will be necessary for access of multiple floors, or if transmission power is reduced. Data communication is bidirectional between the motes and the gateway. Time-stamping is done by the PC when motion events are received. Figure 4.12 shows the current acquisition chain.

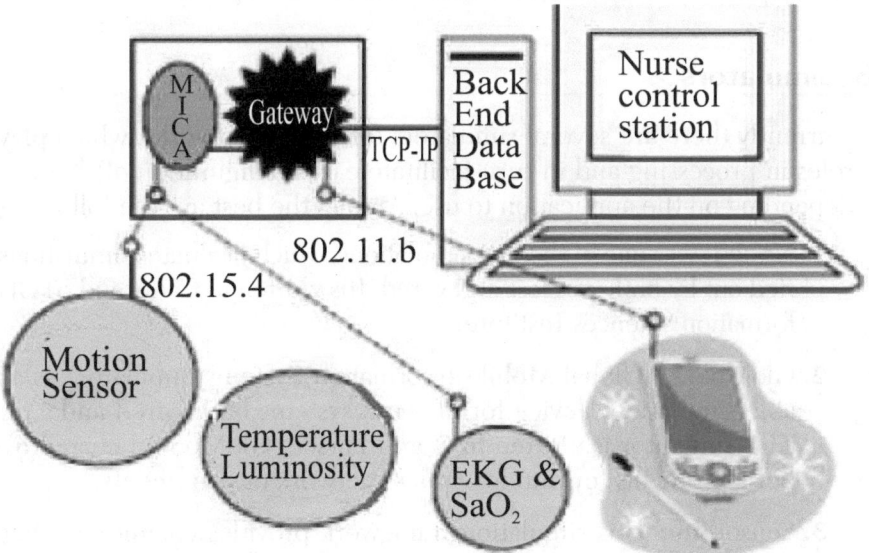

FIGURE 4.12 Current configuration of the medical test bed.

4.7 Wearable Computing

"Wearable Computing" is a technology dealing with computer systems integrated in clothing. One of the possible applications of the project is the rapid availability of patient medical information at any time; this may mean an interesting reduction in medical examination fees, the power to perform medical reviews in the daily circumstances of patients, and in extreme cases could save the life of a patient.

A wearable audio navigation system is a portable device whose characteristics are in navigation software for people with vision loss or even for places in which the vision of the place is limited, and this emphasizes the need to avoid obstacles or to obtain characteristics of the environment quickly. This device consists of a small computer which contains various guidance devices such as GPS, inertial sensors, RFID antennas, and RF sensors, among others. When all devices are synchronized and identify the exact location, sound guidance is delivered through an audio device to the person using the device, which also indicates in real time the location of other characteristics of the sensing environment.

The advanced soldier sensor information system and technology is a well-known program that integrates information on the battlefield (location, time, group activities, among others) on the soldier to collect, disseminate, and display key information, without risking life or physical integrity.

4.8 Simulators

Currently there are several simulators for sensor networks, which play key roles in processing and in turn facilitate easy configuration of the network depending on the application to use. Among the best are the following:

1. NS-2: It was one of the first simulations, which facilitates simulations carried out by both wireless and wired. It is written in C + + and oTCL (Information Sciences Institute).

2. GloMoSim (Global Mobile Information Systems Simulator): is a scalable simulation: device for network systems both wired and wireless. This simulator is written in C and Parsec. GloMoSom currently supports protocols for purely wireless network environments.

3. SensorSim: This simulation framework provides channel sensing and sensor models as models of battery, battery light wide protocols for wireless microsensors.

4.9 Ongoing Research Ideas in Medical

1. **Multimodal data association and multiple residents.** Data association is a way to know "who is doing what?" in a system without biometric identification and with multiple actors present, such as in an assisted living community. It permits us to recognize the right person among others when he is responsible for a triggered event. This is indispensable for avoiding medical errors in the future and properly attributing diagnostics. Consequently, dedicated sensors and data association algorithms must be developed to increase quality of data.

2. **Data integrity.** When the data association mechanisms are not sufficient, or integrity is considered critically important, some functionalities of the system can be disabled. This preserves only the data which can claim a high degree of confidence. In an environment where false alarms cannot be tolerated, there is a tradeoff between accuracy and availability.

3. **Security and privacy.** The system is monitoring and collecting patient data that is subject to privacy policies. For example, the patient may decide not to reveal the monitored data of certain sensors until it is vital to determine a diagnosis and therefore authorized by the patient at the time of a visit to a doctor. Security and privacy mechanisms must be throughout the system.

Summary

- A smart medical home is a system of room labs outfitted with infrared sensors, computers, biosensors, and video cameras.

- The sensors, such as a portable 2-lead ECG, pulse oximeter, wearable Pluto mote with built-in accelerometer, module with accelerometer, gyroscope, and electromyogram sensor for monitoring of stroke patients are included in the medical node.

- Body network and subsystems, emplaced sensor network, back end, backend databases, and human interfaces are the different parts of the body sensor network architecture.

- The intelligent wireless patient-monitoring system framework includes real-time sensing of the patient's vital parameters using the motes, and wireless transmission of such critical information over radio frequencies to the base station.

- A medical node consists of a computing subsystem, communication subsystem, sensing subsystem, and energy storage subsystem.

- BSN technology offers the possibility of developing a detailed diagnosis of the patient, because the network would be able to monitor all vital signs and synthesize all relevant information for more effective patient care.

- Wearable computing is a technology dealing with computer systems integrated in clothing.

Questions

1. What do you mean by body sensor networks?

2. Explain the architecture of body sensor networks.

3. What are the blocks in body sensor network software architecture?

4. Draw the overall framework for patient monitoring. Explain.

5. What are the blocks available in the wireless medical node?

6. Define a homogeneous node.

7. What is meant by an autonomous node?

8. What are the classifications of WSNs?

9. How many classifications are in the functional level of WSNs?

10. What are the characteristics of the layers used for medical?

11. What are the different topologies available for medical? Which is best suited? Justify the answer.

12. Differentiate WSNs and BSNs.

13. What do you mean by wearable computing?

14. Explain the three-phase methodology for the development of applications of biomedical signals acquisition.

15. What are the different types of sensors interfaced for medical monitoring?

16. Explain the WSN system for health monitoring.

Further Reading

1. *Body Sensor Networks* by Guang-Zhong Yang

2. *Wireless Body Area Networks: Technology, Implementation, and Applications* by Mehmet R. Yuce and Jamil Khan

3. *Ultra Wideband Wireless Body Area Networks* by Kasun Maduranga Silva Thotahewa, Jean-Michel Redouté, and Mehmet Rasit Yuce

References

1. *http://cdn.intechweb.org/pdfs/12898.pdf*

2. *http://spriyansasi.blogspot.in/2013/08/wireless-sensor-networking. html*

CHAPTER 5

UBIQUITOUS SENSOR NETWORKS

This chapter discusses the wireless sensor networks for environmental, industrial monitoring, and Ubiquitous Sensor Networks (USN).

5.1 Ubiquitous Sensor Networks (USN)

When you enter a modern office building, it is quite common for the glass doors to open automatically and for lights to come on as you enter a darkened room. This "magic" is achieved by motion sensors. When entering a building of the future, you might be welcomed by name with a personal greeting and given security access suitable to your status (e.g., employee, delegate, newcomer). To do this without human intervention would require not only intelligent sensors but also perhaps ID tags, readers, and interaction with one or more databases containing your profile.

The three elements of sensors, tags, and communication/processing capacity together make up a future network vision identified by a number of different names. Some use the terms "invisible," "pervasive," or "ubiquitous" computing, while others prefer to refer to "ambient intelligence" or to describe a future "Internet of Things." The term "Ubiquitous Sensor Networks" (USN) is used to describe a network of intelligent sensors that could, one day, become ubiquitous. The technology has enormous potential, as it could facilitate new applications and services in a wide range of fields, from ensuring security and environmental monitoring to promoting personal productivity and enhancing national competitiveness. But USNs will also require huge investments and a large degree of customization. As such, it presents a standardization challenge with an unusually high degree of complexity.

Anywhere, Anytime, by Anyone and Anything

The term "ubiquitous" is derived from the Latin word ubique meaning "everywhere." But the literal interpretation of a USN as meaning sensors on every single part of the globe, however remote, is not a realistic aim. Instead, a more reasonable definition is based on socio economic, rather than geographical, lines, and describes a technology which can be available "anywhere" (i.e., anywhere that it is useful and economically viable to expect to find a sensor), rather than "everywhere." The concept of availability is wider than just a geographical measure, and the expression "anywhere, anytime, by anyone and anything" (the "4A vision") has come to be used to illustrate the trend toward a ubiquitous network society. USNs have applications in both civilian and military fields. For civilian applications, these include environment and habitat monitoring, healthcare, home automation, and intelligent transport systems. The main components of a USN, as described in Figure 5.1, are:

Sensor Network: Comprising sensors and an independent power source (e.g., battery, solar power). The sensors can then be used for collecting and transmitting information about their surrounding environment.

USN Access Network: Intermediary or "sink nodes" collecting information from a group of sensors and facilitating communication with a control center or with external entities.

Network Infrastructure: likely to be based on a next generation network (NGN).

USN Middleware: Software for the collection and processing of large volumes of data.

USN Applications Platform: A technology platform to enable the effective use of a USN in a particular industrial sector or application. The nodes may vary enormously in size and in cost and complexity. The medium that nodes use to communicate with the sink would vary according to the characteristics of the application. Depending on the sensor type, the links between sensors may be provided by either wired or wireless communication. The transmission of sensor data using radio frequency might be used, for instance, in the tracking of goods in supply chain management. This application of radio frequency identification (including RFID tags with sensors) corresponds to the lower layers in the schematic model for a USN as follows:

RFID Tags: An RFID processor that may be either passive or active (with potential read/write functions, wider communication ranges, and independent power supplies). An active RFID chip is capable of two-way communication, whereas a passive tag is read only.

RFID Reader: The reader senses and "reads" the information on the tag and passes it on for analysis.

RFID Middleware: Like the USN, the RFID network may have its own software for the collection and processing of data.

As illustrated in Figure 5.1, a USN is not simply a network but can be an intelligent information infrastructure used to support a multitude of different applications. USNs can deliver information to "anywhere, anytime, by anyone." But it is the ability to deliver the information also to "anything" which is groundbreaking. Value is added to the information by using "context awareness," which comes from detecting, storing, processing, and integrating situational and environmental information gathered from sensor tags and/or sensor nodes affixed to any object. For instance, context awareness may relate to where the object is located, whether it is moving or stationary, whether it is hot or cold, and so on.

WPAN (Wireless Personal Area Network)

A wireless personal area network (WPAN) is a personal, short distance area wireless network (typically extending up to 10 meters in all directions) for interconnecting devices centered around an individual person's workspace. WPANs address wireless networking and mobile computing devices such as PCs, PDAs, peripherals, cell phones, pagers, and consumer electronics.

Characteristics of a USN

- Small
- Scale sensor nodes
- Limited power requirements that can be harvested (e.g., solar power) or stored (e.g., battery)
- Able to withstand harsh environmental conditions
- Fault tolerant and designed to cope with high possibility of node failures
- Support for mobility
- Dynamic network topology
- Able to withstand communication failures
- Heterogeneity of nodes
- Large scale of deployment

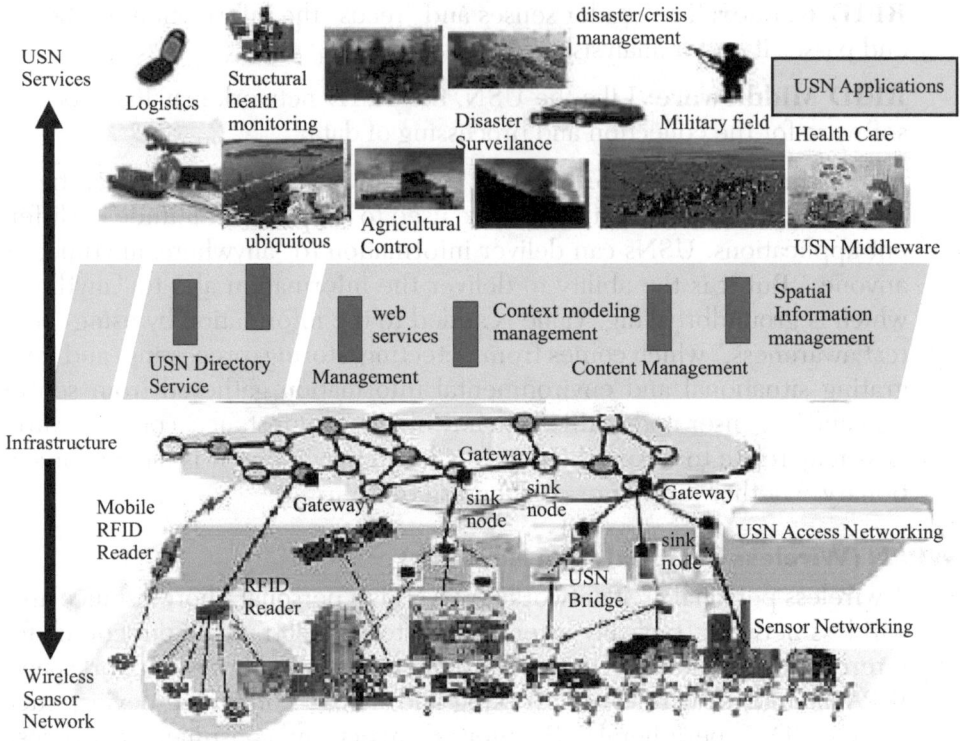

FIGURE 5.1 Schematic layers of Ubiquitous Sensor Networks.

5.2 Applications of USNs

Ubiquitous Sensor Networks provide potentially endless opportunities in a diverse number of different applications. Some of them are listed as follow:

1. *Intelligent Transportation Systems (ITS)*

A network of sensors set up throughout a vehicle can interact with its surroundings to provide valuable feedback on local roads, weather, and traffic conditions to the car driver, enabling adaptive drive systems to respond accordingly. For instance, this may involve automatic activation of braking systems or speed control via fuel management systems. Condition and event detection sensors can activate systems to maintain driver and passenger comfort and safety through the use of airbags and seatbelt pre-tensioning. Sensors for fatigue and mood monitoring based on driving conditions, driver behavior, and facial indicators can interact to ensure safe driving by activating warning systems or directly controlling the vehicle. A broad citywide, distributed sensor network could be accessed to indicate traffic flows, administer tolls, or provide continually updated destination routing feedback to individual vehicles. The feedback may be based on global and local information, combining GPS information with cellular networks.

2. *Robotic Landmine Detection*

A sensor network could be used for the detection and removal or deactivation of landmines. The USN enables the safe removal of landmines in former war zones, reducing the risk to those involved in the removal process. The utilization of advanced sensor technology to detect explosives may help overcome difficulties in detection of unencased landmines.

3. *Water Catchment and Eco System Monitoring*

A network of sensors can be utilized to monitor water flows into catchment areas and areas where access is difficult or expensive. This information can be combined with other sensor networks providing information on water quality and soil condition, together with long-term weather forecasting, to assist with the equitable and efficient distribution of water for irrigation and environmental purposes. Similar technology can be utilized to provide an early warning system for flood-prone regions, particularly for flash flooding.

4. *Real-Time Health Monitoring*

A network of advanced biosensors can be developed using nanotechnology to conduct point-of-care testing and diagnosis for a broad variety of conditions. This technology will reduce delays in obtaining test results, thus having a direct bearing on patient recovery rates or even survival rates. On the basis of the sensed data, physicians can make a more rapid and accurate diagnosis and recommend the appropriate treatment. USNs may also enable testing and early treatment in remote locations, as well as assist triage on location at accident or disaster sites.

5. *Bushfire Response*

A low-cost distributed sensor network for environmental monitoring and disaster response could assist in responding to bushfires by using an integrated network of sensors combining on-the-ground sensors, monitoring local moisture levels, humidity, wind speed, and direction with satellite imagery to determine fire risk levels in targeted regions, and offering valuable information on the probable direction in which fires may spread. This type of USN can provide valuable understanding of bushfire development and assist authorities in organizing a coordinated disaster response by providing early warning for high-risk areas.

6. *Remote Sensing in Disaster Management*

Remote sensing systems have proven to be invaluable sources of information that enable the disaster management community to make critical decisions based on information obtained from study of satellite imagery for better preparedness and initial assessments of the nature and magnitude of damage and destruction. Information derived from satellites can be combined with on-the-ground data from a USN. High-resolution remote sensing data is especially useful for documenting certain hazards, for determining where to locate response facilities and supplies, and for planning related facilities for reconstruction and relocation activities. Data availability and its timely delivery are crucial to saving lives and property during disasters, and technological developments are making positive contributions in this area. Some of the most significant progress in disaster reduction is being made in mitigation, using historical and contemporary remote sensing data in combination with other geospatial data sets as input to compute predictive models and early warning systems.

7. *Detecting, Tracking, Monitoring*

The unique potential and particular characteristics of sensor nodes and network infrastructure have encouraged researchers to identify potential applications in a diverse range of domains. Nevertheless, in most cases, applications can be assigned to one of the following three broad categories:

1) Detection – for example, of temperatures passing a particular threshold, of intruders, of bushfires, of landmines in former war zones, and so on;

2) Tracking – for example, of items in supply chain management, of vehicles in intelligent transport systems, of cattle/beef in the food chain, of workers in dangerous work environments such as mines or offshore platforms, and so forth,

3) Monitoring – for example, of a patient's blood pressure, of inhospitable environments such as volcanoes or hurricanes, of the structural health of bridges or buildings, or of the behavior of animals in their indigenous habitats and so on.

5.3 Monitoring Volcanic Eruptions with a USN

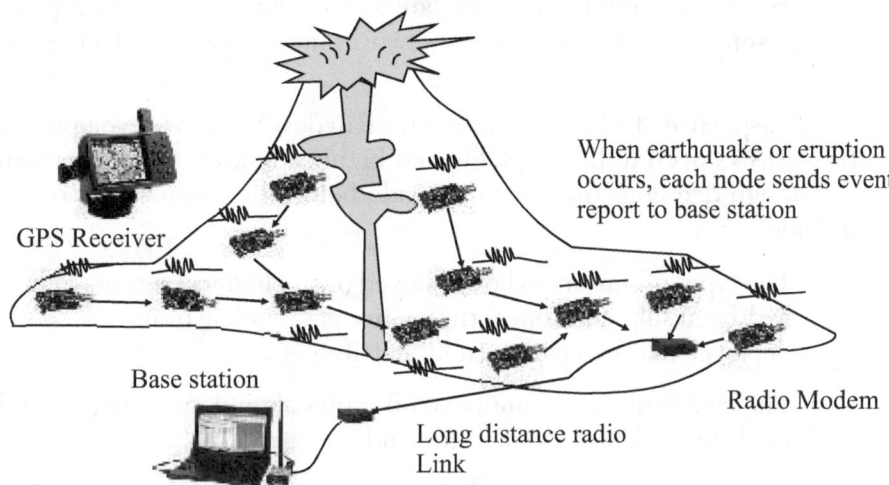

FIGURE 5.2 Wireless seismic and acoustic sensor node with volcano monitoring network architecture.

Studying active volcanoes typically involves sensor arrays built to collect seismic and infrasonic (low-frequency acoustic) signals. This was to study the use of tiny, low-power wireless sensor nodes for geophysical studies, which have advantages in cost, size, weight, and power supply over the traditionally used sensors. Sensor nodes driven by conventional cell batteries were deployed over a 3 km² aperture on the upper flanks of the volcano. The system routed the collected data through a multi-hop network and over a long distance radio link to an observatory, where a laptop logged the collected data, as can be seen in Figure 5.2. Volcano research requires extremely high data quality and reliability. Only one missed or corrupt sample can invalidate an entire record. Similar architectures of sensor networks may be used to monitor and research different natural phenomena, as well as in the field of disaster detection and prevention.

Domains in which USNs are used include civil engineering, education, healthcare, agriculture, environmental monitoring, military, transport, disaster response, and many more. In developing countries, specific applications could cover domains where network engineers face particular challenges such as unreliable power supply, reduced budgets, or the danger of theft. The falling prices of sensor units and RFID tags is greatly increasing a range of potential applications. Furthermore, the possibility of operating independently from electricity networks, by using conventional batteries, or depending upon availability of solar or geothermal power as energy supplies, can make sensor networks more widely available in different environments.

Irrespective of whether they are used in developed or developing countries, USNs need to be adaptable and highly application specific. Some of the design decisions that must be made before the deployment of a USN include:

- The types of sensors to be employed (e.g., chemical sensors to monitor hydrogen sulphide concentration in a gas pipeline or motion sensors deployed in a area with seismic activity);

- The choice of the communication protocols and medium (depending on distance, transmission rate); and

- The energy supply of the nodes (e.g., can batteries be easily replaced? This might be possible in a light sensor in the house, but not if sensors are deployed in a mine field).

5.4 WSNs on Regional Environmental Protection

Regional Environment refers to a particular geographical space of the natural environment or social environment. Different regions face different environmental problems. Wireless sensor networks are widely used in regional environmental protection. By dropping a large number of sensor nodes in a target area through aircraft, the nodes will monitor the changes in the surrounding environment and send data sent back to the base. In this way, one can easily monitor the environment changes. WSNs in regional environmental protection have very good prospects.

Regional environments can be divided into natural regional areas, social regional areas, agricultural regional environments, and tourism regional environments by function. They all have unique structures and characteristics. The divisions of regional environment are designed to contrast with each other, and one can study and solve environmental problems by their particular characteristics. With the rapid development of economies, environmental problems have become increasingly prominent. As the development of a region's industrial production and some other factors, regional environmental water pollution, atmospheric pollution, land pollution, and so on become increasingly serious. Wireless sensor network design meets the needs of a particular application, as it is an application-based network. Its features are particularly applicable to regional environmental protection. The application of sensor networks for environmental monitoring changes the model of single fixed-point sensors in fixed time and a fixed area as in the past. It can attach the targets of multi-angle observations (more types of sensors), synchronization, and continuous measurement. So the data are more comprehensive and more representative. It is easier to describe time and space changes and find the internal relations of observation targets. The main applications of WSNs on regional environmental protection can be divided into information collection, environmental monitoring, security warning, and so on.

Information Collection

Sensors can collect light, temperature, sound, humidity, and other environmental data. In a regional environment, on the basis of these data, one can make some decisions. In the agricultural regional environment, the most common application of WSNs is farmland information collection. The growth of crops is impacted by natural conditions, such as light, temperature, and humidity. Information about crop water demand is the basis for adequate irrigation. We also can use sensor nodes to track crop growth conditions and research the impact of environment change on crops to provide real-time decision making. The volume of data on farmland is very large, and using the network can be a lot more convenient to achieve long-distance data transmission, but laying cable networks in the farmland on the one hand is not easy for farming, and on the other hand the costs are higher. So WSNs are a good option.

Environment Monitoring

The pollution areas that need to be monitored are always places that are hard for humans as well as big devices to reach. But wireless sensor networks can be quickly deployed, and self organization and high reliability are applicable in such an environment. For example, to monitor the atmosphere, traditional manual sampling can only monitor the average concentration of gas at the scene within a certain period of time, cannot provide real-time values, and the results of monitoring are greatly impacted by man-made factors. The use of large-scale automation equipment is expensive. The device is too complex and difficult to maintain. It is very hard to use in the country scale. The atmospheric monitoring system based on wireless sensor networks can effectively meet the needs of real-time monitoring.

The atmospheric monitoring system based on wireless sensor networks has the features of low cost, easy implementation, and high reliability. Similarly, this application can be used in the areas of water pollution, soil pollution, and so on.

Security Warning

There are many potential safety problems hidden in regional natural environments and industrial production. These problems are always hidden, unexpected, and unpredictable; they often lead to some unpredictable consequences, resulting in huge losses. Wireless sensor networks are an important means of early warning. WSNs are widely use in industrial production, construction, transport, medical, and other fields for security warning. For example, forest fires are one of the most serious disasters that make destruction of forest resources and threats to human living environments. Prevention and monitoring of forest fires has become a major research focus for forest fire departments around the world. One can build and import wireless sensor networks to monitor forest fires. The system is capable of real-time monitoring of the measured parameters (such as temperature, relative humidity, etc.) and can send the information to the monitoring center computer, center analyzing the data, compare it with forest resource base data, and then determine whether there is a potential forest fire in the field. This will be an effective basis for the departments to make a decision.

Main parameters of regional agriculture environments

The typical application for agriculture regional environments is collecting information of farmland in a large area as a basis for decision making. Growth of crops is influenced by various factors, mainly the water content in soil, soil temperature, air temperature, air humidity, light intensity, and other factors. A regional agriculture environment has the following characteristics:

1. The coverage of farmland in the region can be significant and broad.

2. Farmland environments are often not available for laying a cable network, so we can't communicate through a cable network.

3. The farmland is accessible, and not a very dangerous environment.

However, the equipment needs to work in the open environment, so it requires high equipment reliability. The characteristics of the agricultural environment ask for certain requirements for a sensor network node. The nodes working in these areas need to be small in size, pollution free, with environmental adaptability and low power consumption, and so on.

Network Structure

A sensor network is mainly composed of the communications nodes, sink nodes, and the application server. Wireless sensor network nodes complete the data collection, preparation, and communication work. Clusters are formed by adjacent network nodes, and each node sends the data to the head node of the cluster. The head node compresses the data and sends it to the sink node. In fact, the sink node is the gateway node, and is responsible for the network initiation, maintenance, and data collection, and for sending data to the control center. The monitor center is responsible for data processing and network management. There is some specific software in the control center that does the job of data processing and makes decisions.

As farmland is a place that human can easily to reach, one can artificially divide farmland into multiple regions, and each region forms a cluster of network topology. Inside each cluster a head node is assigned that is responsible for communication with the gateway. And since the agricultural environment may not have an off-the-shelf cable network, one can consider two communications structures:

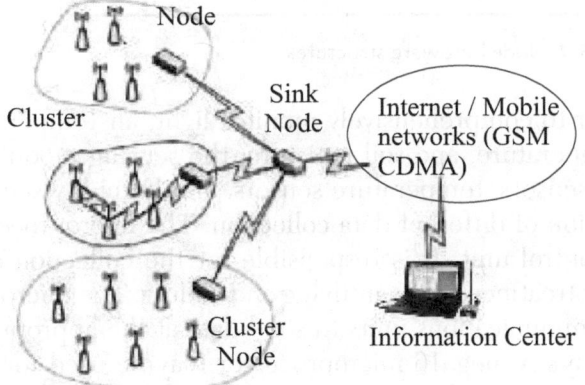

FIGURE 5.3 Network structure.

1. One in which the gateway communicates with the server control center through the cable network.

2. In the mobile networks such as the GSM or CDMA coverage area.

One can use mobile base stations as a transmission medium. The sink node sends the data to base stations, and the base station data is then transmitted to the monitoring center. The structure of this network can be shown as follows as in Figure 5.3.

Node hardware design

The communications node is composed of the sensing element, the central processing unit, the wireless transceiver, and the power components. Based on agricultural environmental monitoring, one needs to use light, temperature, and humidity sensors. The node structure is as follows in Figure 5.4.

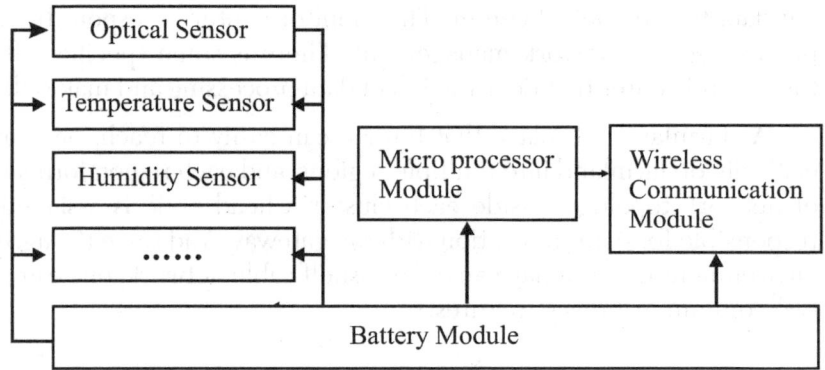

FIGURE 5.4 Node hardware structure.

In order to comprehensively monitor light, air temperature, air humidity, soil temperature, and soil moisture, the sensing modules need to use the optical sensors, temperature sensors, and humidity sensors to achieve the perception of different data collection. The microprocessor module is the main control unit. It is responsible for the collection of sensor data and does pretreatment of quantifying and coding. The microprocessor also controls communications units to send data at the appropriate time. ATMEL Company's Atmega16 microprocessor may be used for this. This MCU is chosen because of having faster, stronger, anti-interference, and C language programming. The wireless communications module uses an XBeePro radio chip. The chip uses IEE802.15.4 specifications. The available power modules are Zn/MnO_2 alkaline batteries, nickel metal hydride batteries, lithium ion batteries, and so on. Each battery has its own characteristics. Lithium ion batteries have the advantages of light weight, and large capacity and energy density. Lithium ion batteries can meet the requirements for volume, environmental adaptation, discharge stability, cost, and so on.

Software Design

Software systems can be divided into node software modules and control center software modules. Node software needs a serial communication module, and a data conversion and an amendment module. The serial communication module is responsible for communicating with gateway nodes, extracting sensors' collected information, and communicating with nodes to address information from relevant data frames. The serial data conversion module is responsible for collecting information from the sensors into digital information. The monitoring center software is the information processing center. The software must receive and store the data from the network. To provide a user-friendly interface, the software still needs to show the data in a variety of ways and make relevant decisions.

This is a typical scheme of a WSN in a regional agricultural environment. Different applications have different characteristics and parameters. They need to use different network topologies and different hardware. Compared with traditional methods for regional environmental protection, wireless sensor networks are a more convenient, real time, reliable, and effective means. They are a good option for regional environment protection.

5.5 The Development of USNs for a Rice Paddy Crop Monitoring Application

The advantages of WSNs in agriculture are as follows: suitability for distributed data collecting and monitoring in tough environments, capablility to handle an economical climate and to control irrigation and nutrient supply to produce the best crop condition, increasing production efficiency while decreasing cost, and providing real-time information about the fields that enable farmers to adjust strategies at any time. The parts of the WSN system will be able to communicate with lower power consumption in order to deliver real data.

Wireless transmission can reduce and simplify wiring, allow the sensor to be deployed at remote, dangerous, and hazardous locations, is easily installed, and can be integrated for extremely low cost, small size, and low power requirement and mobility.

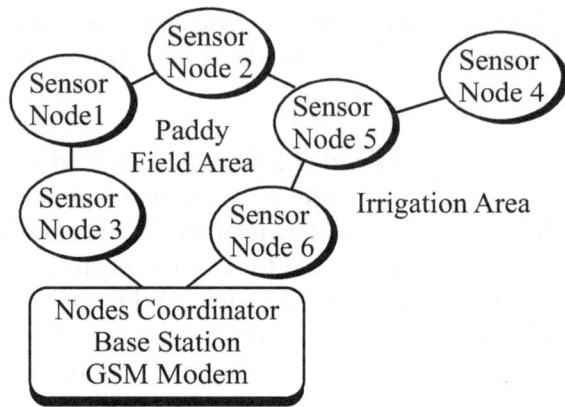

FIGURE 5.5 Wireless Sensor Network routing via PAN (both ad hoc or planned network).

Figure 5.5 describes the overview of the system consisting of a node coordinator/gateway and a few sensor nodes routed via a PAN (personal area network) that can be added to or reduced depending on the requirement of the farming areas. A general-purpose node, which has standard measurement parameter sensors for factors such as ambient air temperature and humidity, soil pH and moisture, are integrated in all nodes. These nodes will route to their neighborhood node(s) until the beacons reach a coordinator destination. The coordinator will coordinate the data collection from each existing PAN address within the network system. Then there are two directions the data will go; it is first linked to the server data base system to be recorded and revealed on an Internet web page, and a real-time alert system using a SMS system via a GSM modem to the person in charge of a cell phone.

The node architecture and hardware is driven by an MCU produced by Microchip, which is a PIC16F684 NanoWatt processor core, and features low current and voltage consumption. This chip is a 14 pin 8-Bit FLASH CMOS MCUs, with 12 I/O and 8 ADC channels at 10-bit maximum resolution. It consumes a current of less than 500µA at 2.0V and a 20 MHz clock cycle. Another series of PIC16F88X chip also can be used as an alternative for more I/O pins (28/40) with a dynamic CPU clock start-up and ultra low power at sleeping mode (approximately 50 nA). Figure 5.6 (a) describes the system architecture that inserts the contribution of a power management system that will utilize two batteries for night time and day time operation. The power management system will manage solar power direction for charging the secondary battery, while the primary battery will remain working until at a certain level of voltage drop, it will trigger an alternate charging-working process for these two batteries. This method of power management system claims a better power life for nodes up to 25% due to the improved charging concept. Sharing the solar energy for charging a battery while at the same time drawing current from the same battery will lose much efficiency of the battery life, and will lead to insufficient charging and a hassle during night operation or dim daylight. The charging engine for the lithium-ion 3.7V cell is driven by a MCP73832 charge pump chip with programmable charging current at 15mA-500mA. The power supply of the sensor node also will apply the low-drop out (LDO) regulator instead of a linear regulator that creates a lot of voltage drops and higher noise for the switching regulator. The system has the capability to attach more sensors such as the wind speed, dissolved oxygen, water pH, solar radiation, and other analog or digital interfaces to the MCUs. Figure 5.6 (b) describes the algorithm sequence for working-charging batteries in the node power management system.

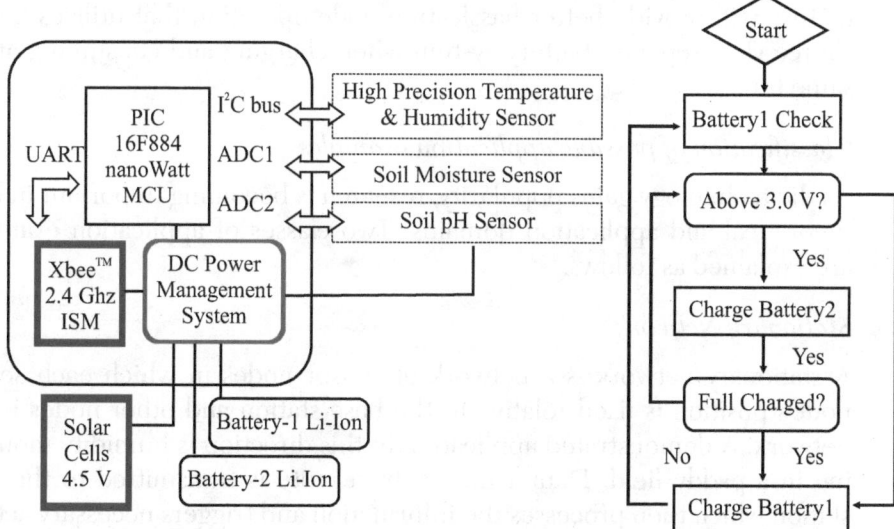

FIGURE 5.6 (a) Sensor node architecture system that features optimal power management system and (b) Battery management algorithm.

The wireless and networking activities will take place after the MCU reads all the ADC values from the sensor output voltage and sends it to FLASH memory. The digital sensors like the I2C bus type will send the readings after the acknowledgment bits sent by the MCU. This IEEE 802.15.4

compatible module has wide range of 65,000 unique networking addresses, and has the capablility to build peer to peer, point to point, and point to multi-point topologies. This module is programmed using both AT command and X-CTU software provided as free software.

A network coordinator as the gateway of accessing the outside world will manage the operation of the wireless networking sensor nodes. This board has the same architecture as the node type, except the MCU has external UART communication with the GSM/GPRS 900/1800MHz modem to send the alert massages via SMS and RS232 serial communication to the server PC. PIC MCU establishes communication together with the GSM modem using AT command to give instant alerts via sms, like the over limit or under limit condition for paddy field environments, such as floods. The node coordinator also becomes a base station that is linked with the LAN and TCP/IP for data storage and a web-based server.

The power management system consists of dual Li-ion/Li-Po batteries that will support the node's life for 24 hour a day operation. This architecture will provide better hassle-free node operation that utilizes a solar source via a separate battery system when charging and consuming at the same time.

Classification of possible application examples

As the technology gains popularity, research is becoming important in both theoretical and application domains. Two classes of application examples are explained as follows.

Stationary Network

A stationary network is a network of sensor nodes in which each sensor node's position is fixed relative to the base station and other nodes in the network. A demonstrated application in this direction is humidity monitoring in a paddy field. Data acquired by a mote is transmitted to the base station which then processes the information and triggers necessary actions such as localized watering. There are two possible cases to transmit data. When a node is in direct wireless contact, that is, in the range of the base station, direct communication is possible. When a node is not in range, it transmits data in an ad hoc environment also referred to as multi-hop. Implementation of an efficient multi-hop system requires optimal routing to facilitate the shortest route, reduced power consumption, and improved transmission.

Network in Motion

An example in this category is a herd of animals on an extensive farm, where each animal is equipped with a sensor node. The animals are in constant motion relative to each other as well as the base station. Such complicated mobility management requires an even more sophisticated implementation of routing algorithms. In order to maximally benefit from wireless sensor networks of this type, additional hardware requirements in the form of GPS devices and other forms of mote location are needed.

5.6 WSNs in the Smart Grid

The power grid is not only an important part of the electric power industry, but also an important part of a country's sustainability. With the dependence on electric power gradually increasing, demand for the reliability and quality of the power grid is also increasing in the world. A smart electricity grid opens the door to new applications with far-reaching impacts: providing the capacity to safely integrate more renewable energy sources (RES), electric vehicles, and distributed generators into the network; delivering power more efficiently and reliably through demand response and comprehensive control and monitoring capabilities; using automatic grid reconfiguration to prevent or restore outages (self-healing capabilities); and enabling consumers to have greater control over their electricity consumption and to actively participate in the electricity market. Sensors will be a key enabler for the smart grid to reach its potential. The idea behind the "smart" grid is that the grid will respond to real time demand; in order to do this, it will require sensors to provide this "real time" information. WSNs as "smart sensing peripheral information" can be an important means to promote smart grid technology development. WSN technology in the smart grid will also further promote the industrial development of WSNs.

Online monitoring system for transmission lines

The condition of transmission lines is directly affected by wind, rain, snow, fog, ice, lightning, and other natural forces; at the same time industrial and agricultural pollution are also a direct threat to the safe operation of transmission lines. The operating environment of transmission lines and the operating states are very complex, which requires more automatic monitoring, more control, and protection equipment to automatically send alarms when accidents occur and dispatch strategy adjustment according to the operation mode, so that the faults will be processed at the early phase or be isolated in a small range.

Traditional wired communications cannot meet the communication needs of online monitoring of transmission lines. WSNs have an advantage of the strong ability to adapt to harsh environments, large area coverage, self-organization, self-configuration, and strong utility independence, and are very suitable for data communication monitoring systems for transmission lines.

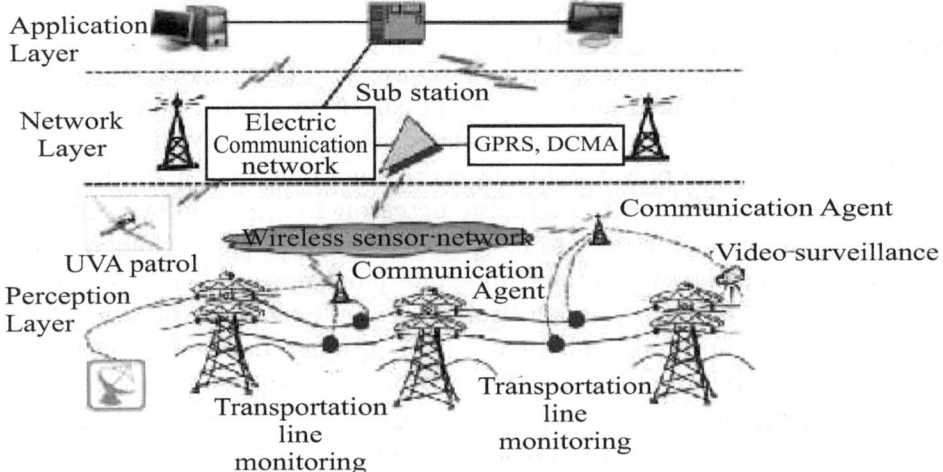

FIGURE 5.7 General architecture of online monitoring systems for a transmission line based on WSNs.

With the technical advantages of WSNs, establishing a full range, multi-element online monitoring system can send timely warnings of disasters, rapidly locate the positions of faults, sense transmission line faults, shorten the time of fault recovery, and thus improve the reliability of the power supply. WSN use can not only effectively prevent and reduce power equipment accidents, when combining with the conductor temperature, environmental, and meteorological real-time online monitoring, but can also provide data to support transmission efficiency improvement and increasing dynamic capacity for transmission lines. The general architecture of online monitoring systems for transmission lines is shown in Figure 5.7.

Intelligent monitoring and early warning system for substations

After decades of development, the domestic substation automation technology has reached the international standards level. Most of the new substations, regardless of voltage level difference, adopt integrated automation systems. Compared with the conventional substations, digital substations focus on the network information digitization, substation information standardization, and networked transmission. For substations in smart grids, more attention is focused on smart power equipment, information exchange, interoperability, and the intelligence application functions of the inner station. Now many smart monitoring functions can be realized and can improve intelligent substation management, including transformer/breaker/temperature monitoring, current leakage monitoring of lightning arresters, electric leakage monitoring equipment, SF6 leakage monitoring of combined electric equipment, secondary equipment environmental monitoring, equipment anti-theft monitoring, and so on.

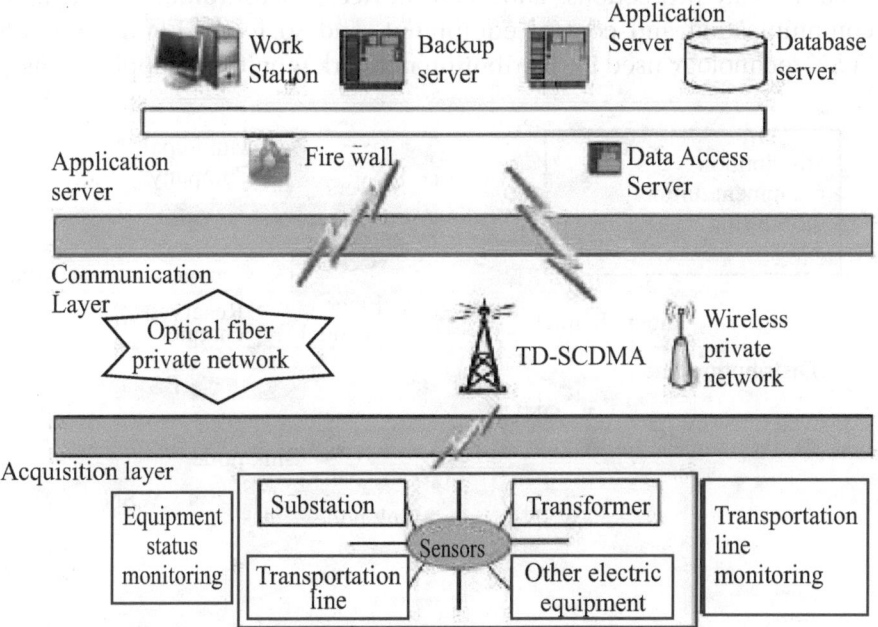

FIGURE 5.8 Architecture of operation status monitoring for equipment.

Applications of WSNs can provide reliable, accurate, real-time, safety, sufficient information for substation management not confined to the traditional electrical quantities information of telemetry, remote communication, remote control, and remote adjustment, but also including equipment information, such as cooling system condition, circuit breaker action times, energy storage state of the transmission mechanism, size of breaking current, and environmental information, video information, and so on, to finally achieve digitalization of information description, integration of data acquisition, data transmission by network, intelligent data processing, data display visualization, and scientific production decision making. Figure 5.8 shows the architecture of operation status monitoring for equipment.

Online Monitoring and Early Warning Systems for Distribution Networks

Distribution networks directly connect the power grid with users, and distribute electrical energy to them. Reliability and quality of the distribution network is an important element for reliable power supplies. The distribution network consists of primary equipment such as feeders, distribution transformers, circuit breakers, switches, and secondary equipment such as relay protections, automatic devices, measurements and meters, communication and control equipment, and so forth. Figure 5.9 shows WSN technology used in distribution network monitoring applications.

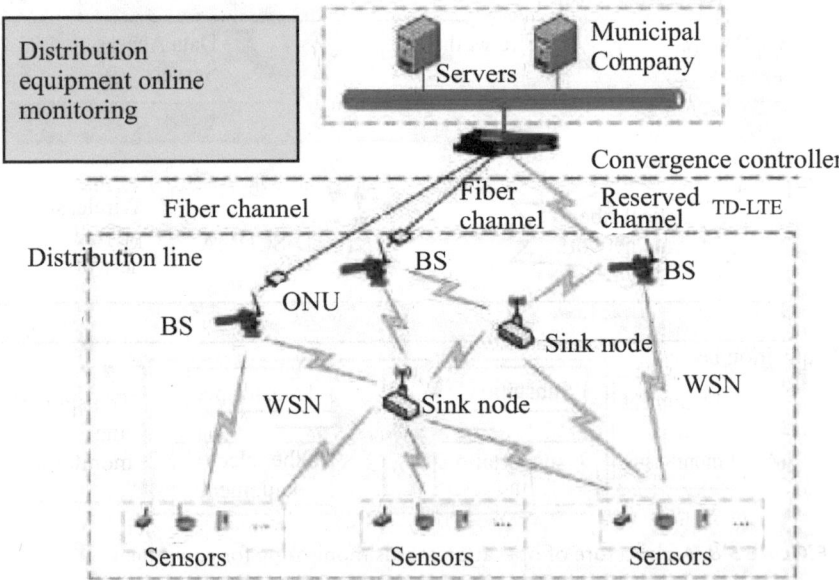

FIGURE 5.9 WSN technology used in distribution network monitoring applications.

Distribution networks have characteristics of a massive number of points, large coverage areas, and long-distance power lines. The application of WSNs in distribution fields can strengthen management, save manpower, improve the reliability of the power supply, and accelerate the recovery efficiency of fault handling. The application of WSN technology in power distribution networks can provide protection and support for the construction of distribution networks in the following aspects:

1. By deploying integrated sensing equipment, power quality variations and load situation of large electricity can be monitored; moreover, the accuracy and timeliness of voltage, current, harmonics, and other information are improved.

2. By utilizing RFID, navigation, video surveillance, and smart wearable technology together, the capability of real-time monitoring of the status of distribution equipment and environmental parameters is strengthened. It can improve the fault location of distribution lines.

3. By monitoring distribution line conditions in underground distribution pipe networks, higher automatic levels of field operation monitoring and anti-theft facilities can be achieved.

Smart electricity consumption services

Intelligent electricity consumption services rely on a strong power grid and the concept of modern management, based on advanced metering, high efficiency control, high-speed communication, and quick energy storage technology, to realize the real-time interaction between power networks, customer energy flow, information flow, and business flow.

WSNs can connect the terminal equipment of the supply side and the user side with sensors to form a complete interactive network for electric energy consumption information and realize electricity information acquisition in a complex environment. Information integration analysis based on WSNs can guide the user or directly adjust the electricity consumption style to achieve the best configuration of power resources and reduce the electrical supply costs, improving reliability and efficiency. WSNs have broad application prospects in intelligent electricity consumption fields such as intelligent communities, intelligent industrial parks, and so forth.

The electric energy data acquisition system is a basis for intelligent electricity consumption services. The system could comprehensively collect several kinds of large user data. This includes data for special transformers, medium and small users of special transformers, three-phase general business users, single-phase general industrial and commercial users, as well as resident users and public distribution transformer data for the examination of metering points. This data can be combined to construct integrated power information platforms. The architecture of a WSN-based electric energy data acquisition system is shown in Figure 5.10.

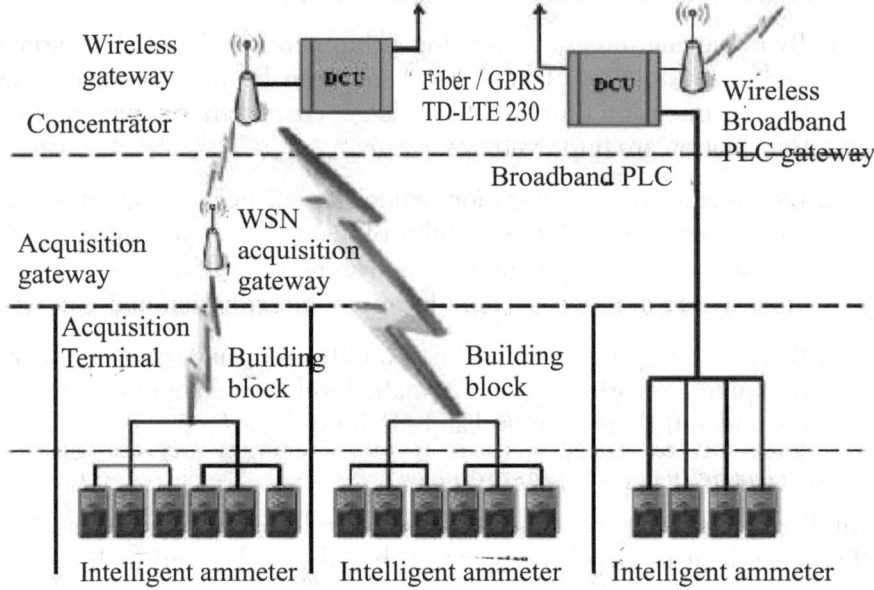

FIGURE 5.10 Architecture of a WSN-based electric energy data acquisition system.

5.7 Smartwater Sensor Networks

Today, the world's water consumption is 300% of what it was in 1950. The strong growth of the world's population combined with a strong growth of what is known as the middle class will continue to create increasing demand for the planet's limited resources. An example of a key resource in this context is the availability of clean water. In addition to the usual governmental regulation and policing of the exploitation of natural resources, many corporations are seeing the impact on the environment. They also see social and commercial advantages in taking steps to ensure the negative impact their operations have on natural resources is minimal.

Sustainability (water resource focus)

There are share price implications as well as regulatory requirements that drive this new green behavior. It can be said that it is generally accepted in modern society that the perceived need to better manage the environmental impact corporations have on scarce resources and the CO_2 footprint as an indication of pollution costs to society versus profit will become increasingly important.

Thus, the trend for corporations to invest in this area in addition to governments creating regulations demanding compliance to new environmental rules and creating new national market entry costs is clear.

When focusing on clean water, a monitoring system has to be built to determine base line quality as well as monitor the various potential sources of contaminants to clean water. Traditional operational technology systems are not usually created to monitor potential pollutants, thus new sensors and actuators need to be used also to monitor airborne pollutants that are usually the most difficult pollutants to track and manage (shown in Figure 5.11). The information gathered can not only be used as a key performance indicator (KPI) dashboard, but can also be used to predict water quality based on real-time monitoring of related events, such as man-made (pollution) or environmental (weather) events. This can be useful for corporations who are always working within an international regulatory framework, and can potentially lead to additional value creation in the form of emission/pollution certificates.

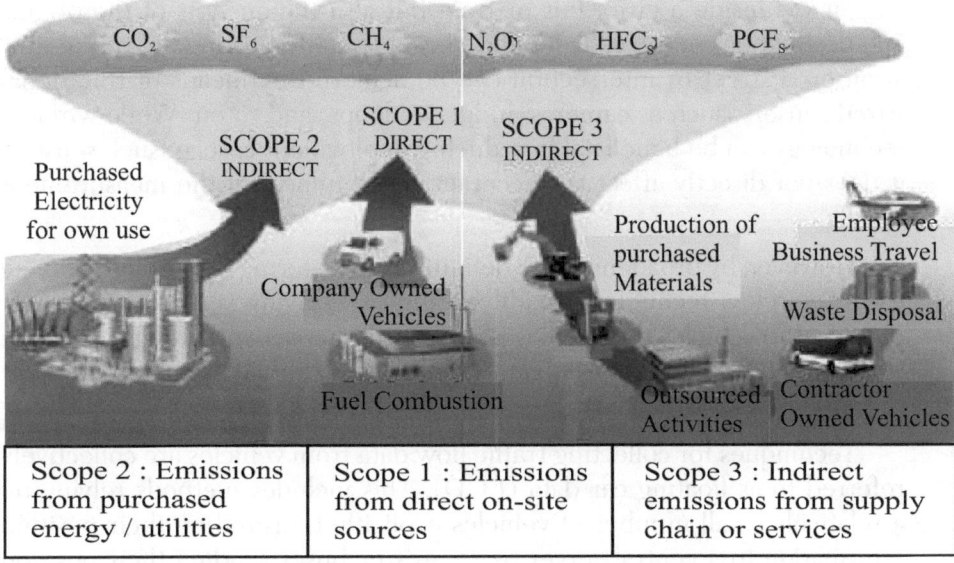

FIGURE 5.11 Airborne pollutants greenhouse gases (GHG) as a threat to water.

5.8 Intelligent Transportation Using WSNs

Wireless sensing in intelligent transportation differs on several points from the traditional concepts and design requirements for WSNs. In most cases, sensors can rely on some sort of infrastructure for power supply, for example, the aspect of energy efficiency is usually of secondary importance in these systems. WSN applications in intelligent transportation can be subdivided into two categories:

1. Stationary sensor networks, either on board a vehicle or as part of a traffic infrastructure.

2. Floating sensor networks, in which individual vehicles or other mobile entities act as the sensors.

The latter category comprises applications related to the tracking and optimization of the flow of goods, vehicles, and people, whereas the former comprises mainly applications that were formerly covered by wired sensors.

Sensing of traffic flows

Intelligent traffic management solutions rely on the accurate measurement and reliable prediction of traffic flows within a city. This includes not only an estimation of the density of cars on a given street or the number of passengers inside a given bus or train but also the analysis of the origins and destinations of the vehicles and passengers. Monitoring the traffic situation on a street or intersection can be achieved by means of traditional wired sensors, such as cameras, inductive loops, and so on. While wireless technology can be beneficial in reducing deployment costs of such sensors, it does not directly affect the accuracy or usefulness of the measurement results.

However, by broadening the definition of the term "sensor" and making use of wireless technology readily available in many vehicles and smart phones, the vehicles themselves as well as the passengers using the public transportation systems can become "sensors" for the accurate measurement of traffic flows within a city.

Techniques for collecting traffic flow data from vehicles are collectively referred to as floating car data (FCD). This includes methods relying on a relatively small number of vehicles explicitly transmitting their position information to a central server (e.g., taxis or buses sending their position obtained via GPS) as well as approaches relying on location information of mobile phones obtained from real-time location databases of cellular network operators. The latter approach does not actually involve any sensing by the vehicle itself, but still makes use of a wireless network (i.e., the existing cellular network) to sense or rather infer the current characteristics of traffic flows. The technical challenges lie particularly in the processing of the potentially large amounts of data, the distinction between useful and non-useful data, and the extrapolation of the actual traffic flow data from the observation of only a subset of all vehicles. Extensions of the FCD idea involving information gathered from the on-board electronics of the vehicles have been proposed under the term extended floating car data (XFCD). Collecting and evaluating data from temperature sensors, rain sensors, anti-lock braking system (ABS), Electronic Stability Control (ESC), and traction control systems of even a relatively small number of cars can be used to derive real-time information about road conditions which can be made available to the public and/or used for an improved prediction of traffic flows based on anticipated behavior of drivers in response to the road conditions.

Privacy issues must be taken into consideration whenever location or sensor data is collected from private vehicles. However, this is a general concern related to the monitoring of traffic flows, and schemes that don't make use of wireless technology (e.g., relying on license-plate recognition) also have to consider the car owners' privacy.

Equivalent to the measurement of vehicle movement by FCD, passenger behaviour in public transportation systems can be analyzed with the help of wireless technology. For example, electronic tickets, which typically employ RFID technology for registering the access to a subway station, bus, or train, effectively turn the passenger into a part of a sensor network, as shown in Figure 5.12.

The possibilities for gathering information about passenger movement and behavior can be further increased if smart phones are used to store electronic tickets. Especially for gathering information about intermodal transportation habits of passengers, electronic ticket applications for smart phones offer possibilities that conventional electronic tickets cannot provide. It remains to be seen, however, to which extent users will be willing to share position data in exchange for the convenience of using their mobile phone as a bus or metro ticket.

City logistics

Urbanization is posing a lot of challenges, especially in rapidly developing countries where already-huge cities are still growing and the increasingly wealthy population leads to a constantly rising flow of goods into and out of the city centers. Delivery vehicles account for a large portion of the air pollution in the cities, and streamlining the flow of goods between the city and its surroundings is the key to solving a lot of the traffic problems and improving the air quality.

FIGURE 5.12 Electronic tickets for smarter travel.

A promising approach toward reducing the traffic load caused by delivery vehicles is the introduction of urban consolidation centers (UCCs), that is, warehouses just outside the city where all the goods destined for retailers in a city are first consolidated and then shipped with an optimized routing, making the best possible use of truck capacity and reducing the total number of vehicles needed and the total distance traveled for delivering all goods to their destinations.

To achieve such optimization, careful analysis and planning of traffic flows in the city as well as monitoring of the actual flow of the goods are needed. Rather than just tracking a subset of vehicles as they move through the city, tracking of goods at least at a pallet level is required. The pallet (or other packaging unit) thus becomes the "sensor" for measuring the flow of goods, and a combination of multiple wireless technologies (GPS, RFID, WLAN, cellular) in combination with sophisticated data analysis techniques are applied to obtain the required data for optimizing the scheduling and routing of the deliveries and ensure timely arrival while minimizing the environmental impact of the transportation.

Vehicles of all kinds rely on an increasingly large number of sensors to ensure safe and smooth operation. This includes sensors primarily providing information to the driver as well as sensors that are part of the propulsion or vehicle dynamics systems. Due to the safety-critical nature of those subsystems, wireless technology is not usually a feasible option for these applications.

However, especially in large vehicles such as buses, trains, and airplanes, a lot of sensors and actuators serve non-safety-critical purposes, for example, monitoring cabin temperature, collecting data used in preventive maintenance of the vehicle, or monitoring the status of transported goods. In railway applications, WSNs can play an important role in the refurbishment of old carriages with state-of-the-art electrical systems. In airplanes, saving the weight of copper or aluminium cables by applying wireless sensors for non-critical applications is an important consideration. Wireless sensors employing energy harvesting techniques have been discussed even for monitoring the mechanical stress on composite materials forming part of the aircraft structure. Wiring the sensors in such "smart materials" would increase the weight of the structure and therefore significantly reduce the advantages of the composite material over conventional metal structures.

5.9 WSNs in Traffic Infrastructures

Traffic lights at intersections are usually controlled by units located close to the intersection, taking inputs from a set of sensors (e.g., inductive loops) as well as commands from a centralized control unit and switching the individual lights (also known as signal heads) according to the traffic rules and situational requirements.

With the number and complexity of sensors and display elements increasing, the task of a traffic controller today is really based on communication rather than a pure switching of the connected components. Traffic lights may be equipped with countdown timer displays, variable message signs display updated speed limits, and optical or radar-based sensors deliver information about the occupancy of individual lanes or the speed of vehicles passing the intersection. Upgrading the infrastructure of an existing intersection with state-of-the-art technology requires also providing the necessary communication links between sensors, signal heads, variable message signs, traffic controllers, and other components.

Wireless technology can help reduce the cost by eliminating the need to route communication cables (e.g., Ethernet) to all devices in an intersection. Such an installation will in most cases not be a pure sensor network, as it will usually also include display components or actuators.

Interaction of the traffic infrastructure with vehicles through wireless communication (e.g., granting priority to buses or emergency vehicles at intersections) is another promising application for wireless technology in traffic infrastructure. Though not all possible applications actually involve the exchange of sensor data over the wireless communication links, there are also a number of scenarios in which either vehicles share their sensor data with the infrastructure elements (e.g., regarding speed when approaching the intersection) or where the infrastructure provides sensor data to the vehicles (e.g., regarding road congestion on the other side of the intersection).

5.10 WSNs in Smart Homes

Faced with growing consumption and high energy costs, as well as the scarcity of fossil fuels, all of the scenarios developed by public institutions and experts to curb energy demand and our CO_2 emissions at the same time converge on energy efficiency being an absolute priority.

Based on this vision, the protocol of active control is articulated around the three strategies to maximize building performance while making it smart grid-compatible.

1. Act room by room: for maximizing the energy performance of a building, it is necessary to optimize the services rendered to the occupant, which is to say at the level of a room or a zone in a tertiary building. Thanks to the zone control, the occupant can adapt the environment to his or her activities and comfort.

2. Optimize energy supplies: to serve the needs of the occupants of a building, it is necessary to optimize the supply of energy based on the economic and carbon costs. The supply and distribution of energy are then managed as a function of the sum of the needs of each location. It enables control of the energy sources and the relationship with the upstream ecosystem consisting of the district, city, and so on. This strategy facilitates anticipation of the development of smart grids. It creates a system where each level contributes to optimization at a higher level. It also participates in developing the demand manage-

ment potential for electricity in buildings. Therefore, it is necessary to move from vertical independent application control to a multi-application control by zone.

3. Act on the engagement of the stakeholders: to improve the energy performance of a building, it is necessary to establish an incremental action plan to progressively look for sources of savings. However, the needs differ depending on the stakeholders involved.

Information strategies must be implemented that are tailored to the specific needs of each stakeholder and their areas of responsibility for helping them to make energy efficient decisions. Space and time fragmentation of the building and its technical systems have a strong impact on the efficiency of monitoring and energy savings through active control. Therefore, the implementation of active control strategies modify the sensing and control command architectures of active control solutions, to be based on a zone control ecosystem as shown in Figure 5.14, where the comfort sensor is one of the key elements. Further, this has been assessed on five pilot sites representing different climatic zones, sectors, constructive age, heating energy, hot water energy, and owner type. The savings went from 25% for areas residential areas up to 56% for schools. This is achieved as shown in Figure 5.13.

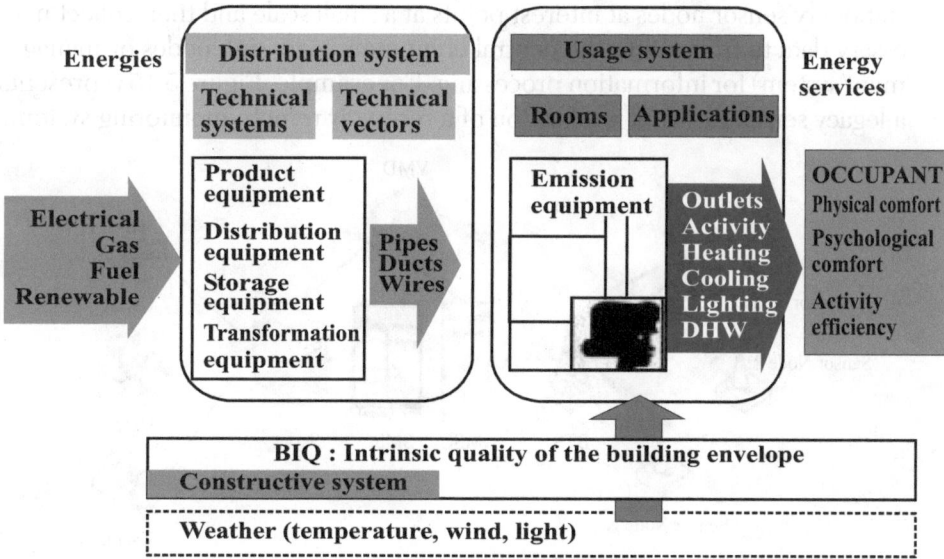

FIGURE 5.13 Systemic approach of energy in buildings.

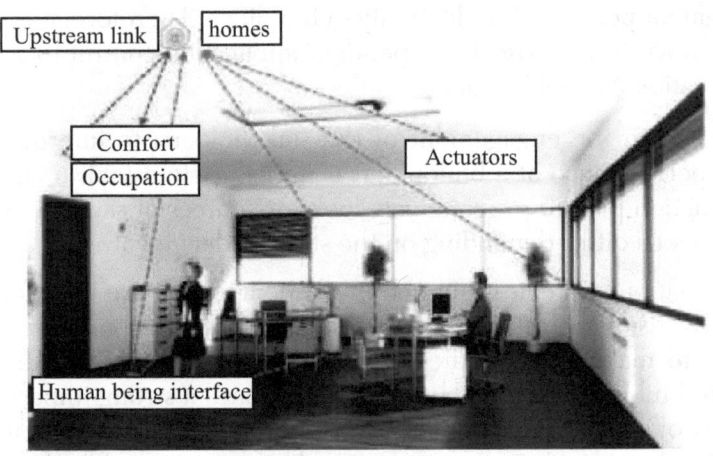

FIGURE 5.14 Zone control.

5.11 Monitoring Systems for Structure

Sensor network monitoring applications include structural health (bridge, building, dam, etc.) monitoring, home safety and intrusion detection, industrial distribution management, and critical resource and environment (fire, flooding, earthquake, etc.) surveillance. Most monitoring systems install stationary sensor nodes at interest points at a small scale and then collect necessary data to transmit to the central computers (e.g., sink nodes or management system) for information processing. For example, Figure 5.15 represents a legacy sensor network application of a reservoir remote monitoring system.

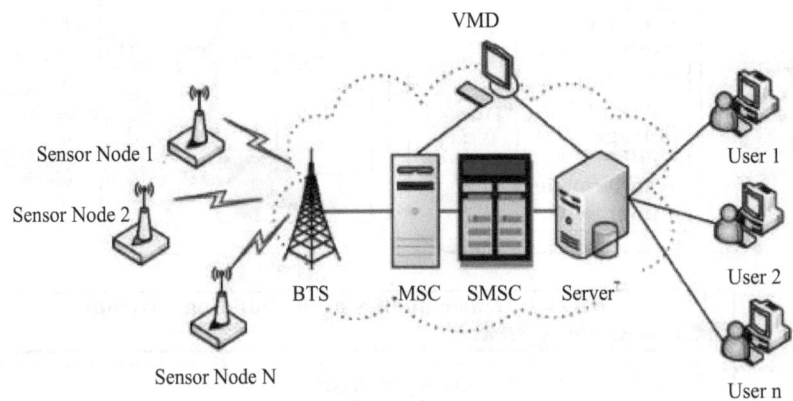

FIGURE 5.15 Legacy sensor network application.

The system collects information related to an open and close degree of sluice gate, water quantity, and BOD (biochemical oxygen demand) from sensors (sensor devices of Figure 5.15) installed around sluice gates. And then it transmits collected information to the central monitoring system (server of Figure 5.15) through BTS (base transceiver station) using wireless communication technologies such as HSDPA. The system analyzes the information taken from plenty of places and then displays it through a graphic user interface. Users can get analyzed results through SMS (short message service) from the monitoring system. A building monitoring system observes environmental information of temperature, humanity, CO_2, and crack information of buildings in real-time, and stores the information into databases using the Internet. This system can analyze their variance in sensing values and environmental information of the special zone so that it can estimate energy consumption of buildings. Analyzed and estimated information is displayed on a PDA or website via the Internet.

Representative WSN applications regarding ground environmental monitoring are a landslide precaution system and geological structure monitoring system. These systems monitor a fault plane and a landslide around primary national facilities. For example, Figure 5.16 illustrates a WSN-based, real-time landslide monitoring system. The system has been developed to forecast landslides, rock falls, and soil flow. In order to make predictions, the system keeps collecting monitoring data using several different sensors such as inclinometers, its chain, tachymetry, and GPS. The data are visualized in the web-based geographical information system WebGIS. With the analyzed data, preliminary warnings are offered via WebGIS and then users are informed of the warnings using SMS if the factor of safety gets lower than a specified threshold value.

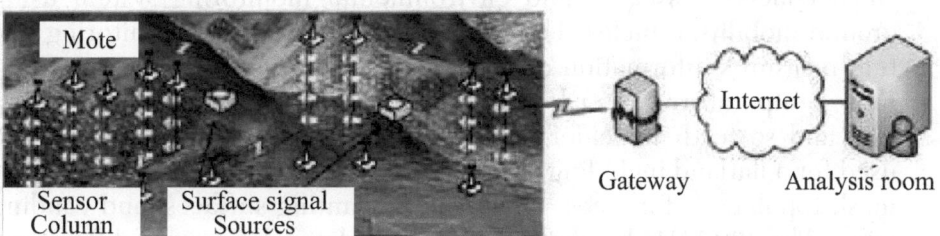

FIGURE 5.16 WSN-based real-time landslide monitoring system.

When collecting data from sensor nodes installed on the bridge, the system monitors ambient vibration of the structure. The sensor nodes have two types, an accelerometer sensor and a thermometer sensor connected to a patch antenna; Data are collected in a PC station by wireless network provided by the patch antenna. The collected data are used to analyze the structural dynamic of the bridge by estimating its modal properties. Figure 5.17 shows the hardware block diagram of the sensor nodes. Most of these systems consist of a small-scale network connected to the sink node directly, so that their network expansion capability is a limitation. Also, these systems have used a self-designed sensor format and network.

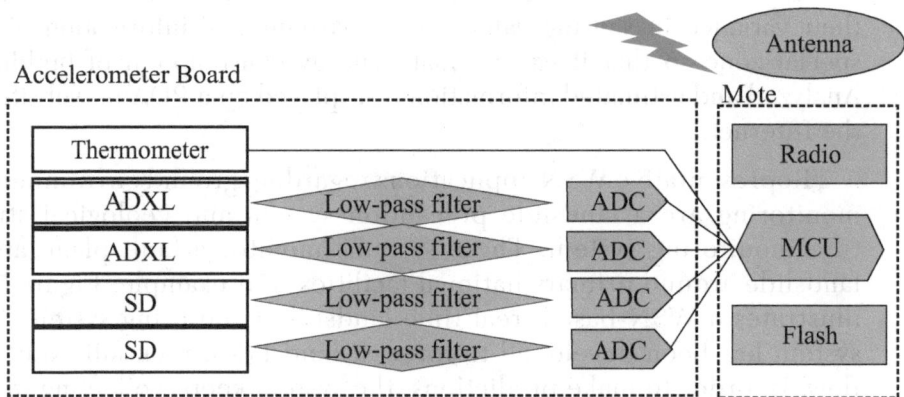

FIGURE 5.17 A structural health monitoring system.

5.12 IP-WSN Based Integration System

Figure 5.18 shows the real-time global monitoring integration system. The system consists of the ground/environment monitoring system, video collection system, and environmental monitoring system using ground mobility vehicles. The ground and environment monitoring system measures information on microseism, minute displacement, strain ratio, temperature, water level, water quality, exhaust/atmospheric gas, soil, and so forth for chief national facilities. A star topology may be used for a flatland including several types of farm fields, and a multi-hop mesh topology is for trees, forests, hills, slanting surfaces, and winding areas. The 400 MHz band may be a better choice because of diffraction. The video collection system provides video information required for the global monitoring system in liaison with closed-circuit television (CCTV) systems built in a variety of public/private institutions. A star topology and peer-to-peer topology may be used for network cameras for the system. Also, it supports streaming services for real-time transport of video data and transformation services for streaming the types of existing CCTV videos.

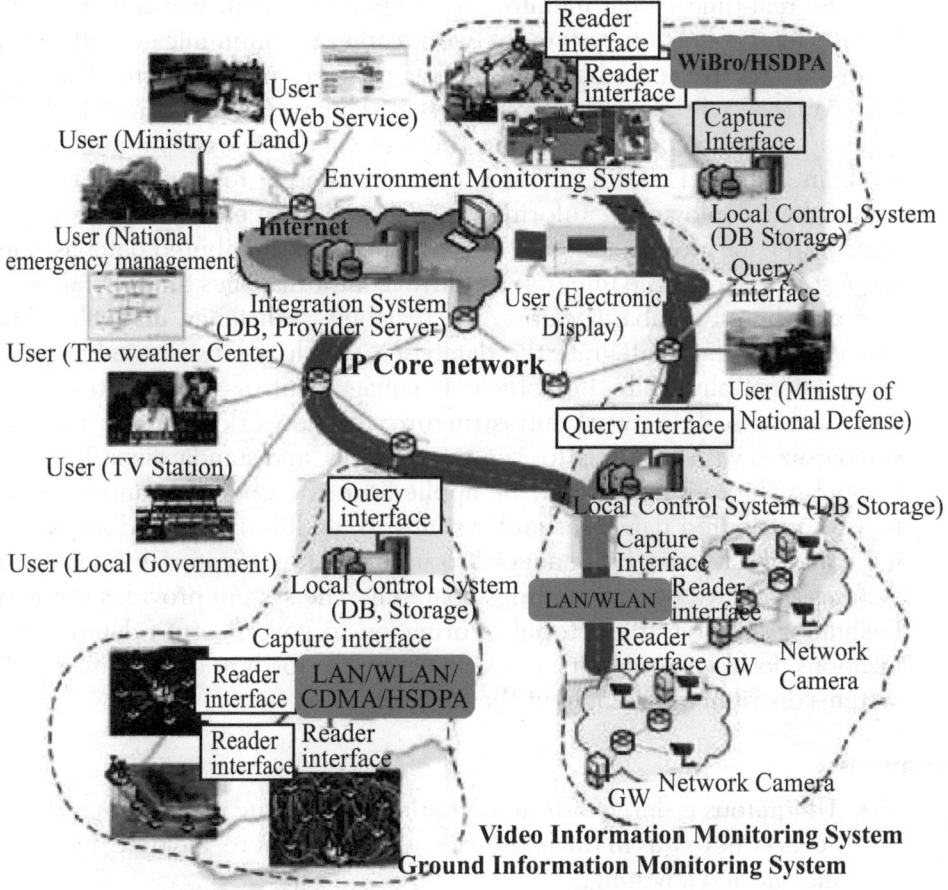

FIGURE 5.18 An overall system designed for real-time global monitoring.

The environmental monitoring system using ground mobility vehicles monitors a wide scope of areas at runtime by compensating the limit of fixed monitoring systems using special purpose vehicles or public transportation equipped with mobile IP-WSN gateways and sensors. Its targets to monitor include road conditions, road materials, road information, facilities near roads, and information on the yellow dust. A star topology may be mainly used in/on a vehicle, but a tree topology may be more useful due to the limit of communication coverage caused by the deployment of sensor nodes inside/outside the vehicle. The 2.4 GHz band may be a good choice for this system.

The real-time global monitoring integration system works as follows. In the video collection system, network cameras communicate with each other via TCP/IP, and a base station is connected to the Internet through LAN/WLAN, whereas other systems that deploy IP-WSNs and IP-WSN gateways may be connected to the Internet via a wireless communication medium (Wibro, HSDPA, CDMA, etc.) considering regional characteristics of their deployment. Information gathered from each system is managed by the local control system and it is stored in distributed databases in request to a query. The integrated control center manages those databases. The distributed databases can secure scalability, efficiency, and reliability since they logically integrate the databases, which are distributed in multiple systems physically. For efficient management of system integration and national-scale network infrastructure, the network protocols may be standardized with IP-WSN for sensor networks, and standardized Reader/Capture/Query interfaces may be applied for interoperation and integration of distributed databases and systems. In addition, standardized sensing data formats may be beneficial to unify existing territorial monitoring systems that have been operating separately. The system provides services facilitating integrated territorial information to government-related organizations, local governments, individuals, and so forth in close liaison with systems distributed throughout the country.

Summary

- Ubiquitous is derived from the Latin word ubique meaning "everywhere," as in the expression of 4A, anywhere, anytime, by anyone and anything.

- The main components of a USN are the sensor network, access network, middleware, applications platform, RFID tags, and so forth.

- A wireless personal area network (WPAN) is a personal, short distance area wireless network for interconnecting devices centered around an individual person's workspace.

- Intelligent transportation, robotic landmine detection, water catchment and ecosystem monitoring, bushfire response, detecting tracking, and monitoring of the events are some of the applications of USNs.

- A smart grid is a grid that will respond to real-time demand.

- Intelligent transportation is divided into a stationary sensor network and a floating sensor network.

Questions

1. Define a ubiquitous sensor network.

2. What are the main components of USNs?

3. List the characteristics of a USN.

4. Define the term WPAN.

5. List the application area of USNs.

6. Write about volcanic eruptions with WSN monitoring.

7. With the help of diagram write about the main parameters of a regional agriculture environment.

8. Explain the design of USNs for rice paddy crop monitoring.

9. Draw the general architecture of online monitoring systems for transmission lines based on WSNs. Explain.

10. Write about intelligent monitoring and early warning systems for substations.

11. Draw the architecture of a WSN-based electric energy data acquisition system.

12. Write in detail about WSN applications for smart water networks.

13. In detail write about intelligent transportation.

14. Explain WSN application for smart homes.

Further Reading

1. *Attacks and Defenses of Ubiquitous Sensor Networks: A Systematic Approach to Sensor Network Security* by Tanya G. Roosta

2. *Industrial Wireless Sensor Networks: Monitoring, Control, and Applications* by R. Budampati and S. Kolavennu

3. *Industrial Wireless Sensor Networks: Applications, Protocols, and Standards* by V. Cagri Gungor and Gerhard P. Hancke

References

1. *http://www.itu.int/dms_pub/itu-t/oth/23/01/T23010000040001PDFE.pdf*

2. *http://www.technicaljournalsonline.com/ijaers/20SEPTEMBER13/350.pdf*

3. *http://www.iec.ch/whitepaper/pdf/iecWP-internetofthings-LR-en.pdf*

CHAPTER 6

UNDERWATER WIRELESS SENSOR NETWORKS (UWSNS)

This chapter discusses the requirements of wireless sensor networks for oceanographic and water monitoring.

6.1 Wireless Sensor Networks for Oceanographic Monitoring

Coastal marine systems are particularly vulnerable to the effects of human activity attendant on industrial, tourist, and urban development. Information and communications technologies offer new solutions for monitoring, such ecosystems in real time. These systems are composed of sensor nodes, frequently wireless, which transmit data to a sink node, in real time, on a number of physical, chemical, and/or biological measurements (temperature, pH, dissolved oxygen, salinity, turbidity, phosphates, chlorophyll, etc.).

The design, implementation and deployment of a WSN for oceanographic applications poses new challenges different from the ones that arise on land, as the impact of the marine environment on the sensor network limits and affects its development. The following are some of the most important differences:

- The marine environment is an aggressive one which requires greater levels of device protection.

- Allowances must be made for movement of nodes caused by tides, waves, vessels, and so on.

- Energy consumption is high since it is generally necessary to cover large distances, while communications signals are attenuated due to the fact that the sea is an environment in constant motion.

- The price of the instrumentation is significantly higher than in the case of a land-sited WSN.

- There are added problems in deployment of and access to motes, the need for flotation and mooring devices, possible acts of vandalism, and others.

WSNs are largely designed and implemented ad hoc (buoys, electronics, and software) and oceanographic sensors. The two broad categories of marine wireless networks, depending on the data transmission medium that they use, are as follows:

1. WSNs based on radio frequency (RF) aerial communications (hereafter called Aerial WSNs or A-WSNs) and

2. Under Water Acoustic Sensor Networks (UW-ASNs).

In underwater conditions RF does not work well because radio waves propagate only at very low frequencies (30–300 Hz), and special antennas and a bigger power supply are required.

The limitations of UW-ASNs are

1. Bandwidth is severely limited

2. Propagation delays are five orders of magnitude greater than in terrestrial radio frequency channels

3. Higher bit error rates

4. Temporary losses of connectivity

5. Limited battery power because solar energy cannot be used

UW-ASNs are the best solution for viable oceanographic monitoring at great depths entailing the use of Autonomous Underwater Vehicles (AUVs) equipped with underwater sensors.

A-WSNs consist of a set of nodes with scanty power supplies, which moreover communicate with one another by way of low consumption radio modules. In addition, they have one or more nodes with bigger power supplies which act as sinks. These communicate with a remote station using longer range connections (via satellite, GPRS, etc.). This type of network should not be confused with the ones in which each node has a large power supply and connects directly to the base station. These are isolated buoys linked to a data collection center using satellite communications.

A-WSNs do not present the problems described previously in the case of UW-ASNs, but they do pose other problems; for example, they have to transmit via data cables running from underwater sensors to buoys on the surface. Where these sensors are located at great depths, the problems that arise can also be serious. In short, there is no ideal solution, and the most suitable technology will depend on the particulars of each case.

6.2 Aerial Wireless Sensor Networks for Oceanographic Monitoring

The various components of an A-WSN network and the resources needed to deploy it for oceanographic monitoring are listed as follows.

Sensor Nodes

Figure 6.1 details the elements commonly used in the design and implementation of a sensor node. It is normal to include a flotation device such as a buoy to keep part of the node out of the water. This out of the water part always includes an antenna for RF transmission, optionally a harvesting system (solar panel, generator, etc.) to supplement the power source, and in some cases one or more external sensors essentially to monitor meteorological data (wind speed, air temperature, atmospheric humidity, etc.). The submerged part of the node is composed of one or more sensors, which may be placed at different depths (sensor strings), a sonde to transmit the data collected to the buoy, and finally some means of anchoring the buoy to the seabed in order to prevent it from moving (due to marine currents, wind, waves, etc.).

The mote's electronics include: a module for RF transmissions, a power supply regulation and management system, a set of interfaces for accessing the sensors, a module for amplification, conversion (analog to digital) and multiplexing of the data read from the sensors (surface and underwater), a flash type permanent read/write memory, a clock to act as a timer, scientific instruments (e.g., improved meteorological packages, acoustic recording packages, biological samplers, etc.), and lastly a CPU (microprocessor) to centralize the whole process and implement the user-defined monitoring functions.

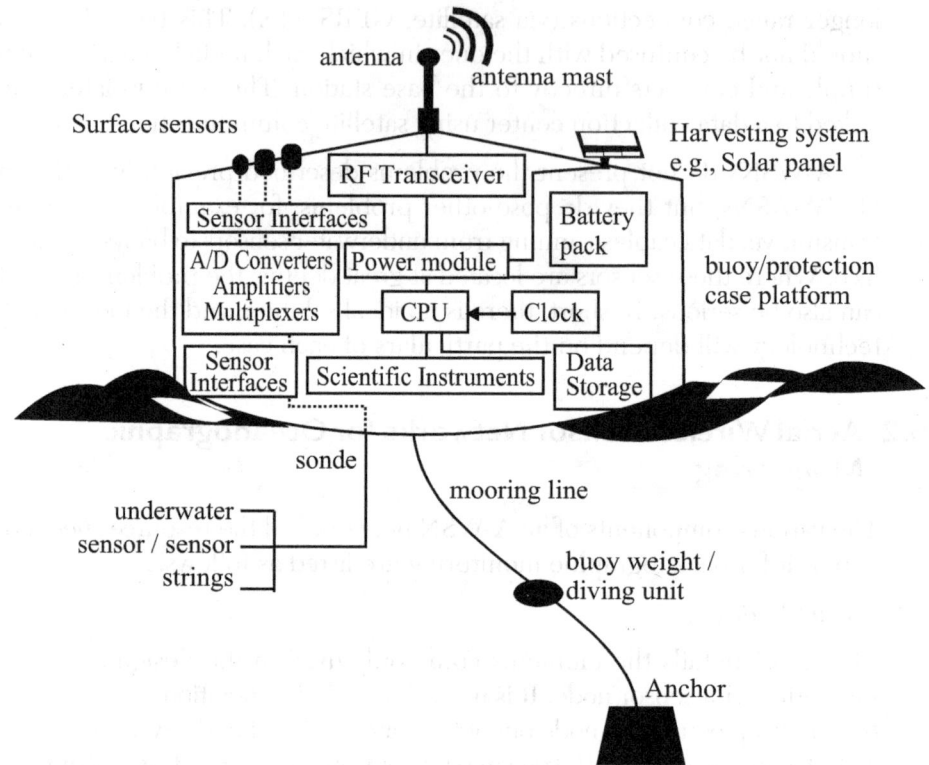

FIGURE 6.1 General scheme of a sensor node for oceanographic monitoring.

A-WSN General Architecture

Figure 6.2 shows a general architecture of an A-WSN for oceanographic monitoring. There are two main types for inter node communication: point to point vs. multi-hop. There are almost always one or more nodes that communicate directly with the base station. These act as sinks and may not have a role in the monitoring process. The differences between deployments are essentially determined by decisions concerning:

1. The network topology;

2. The dimensions of the area to be monitored and the number of nodes used in the deployment;

3. The communication devices/protocols used and the radio frequencies chosen;

4. Facilities for accessing the nodes for repair or removal (maintenance);

5. The flotation and mooring systems used;

6. The types of oceanographic sensors considered;

7. The tools for monitoring the network developed for real-time visualization of the data gathered; and

8. The electronics used for autonomous sampling of the requisite parameters and for wireless transmission to a data server.

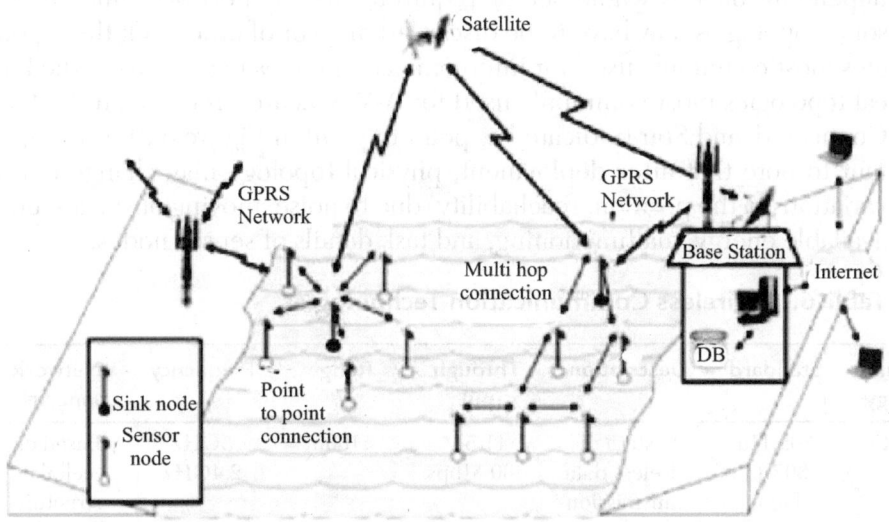

FIGURE 6.2 General structure of an A-WSN for oceanographic monitoring.

The component elements of an A-WSN are generally common, irrespective of where they are used (wireless sensor nodes, a communications protocol, a monitoring application, etc.). But even so, there are major differences depending on the characteristics of the deployment. One of the most obvious differences is the type of sensors that are to be used, which will be very specific to the environment it is proposed to monitor.

Wireless Communications

Network physical topology and density are entirely application dependant, so before deploying an A-WSN it is necessary to understand the environment in which it will be installed. This implies choosing the most suitable number of nodes and their absolute position inside the area to be monitored. Denser deployments improve data accuracy and provide sensor networks with more energy and better connectivity. However, at the same time a denser infrastructure can negatively affect network performance (data collisions, interferences, etc.). Thus, network density, physical topology, and communication type all determine the logical choice of topology.

Every topology has its own characteristics, which determine whether or not it is more suitable than others in terms of attributes of network functionality such as fault tolerance, connectivity, and so on. This means that depending on the whole set of requirements for network functionality, some topologies may have to be discarded in favor of others. Of the topologies most commonly used for interconnecting nodes in a network, the logical topologies most commonly used for A-WSNs are Tree, Chain, Partially Connected, and Star (indicated as point to point in Figure 6.2). It is important to note that after deployment, physical topology may change due to variations in the position, reachability (due to noise, moving obstacles, etc.), available energy, malfunctioning, and task details of sensor nodes.

Table 6.1 Wireless Communication Technologies

Techno-logy	Standard	Description	Through put	Range	Frequency	Network connectivity
Wi-Fi	802.11a 802.11 b/g/n	System of wireless data transmission over computational networks	11/54/ 300 Mbps	<100m	5GHz 2.4GHz	Based on cellular structure WLAN/ WiFi
WiMAX	IEEE 802.16	Standard for data transmission using radio waves	<75 Mbps	<10m	2-11GHz 3.5GHz : Europe	Point-to-point mobile cellular
Blue tooth	IEEE 802.15.1	Individual specification for WPAN which enables voice and data transmission between different devices by means of a secure, globally free radio link (2.4GHz)	v 1.2:1 Mbps v 2.0:3 Mbps UWB: 53–480 Mbps	Class 1: 100m Class2: 15–20m Class 3:1m	2.4GHz	Star (up to 7 nodes)

Techno-logy	Standard	Description	Through put	Range	Frequency	Network connectivity
GSM		Standard system for communica-tion via mobile telephones incorporat-ing digital technology	9.6 Kpbs	Dependent on cellular network service provider	900/ 1800 MHz: Europe 1900 MHz: USA	Point-to-point mobile cellular
GPRS		GSM extension for unswitched (or packaged) data transmission	56–144 Kbps	Dependent on cellular network service provider	2.5 GHz	Point-to-point mobile cellular
	IEEE 802.15.4	Standard defining the physical level and control of medium access of WPANs with low data transmission rates	20Kbps: 868MHz: Europe 40Kbps: 915MHz: America 250 Kbps 2.4 GHz: world wide	<100m	868/915 MHz and 2.4GHz	Star/Mesh Peer-to-Peer
ZigBee	IEEE 802.15.4	Specification of a set of high level wireless communica-tion protocols for use with low consump-tion digital radios, based on WPAN standard IEEE:802.15.4	250 Kbps –2.4GHz world wide	<75m	2.4GHz	Star/Mesh Peer-to-Peer/ Tree

For wireless communication, the sensor node incorporates a radio module, which is chosen to suit the desired range. Sometimes, in order to increase the range, range extenders for RF transceivers are incorporated, thus providing amplification to improve both output power and LNA (Low Noise Amplification). Another option, where such devices are insufficient to cover the distance, is to include a GSM/GPRS module.

For communication between sensor nodes it is possible either to develop communications protocols on the data linking layer using different medium access mechanisms (such as TDMA, FDMA, and CSMA), or else to use different wireless communication standards and technologies (Table 6.1) in which the technology chosen will depend on the requirements of the A-WSN it is proposed to implement, which in turn will be determined chiefly by the amount of information that has to be sent and whether images are to be sent in real time. Another requirement that has to be considered are the maximum distances that a communications link will have to cover, as this will determine the choice of RF antenna.

There are several types of antenna (omnidirectional, sector type, etc.) which are chosen on the basis of characteristics such as the radiation diagram, the bandwidth needed, directionality, gain, efficiency, beam width, and the desired polarization. In the case of sensor nodes, communication is more effective with omnidirectional antennas so that the radiated power is the same in all directions. This is necessary in that the movement of the sea can cause the sensor node to move rotationally, vertically, or horizontally, thus altering the original position of the buoy. The drawback of this kind of antenna is that the radiated power is more dispersed and hence the range is smaller than with more directional antennas. Directional antennas need to be properly aligned and the power channeled in a single direction; this assures more range in that direction and in some cases avoids interference with other services. One important factor that must be taken into account is the height of the antenna with respect to the flotation device supporting the node, since over long distances, visual line of sight is not sufficient for propagation due to attenuation and so RF line of sight is required.

Oceanographic Sensors

There are many types of sensors for monitoring oceanographic parameters (physical, chemical, and biological). The right choice of sensor depends on the requirements defined by the user and the requirements imposed by the characteristics of the area where they are to be deployed. These requirements include the measurement range within which the parameter is to be measured, the place where the sensor is to be deployed, sensitivity, linearity, accuracy, precision, resolution, measurement rate, power consumption, and deployment time. The parameters most commonly measured in a marine environment and the measurement units used are shown in Table 6.2. In addition, depending on what sensor is used, it is essential to consider its position within the node and the depth at which it will be working. For example, to determine the temperature profile of a water column, several sensors will have to be placed at different depths on the same vertical line.

Table 6.2 Common Oceanographic Sensors

Measured Parameter	Unit
Temperature	°C, °F
Pressure	mmHg
Salinity (Conductivity)	g/L
Water speed	m/s
Turbidity	FTU (Formazin Turbidity Unit) NTU (Nephelometric Trubidity Units) JTU (Jackson Turbidity Unit) mg/LSi02
Chlorophyll	µlL
Dissolved oxygen	mg/L
Nitrate	mg/L
pH	pKa
Swell	Height: (meters) Direction: (degrees)
Blue-Green Algae Phycocyanin	Relative Fluorescence Units
Ammonium/ammonia	mg/l-N
Chloride	mg/L
Rhodamine	µlL
Hydrocarbons	ppm

On the other hand, the sensor node may be equipped with surface sensors (Table 6.3), which are normally used to determine the state of the water surface or the atmosphere. These conditions may be important when setting up a sampling strategy. For example, in the event of bad atmospheric conditions, the sensor node may decide to raise the sampling frequency to as sure more precise monitoring of the environment.

Hardware/Software Solutions for Node Implementation

Some sensor node implementations reuse commercial solutions (MicaZ®, TelosB®, Mica2®, etc.) which come with an incorporated microprocessor, and communications electronics (radio modules, antennas, etc.). These motes normally come with a set of software development tools (operating system, programming languages, reusable components, etc.). When the characteristics of such commercial motes are inadequate or unsuitable, sensor nodes are commonly developed from scratch using the electronic components shown in Figure 6.1.

Table 6.3 Surface Sensors

Measured Parameter	Unit
Air temperature	°C, °F
Air pressure	mb
Wind speed	m/s
Wind direction	degrees
Precipitation	mm, inch
Atmospheric pressure	mmHg
Relative humidity	%RH
Solar radiation	W/m^2
Surface salinity	Ppt
Surface conductivity	S/m

The main component is a low power microprocessor, which is the core of the platform and is responsible for managing node operation. This microprocessor must possess certain features if it is to be suitable for use with an A-WSN: its architecture, combined with some low power modes, has to be optimized to achieve extended battery life in portable measurement applications. Also, it must include several universal serial synchronous/asynchronous communication interfaces (such as UART, I^2C, SPI, etc.) so that the sensors can be integrated with different types of electrical signals.

The lifetime of the network depends on the autonomy of the sensor nodes. Power is normally supplied by batteries (commonly D cell, Lithium ion, AA, or AAA batteries), which may be supplemented by harvesting systems (solar panels, generators, etc.) to prolong the useful life of the sensor node. It is sometimes necessary to adapt the voltage between the node's power supply and the rest of the components by means of DC/DC converters. Inclusion of a flash read/write permanent memory (SD, MMC, etc.) enhances the robustness of the mote by allowing data to be stored and transmitted later on when conditions permit, thus avoiding loss of information. Another important component is a low consumption clock operating in real time. With a clock synchronized with all the other motes, when a reading is taken from a sensor it can be stored along with the exact time of the reading. Later on, the information can be relayed to the data server, which is important when it comes to analyzing the resulting data.

Monitoring Application

The information gathered by the sensor nodes has to be transmitted to a base station or monitoring station or PC or Laptop with a massive data storage system (relational databases are the commonest solution such as SQL, etc.) which can also be used for the necessary studies using the existing oceanographic theoretical models. Having integrated monitoring tools makes it possible to maintain permanent communication with the sensor network deployed and access to the stored data via the Internet. The information displayed by these tools usually consists of the number of nodes deployed, the parameters analyzed, the geographical location of each node, the most recent data gathered by the sensor nodes, and visualization of a data historical table. The number of buoy implementations on the basis of the location of their components (electronics, radio, batteries, sensors) and mooring system are shown in Figure 6.3.

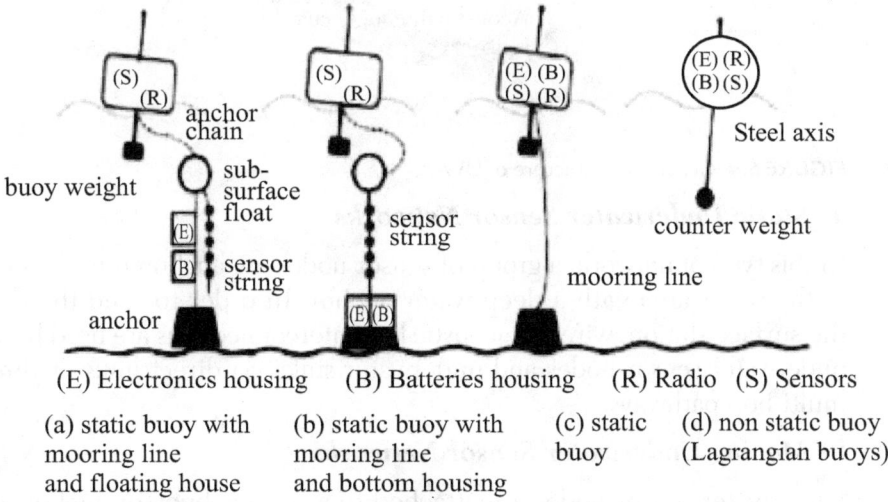

(E) Electronics housing   (B) Batteries housing   (R) Radio   (S) Sensors

(a) static buoy with mooring line and floating house

(b) static buoy with mooring line and bottom housing

(c) static buoy

(d) non static buoy (Lagrangian buoys)

FIGURE 6.3 Most representative configurations of buoys used in A-WSNs.

6.3 Underwater Acoustic Sensor Networks (UW-ASNs)

UW-ASNs consist of underwater sensor nodes and autonomous underwater vehicles, which are deployed to carry out cooperative surveillance in a given area as shown in Figure 6.4. The typical physical layer technology in underwater networks involves acoustic communication. The architectures of underwater acoustic sensor networks can be categorized depending on the network topology used. Thus, network topology is considered as a crucial factor in terms of the capacity of the network, as well as the energy consumption requirements.

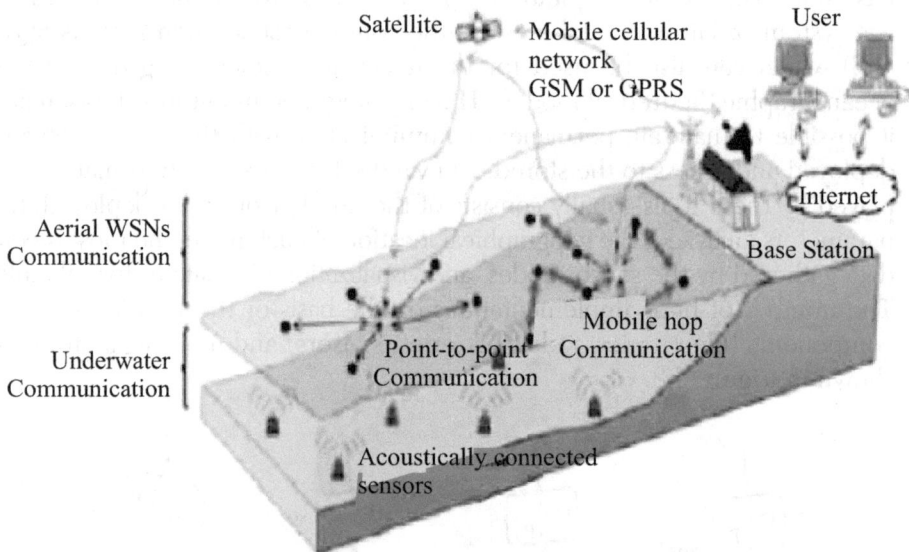

FIGURE 6.4 General architecture of UW-ASNs.

a) Static Underwater Sensor Networks

In this type of network, a group of sensor nodes are anchored to the bottom of the water area with a deep water anchor. In order to send the data to the surface station, wireless acoustic link interconnections are used between underwater sensor nodes and underwater sinks via direct links or through multi-hop pathways.

b) Moving Underwater Sensor Networks

Underwater sensor nodes are attached to a surface buoy or anchor to the bottom of the water area, with flexibility of movement in a specific area.

c) Underwater Sensor Networks with Autonomous Underwater Vehicles

Autonomous Underwater Vehicles (AUVs) are used to enhance the capabilities of underwater sensor networks in terms of self configuration of sensor nodes (e.g. maintenance of underwater network infrastructure), adaptive sampling, power supply issues, and depth capability, which can reach up to 1500m. Furthermore, global position satellite (GPS) technology can be used to track the location of the vehicles on or near the surface.

6.4 Underwater Wireless Sensor Networks (UWSNs)

A scalable UWSN provides a promising solution for efficiently exploring and observing aqueous environments and operates under the following constraints:

1. Unmanned Underwater Exploration

Underwater conditions are not suitable for human exploration. High water pressure, unpredictable underwater activities, and a vast size of water area are major reasons for unmanned exploration.

2. Localized and Precise Knowledge Acquisition

Localized exploration is more precise and useful than remote exploration because underwater environmental conditions are typically localized at each venue and variable in time. Using long-range SONAR or other remote sensing technology may not acquire adequate knowledge about physical events happening in the volatile underwater environment.

3. Tetherless Underwater Networking

The Internet is expanding to outer space and underwater. The current tethered technology allows constrained communication between an underwater venue and the ground infrastructure, and it incurs significant cost of deployment, maintenance, and device recovery to cope with volatile undersea conditions.

4. Large Scale Underwater Monitoring

Traditional underwater exploration relies on either a single high cost underwater device or a small scale underwater network. Neither existing technology is suitable to applications covering a large area. Enabling a scalable underwater sensor network technology is essential for exploring a huge underwater space.

Underwater sensor networks have many potential applications. The seismic imaging of undersea oilfields are a representative application. Today, most seismic imaging tasks for offshore oilfields are carried out by a ship that tows a large array of hydrophones on the surface. The cost of such technology is very high, and the seismic survey can only be carried out rarely, for example, once every 2–3 years. In comparison, sensor network nodes have very low cost, and can be permanently deployed on the sea floor. Such a system enables frequent seismic imaging of a reservoir (perhaps every few months), and helps to improve resource recovery and oil productivity. The following Figure 6.5 shows the schematic diagram of UWSNs.

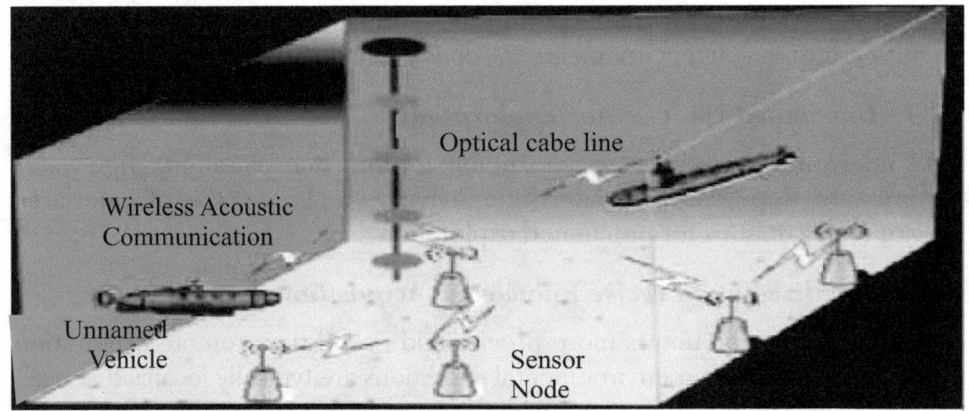

FIGURE 6.5 Schematic of Underwater Wireless Sensor Networks.

Some of the challenges are different. First, radio is not suitable for underwater usage because of extremely limited propagation (current mote radios transmit 50–100 cm). While acoustic telemetry is a promising form of underwater communication, off the shelf acoustic modems are not suitable for underwater sensor nets with hundreds of nodes: their power draws, ranges, and price points are all designed for sparse, long range, expensive systems rather than small, dense, and cheap sensor nets. Second, the shift from RF to acoustics changes the physics of communication from the speed of light (3×10^8 m/s) to the speed of sound (around 1.5×10^3 m/s), a difference of five orders of magnitude. While propagation delay is negligible for short-range RF, it is a central fact of underwater wireless. This has profound implications on localization and time synchronization. Finally, energy conservation of underwater sensor nets will be different than on the ground because the sensors will be larger, and because some important applications require large amounts of data, but very infrequently (once per week or less).

In a UWSN, the sensor mobility can bring two major benefits:

1. Mobile sensors injected in the current in relative large numbers can help to track changes in the water mass, thus providing 4D (space and time) environmental sampling. 4D sampling is required by many aquatic systems studies, such as estuary monitoring; the alternative is to drag the sensors on boats and or on wires and carry out a large number of repeated experiments. This latter approach would take much more time and possibly cost. The multitude of sensors helps to provide extra control on redundancy and granularity.

2. Floating sensors can help to form dynamic monitoring coverage and increase system reusability. In fact, through a "bladder" apparatus one can dynamically control the depth of the sensor deployment, and force resurfacing and recovery when the battery is low or the mission is over. In traditional aquatic monitoring or surveillance applications, sensors are usually fixed to the sea floor or attached to pillars or surface buoys, and sensors with computational power are usually of big size. Thus, the sensor replacement and recovery cost is very high, as also results in low system reusability.

System Architecture of UWSNs

Figure 6.6 shows the general architecture for an underwater sensor network.

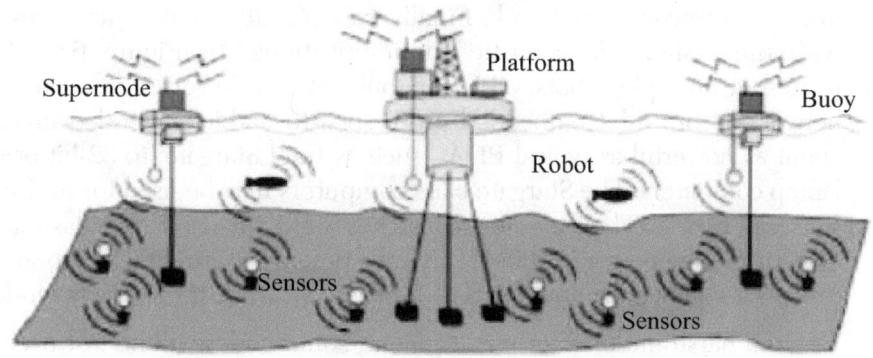

FIGURE 6.6 Underwater network deployment.

In Figure 6.6, four different types of nodes are shown in the system. At the lowest layer, the large number of sensor nodes are deployed on the sea floor. They collect data through attached sensors (e.g., seismic) and communicate with other nodes through short-range acoustic modems. They operate on batteries, and to operate for long periods they spend most of their life asleep. Several deployment strategies of these nodes are possible; they are anchored to the sea floor or buried for protection. Tethers ensure that nodes are positioned roughly where expected and allow optimization of placement for good sensor and communications coverage. Node movement is still possible due to anchor drift or disturbance from external effects. Nodes are able to determine their locations through distributed localization algorithms. At the top layer are one or more control nodes with connections to the Internet. The node shown on the platform in Figure 6.6 is this kind of node. These control nodes may be positioned on an off-shore platform with power, or they may be on shore; these nodes are to have a large storage capacity to buffer data, and access to sufficient electrical power.

Control nodes will communicate with sensor nodes directly, by connecting to an underwater acoustic modem with wires. In large networks, a third type of nodes, called super nodes, can be deployed. Super nodes have access to high speed networks, and can relay data to the base station very efficiently. There are two possible implementations: the first involves attaching regular nodes to tethered buoys that are equipped with high-speed radio communications to the base station, as shown in Figure 6.6. An alternative implementation would place these nodes on the sea floor and connect them to the base station with fiber optic cables. Super nodes allow a much richer network connectivity, creating multiple data collection points for the underwater acoustic network. Finally, robotic submersibles are interacting with the system via acoustic communications. In Figure 6.6, "fishes" represent multiple robots. CPU capability at a node varies greatly in sensor networks, from 8-bit embedded processors, to 32-bit embedded processors about as powerful as typical PDAs, such as Intel Stargate, to 32-bit or 64-bit laptop computers. The Stargate class computers may be used for underwater sensor networks. Their memory capacities (64MB RAM, 32MB flash storage) and computing power (a 400MHz XScale processor) are sufficient to store and process a significant amount of data temporarily, while their cost is moderate.

In a harsh underwater environment, some nodes will be lost over time. Possible risks include fishing trawlers, underwater life, or failure of waterproofing. Therefore, one should expect basic deployments to include some redundancy, so that loss of an individual node will not have wider effects. In addition, one will be able to recover from multiple failures, either with mobile nodes, or with deployment of replacements. Operating on battery power, sensor nodes must carefully monitor their energy consumption. It is essential that all components of the system operate at as low a duty cycle as possible.

In addition, it is necessary to coordinate with the application to entirely shut off the node for very long periods of time, up to days or months, and build on techniques for long duration sleep. Networking protocols that allow underwater nodes to self configure and coordinate with each other are time synchronization, localization, MAC, and routing. First, applications benefit from local processing and temporary data storage. Storage can be used to buffer data to manage low speed communications, "time-shifting" data collection from retrieval. In some cases, nodes benefit from pair-wise communications and computation. Finally, in most sensing applications, the data is to be eventually relayed to the user through the Internet or a dedicated network.

Applications of UWSNs

The application of wireless sensor networks to the underwater domain has huge potential for monitoring the health of river and marine environments. Monitoring these environments is difficult and costly for humans: divers are regulated in the hours and depths at which they can work, and require a boat on the surface that is costly to operate and subject to weather conditions. Figure 6.7 shows anti-submarine warfare, a branch of naval warfare that uses surface warships, aircraft, or other submarines to find, track, and deter, damage, or destroy enemy submarines.

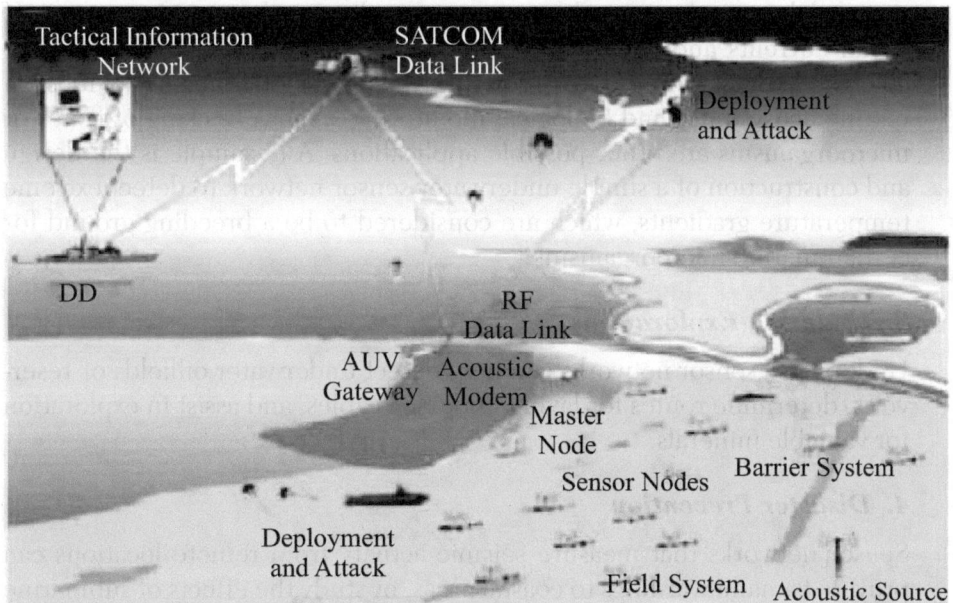

FIGURE 6.7 Anti-submarine warfare.

A sensor network deployed underwater could monitor physical variables such as water temperature and pressure as well as variables such as conductivity, turbidity, and certain pollutants. The network could track plumes of silt due to dredging operations or pollutants flowing in from land, and it could monitor and model the behavior of underwater ecosystems. Imaging sensors could be used to measure visible change in the environment or count, and perhaps even classify species, and is also useful for disaster prevention. The applications of underwater sensor networks are classified as follows.

1. Ocean Sampling

Networks of sensors and AUVs can perform synoptic, cooperative adaptive sampling of the 3D coastal ocean environment and advanced ocean models to improve the ability to observe and predict the characteristics of the oceanic environment.

2. Environmental Monitoring

UW-ASNs can perform pollution monitoring (chemical, biological, and nuclear). For example, it may be possible to detail the chemical slurry of antibiotics, estrogen-type hormones, and insecticides to monitor streams, rivers, lakes, and ocean bays (water quality analysis). Monitoring of ocean currents and winds, improved weather forecast, detecting climate change, understanding and predicting the effect of human activities on marine ecosystems, and biological monitoring such as tracking of fishes or microorganisms are other possible applications. An example is the design and construction of a simple underwater sensor network to detect extreme temperature gradients, which are considered to be a breeding ground for certain marine microorganisms.

3. Undersea Explorations

Underwater sensor networks can help detect underwater oilfields or reservoirs, determine routes for laying undersea cables, and assist in exploration for valuable minerals.

4. Disaster Prevention

Sensor networks that measure seismic activity from remote locations can provide tsunami warnings to coastal areas, or study the effects of submarine earthquakes (seaquakes).

5. Assisted Navigation

Sensors can be used to identify hazards on the seabed, locate dangerous rocks or shoals in shallow waters, mooring positions, submerged wrecks, and perform bathymetry profiling.

6. Distributed Tactical Surveillance

Autonomous Underwater Vehicles (AUV) and fixed underwater sensors can collaboratively monitor areas for surveillance, reconnaissance, targeting, and intrusion detection systems. For example, a 3D underwater sensor network is designed for a tactical surveillance system that is able to detect and classify submarines, small delivery vehicles (SDVs), and divers based on the sensed data from mechanical, radiation, magnetic, and acoustic microsensors. With respect to traditional radar/sonar systems, underwater sensor networks can reach a higher accuracy and enable detection and classification of low signature targets by also combining measures from different types of sensors.

7. Mine Reconnaissance

Simultaneous operation of multiple AUVs with acoustic and optical sensors can be used to perform rapid environmental assessment and detect mine like objects.

8. Flocks of Underwater Robots

A very different application is supporting groups of underwater autonomous robots. Applications include coordinating adaptive sensing of chemical leaks or biological phenomena (for example, oil leaks or phytoplankton concentrations), and also equipment monitoring applications.

9. Oceanography

Oceanography is the study of processes that govern the complex interplay of tides, currents, waves, and seabed and coastal modeling. Oceanography can tell us about coastal deposition and erosion and consequently about flooding and sea defenses. Sensor networks offer a new paradigm for oceanography, and many other scientific, commercial, agricultural, and industrial applications.

Marine Sensor Package Design

The sensor package was designed as a waterproof cylinder (approx. 50 cm long) containing a sensor section, a data logger, and a microprocessor (PIC) running the lightweight device control algorithm, designed to control the measurement rates, data processing, queue management, data aggregation, and data forwarding, together with two alkaline D cells. The sensor package is designed to last several months using these batteries. Attached to this cylinder is a "dongle" which is able to move in the current, and thus provide current velocities. It includes a sensor section with which wave height can be derived, an optical backscatter sensor that measures turbidity, and an electrical conductivity sensor, which is used as a surrogate for salinity. The sensor package as shown in Figure 6.8 is suspended within a pyramidal-shaped cage, designed to remain fixed on the seabed, thereby giving a consistent reference orientation for current velocities and clearance above the sea bed for the optical sensor.

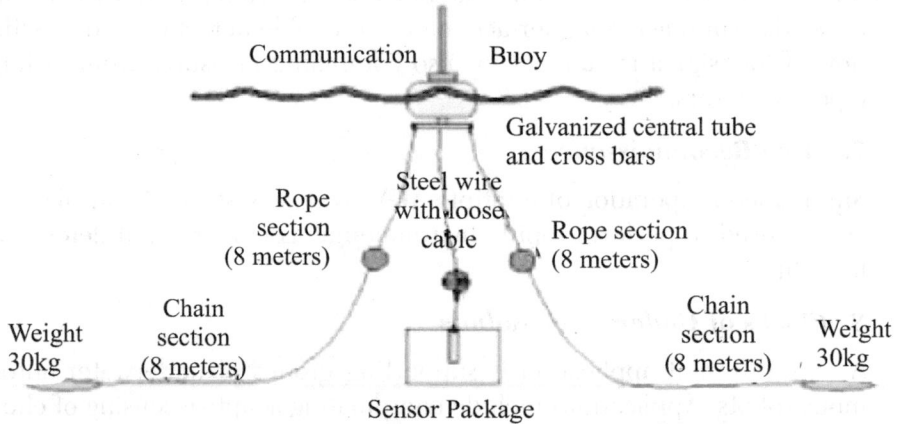

FIGURE 6.8 Sensor node mechanical design.

Sensors transmit the collected data from different forms of natural and human made phenomena such as sound, light, temperature, salinity, and pollution in water areas to a server "sink/gateway" and subsequently to the end user. Sensor nodes are interconnected by point to point and multi-hop communication networks. Wireless sensor network infrastructure requires standards and protocols (zigbee, IEEE 802.15.4, etc.) which take into account the battery life of the node and node cost in addition to the operating environment.

Security Issues in UWSNs

Security is considered as a central issue in WSNs, providing confidentiality, authentication, and the integrity of sensor data transmission. In order to achieve secure data transmission between nodes, complex cryptographic algorithms are required. However, the capabilities and constraints of UWSNs dictate the security services that are needed and the mechanisms that can be used. In particular, with communication between a large number of sensor nodes, power consumption, capability of key storage, and computation of new security keys must be considered.

The services include: confidentiality, authentication, integrity of data, and node/data availability. The WSN world in particular offers many obstructions/difficulties in providing these services. These difficulties are discussed as follows:

1. Confidentiality

Sensor nodes may be attacked in order to reveal the sensor data. Encrypted information with a secret key will maintain data confidentiality. This data should only be exposed to permissible users, who can decrypt the data with the correct key.

2. Authentication

Data transmission between nodes must be trusted. As such the receiver must ensure that any data received is authenticated. This can be provided using resource friendly tools such as hardware-implemented hashing algorithms.

3. Integrity

The same hashing algorithms that can be used to provide source authentication are used to provide data integrity. Hardware implementations of these algorithms can limit their draw on system resources such as power and memory.

4. Availability

Nodes in the network may suffer from Denial of Service (DoS) attacks. Network systems can protect the availability of nodes by enabling them to be self organizing and through the use of suitable rekeying algorithms. This rekeying will enable the network to be self healing while keeping security of data at the fore.

Key Managementin UWSNs

To ensure the security of any application in UWSNs, key management mechanisms are a most critical operation. These include generating, distributing, and revoking cryptographic keys. In UWSNs, there are two kinds of keying schemes generally used: network-wide and node-specific pre-deployed keying. The former supplies the same system-wide master key to each sensor node for the entire network, whereas the latter equips each neighboring node with a unique key to allow communication pairing between neighbor nodes to take place.

1. Key Pre-distribution

Keys are generated and then installed in the memory of each sensor node, which creates a key ring. Furthermore, the key ring identifiers of each sensor node and its associated key ring are kept in a controller node in the network. This phase must be completed before deploying the sensornodes.

2. Discovery of the Common S hared Key

In this step, nodes broadcast their identifier key ring in order to discover a pair-wise key. At this point in the operation, the topology of the network is established by the communication links between the nodes that share a common key.

3. Establishment of Path Keys

In some cases if the node does not discover a shared key with other nodes, and they are connected by a multi-hop path, then it is possible for a path key to be established between the nodes. This key is known as an end-to-end path key.

4. Revocation of Stray Sensor Nodes

During the operation of UWSNs, some nodes may not function as expected due to reasons such as a compromised sensor nodes, or power becoming exhausted. As a result of this, these nodes must be isolated. Revoking the entire key ring of these nodes from the network will remove particular communication links in the network. Revocation messages consist of a set of key identifiers of revoked nodes which are broadcast by controller nodes.

5. Rekeying

This phase occurs after isolating deviant nodes. The rekeying step must take place in sensor nodes in order to generate and replace the expired key rings after employing the revocation algorithm.

The UWSNs which can be subjected to many kinds of node/data attacks are dependent on many factors, such as: environmental and water conditions, energy constrained operation requirements, suitable network communication and software design/topologies, and finally security.

A UWSN architecture can be classified into two classes based on the required coverage area: small coverage area and wide coverage area. Personal area networks (PAN) connect sensor nodes in a wireless communication range up to 10m in the 2.4GHz ISM band. For instance, IEEE 802.15.4 or Bluetooth can be used to connect several wireless sensors inside a circle with a radius of ten meters, with low power consumption and a data rate up to 480 Mbps. Wireless Local Area Networks (WLAN) have a communication range of 250 meters, in some cases up to 600 Mbps. On the other hand, point-to-point mobile cellular networks used in UWSNs offer ranges up to 50km such as with IEEE802.15.1 or GPS/GRPS.

Second, a particularly important factor in the deployment of UWSNs is how to overcome security attacks with dynamically changing network topology. These attacks can consist of node impersonation, denial of service, and data disclosure attacks. Hence, applying different kinds of security mechanisms and management techniques is particularly important in order to prevent and detect attacking attempts and to ensure integrity, confidentiality, and authentication of transmitted data. To facilitate this, key management and encryption schemes are at the core of security communication requirements. Dynamic key management schemes are finding considerable use in UWSNs, where they are capable of adding new nodes and ejecting compromised nodes. In some approaches, to ensure the security of the network, sensor nodes share a single symmetric key, known as a network-wide master key, which is used to facilitate rekeying of a network with session keys used for encrypting and decrypting messages. These session keys can be updated and redistributed when the sensor nodes change, drop out of the network, or are attacked (rekeying).

Another approach is neighborhood key management. In this approach, each sensor node only keeps and shares a symmetric key with its closest nodes (neighbors). The sender encrypts the message key with the neighborhood key and attaches the encrypted key (like a single session key) to the message. Thus, in order to forward the message, the node must re-encrypt the message key. When the node needs to send data to another node for the first time, the received node demands a certificate from the sender, which identifies it as a legitimate node. Once the receiver ensures the authentication of the sender by exchanging certificates, the secret keys can be exchanged to use for encrypting and signing messages.

6.5 Challenges In Network Protocol

The design challenges along the network protocol stack in a top-down manner are given as follows. At each layer, there are many critical problems awaiting the solutions.

1. *Security, Resilience, and Robustness*

A self-organizing sensor network needs more protections than cryptography due to the limited energy, computation, and communication capabilities of sensor nodes. A critical security issue is to defend against denial-of-service attacks, which could be in the form of depleting nodes on device resources (especially draining batteries by incurring extra computation and communication) and disrupting network collaboration (e.g., routing, data aggregation, localization, clock synchronization). Such attacks can disrupt or even disable sensor networks independent of cryptographic protections.

In a UWSNs, due to the unique characteristics of underwater acoustic channels, denial-of-service attacks are lethal. In particular, a wormhole attack (in which an attacker records a packet at one location in the network, tunnels the data to another location, and replays the packet there) and its variants impose a great threat to underwater acoustic communications. Many countermeasures that have been proposed to stop wormhole attacks in radio networks are ineffectual in UWSNs. Thus, to protect against wormhole attacks in UWSNs, new techniques are demanded.

Another problem that may arise in UWSNs is intermittent partitioning due to water turbulence, currents, ships, and so forth. In fact, there may be situations where no connected path exists at any given time between source and destination. This intermittent partitioning situation may be detected through routing and by traffic observations. Disruption Tolerant Networking (DTN) includes the use of intermediate stores and forward proxies. If the data sink (i.e., the command center) suspects the presence of such conditions, it can then take advantage of some of the DTN techniques to reach the data sources.

2. *Reliable and/or Real Time Data Transfer*

Reliable data transfer is of critical importance. There are typically two approaches for reliable data transfer: end-to-end or hop-by-hop. The most common solution at the transport layer is TCP (Transmission Control Protocol), which is an end-to-end approach. TCP performance is problematic because of the high error rates incurred on the links, which were already encountered in wireless radio networks.

Another type of approach for reliable data transfer is hop-by-hop. The hop-to-hop approach is favored in wireless and error-prone networks, and is believe to be more suitable for sensor networks. PSFQ (Pump Slowly and Fetch Quickly) employs the hop-by-hop approach. In this protocol, a sender sends a data packet to its immediate neighbors at a very slow rate. When the receiver detects some packet losses, it has to fetch the lost packets quickly. Hop-by-hop, data packets are finally delivered to the data sink reliably. In PSPQ, ARQ (Automatic Repeat Request) is used for per-hop communication. However, due to the long propagation delay of acoustic signals, in UWSNs, ARQ would cause very low channel utilization. One possible solution to solve the problem is to investigate erasure coding schemes, which, though introducing additional overhead, can effectively avoid retransmission delay. The challenge is to design a tailored, efficient coding scheme for UWSNs. In UWSNs, due to the high error probability of acoustic channels, efficient erasure coding schemes could be utilized to help achieve high reliability and at the same time reduce data transfer time by suppressing retransmission.

3. Traffic Congestion Control

Congestion control is an important while tough issue to study in many types of networks. In UWSNs, high acoustic propagation delay makes congestion control even more difficult. In ground-based sensor networks, the congestion control problem is thoroughly investigated in CODA (Congestion Detection and Avoidance). In CODA, there are two mechanisms for congestion control and avoidance: open loop hop-by-hop backpressure and closed loop multisource regulation. In the open loop hop-by-hop backpressure mode, a node broadcasts a backpressure message as soon as it detects congestion.

For UWSNs, a combination of open and closed loops may apply, since it provides a good compromise between fast reaction (with open) and efficient steady-state regulation (with closed). Considering the poor quality of acoustic channels, one aspect that deserves further investigation is the distinction between loss due to congestion and loss due to external interference. From received packet inter-arrival statistics and from other local measurement, the data sink may be able to infer random loss versus congestion and maintain the rate (and possibly strengthen the channel coding) if loss is not congestion related.

4. Efficient Multi-Hop Acoustic Routing

Like in ground-based sensor networks, saving energy is a major concern in UWSNs (especially for the long-term aquatic monitoring applications). Another challenge for data forwarding in UWSNs is to handle node mobility. This requirement makes most existing energy-efficient data-forwarding protocols unsuitable for UWSNs. There are many routing protocols proposed for ground-based sensor networks.

They are mainly designed for stationary networks and usually employ query flooding as a powerful method to discover data delivery paths. In UWSNs, however, most sensor nodes are mobile, and the "network topology" changes dramatically even with small displacements. Thus, the existing routing algorithms using query flooding designed for ground-based sensor networks are no longer feasible in UWSNs.

With no proactive neighbor detection and with less flooding, it is a big challenge to furnish multi-hop packet delivery service in UWSNs with the node mobility requirement. One possible direction is to utilize location information to do geo-routing, which proves to be very effective in handling mobility. However, how to make geo-routing energy efficient in UWSNs is yet to be answered.

5. Distributed Localization and Time Synchronization

In aquatic applications, it is critical for every underwater node to know its current position and the synchronized time with respect to other coordinating nodes. Due to quick absorption of high frequency radio waves, a Global Positioning System (GPS) does not work well under the water. A low-cost positioning and time-synchronization system with high precision like GPS for ground-based sensor nodes is not yet available to underwater sensor nodes. Thus, it is expected that UWSNs must rely on a distributed GPS-free localization or time synchronization scheme, which is referred to as cooperative localization or time synchronization. To realize this type of approach in a network with node mobility, the key problem is the range and direction measurement process.

Promising approaches may include acoustic only Time-of-Arrival (ToA) approaches (e.g., measuring round trip time by actively bouncing the acoustic signal) as well as deploying many surface-level radio anchor points (via GPS for instant position and time-synchronization information). Moreover, the underwater environment with motion of water and variation in temperature and pressure also affect the speed of the acoustic signal. Sophisticated signal processing will be needed to compensate for these sources of errors due to the water medium itself.

6. Efficient Multiple Access

The characteristics of the underwater acoustic channel, especially limited bandwidth and high propagation delays, pose unique challenges for media access control (MAC) that enables multiple devices to share a common wireless medium in an efficient and fair way. It has been observed that contention-based protocols that rely on carrier sensing and handshaking are not appropriate in underwater communications. One possible direction is to explore ALOHA/slotted ALOHA in UWSNs since satellite networks, which share the feature of long propagation delay, employ these random access approaches. On the other hand, FDMA is not suitable due to the narrow bandwidth of the underwater acoustic channel, and TDMA is not efficient due to the excessive propagation delay.

As a result, CDMA has been highlighted as a promising multiple access technique for underwater acoustic networks. If multiple antenna elements are deployed at certain relay or access points, then spatial division multiple access (SDMA) is a viable choice. Like in CDMA, users can transmit simultaneously over the entire frequency band. With different spatial signature sequences, users are separated at the receiver through interference cancellation techniques. SDMA and CDMA can be further combined, where each user is assigned a signature matrix that spreads over both space and time, extending the concept of temporal or spatial spreading.

7. Acoustic Physical Layer

Compared with the counterpart on radio channels, communications over underwater acoustic channels are severely rate limited and performance limited. That is due to the inherent bandwidth limitation of acoustic links, the large delay spread, and the high time variability due to slow sound propagation in an underwater environment. As a result, unlike the rapid growth of wireless networks over radio channels, the last two decades have only witnessed two fundamental advances in underwater acoustic communications. One is the introduction of digital communication techniques (non-coherent frequency shift keying (FSK)) in the early 1980s, and the other is the application of coherent modulations, including phase shift keying (PSK) and quadrature amplitude modulation (QAM) in the early 1990s.

6.6 Distinctions Between UWSNS And Ground-Based Sensor Networks

A UWSN is significantly different from any ground-based sensor network in terms of the following aspects:

1. Communication Method

Electromagnetic waves cannot propagate over a long distance in underwater environments. Therefore, underwater sensor networks have to rely on other physical means, such as acoustic sounds, to transmit signals. Unlike wireless links among ground-based sensors, each underwater wireless link features large latency and low bandwidth. Due to such distinct network dynamics, communication protocols used in ground-based sensor networks may not be suitable in underwater sensor networks. Specially, low bandwidth and large latency usually result in long end-to-end delay, which brings big challenges in reliable data transfer and traffic congestion control. The large latency also significantly affects multiple access protocols. Traditional random access approaches in RF wireless networks might not work efficiently in underwater scenarios.

2. Node Mobility

Most sensor nodes in ground-based sensor networks are typically static, though it is possible to implement interactions between these static sensor nodes and a limited amount of mobile nodes (e.g., mobile data collecting entities like "mules," which may or may not be sensor nodes). In contrast, the majority of underwater sensor nodes, except some fixed nodes equipped on surface-level buoys, are with low or medium mobility due to water current and other underwater activities. From empirical observations, underwater objects may move at the speed of 23 knots (or 3-6 kilometers per hour) in a typical underwater condition.

6.7 Networking Architectures for UWSNs

In general, depending on the permanent vs. on-demand placement of the sensors, the time constraints imposed by the applications, and the volume of data being retrieved, the aquatic application scenarios could roughly be classified into two broad categories:

1. long-term non-time-critical aquatic monitoring

2. short-term time-critical aquatic exploration

Applications that fall in the first category include oceanography, marine biology, pollution detection, and oil/gas field monitoring, to name a few. The examples for the second category are underwater natural resource discovery, hurricane disaster recovery, antisubmarine military missions, lost treasure discovery, and so on.

UWSNs for Long-Term Non-Time-Critical Aquatic Monitoring

Figure 6.9 illustrates the mobile UWSN architecture for long-term non-time-critical aquatic monitoring applications. In this type of network, sensor nodes are densely deployed to cover a spacial continuous monitoring area. Data are collected by local sensors, related by intermediate sensors, and finally reach the surface nodes (equipped with both acoustic and RF (Radio Frequency) modems), which can transmit data to the on-shore command center by radio. Since this type of network is designed for long-term monitoring tasks, energy saving is a central issue to consider in the design. Among the four types of sensor activities (sensing, transmitting, receiving, and computing), transmitting is the most expensive in terms of energy consumption.

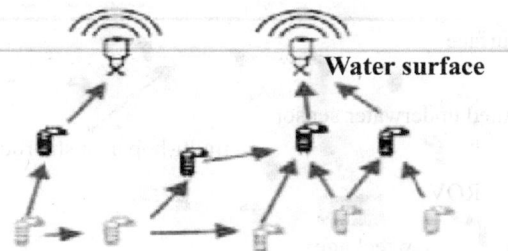

    Distributed underwater sensor nodes
Surface level gateway nodes
communicating with the onshore
command center through radio
⟶ Data path

FIGURE 6.9 The UWSN architecture for long-term non-time-critical aquatic monitoring applications.

In Micro-Modems, the transmit power is 10 Watts, and the receive power is 80 milli watts. Note that the Micro-Modem is designed for medium range (1 to 10 km) acoustic communications. Efficient techniques for multiaccess and data forwarding play a significant role in reducing energy consumption. Moreover, depending on the data sampling frequency, one may need mechanisms to dynamically control the mode of sensors (switching between sleeping mode, wake-up mode, and working mode). In this way, one may save more energy. Further, when sensors are running out of batteries, they should be able to pop up to the water surface for recharge, for which a simple air-bladder-like device would suffice. Clearly, in the UWSNs for long-term aquatic monitoring, localization is a must-do task to locate mobile sensors, since usually only location-aware data is useful in aquatic monitoring. In addition, the sensor location information can be utilized to assist data forwarding, since geo-routing proves to be more efficient than pure flooding. Self-relocation obviously needs some buoyancy control, which is very energy consuming. Thus, a UWSN system design has to deal with the tradeoff between energy efficiency and self-reorganizability.

UWSNs for Short-Term Time-Critical Aquatic Exploration

In Figure 6.10, a civilian scenario of the UWSN architecture for short-term time-critical aquatic exploration applications is shown. Assume a ship wreckage and accident investigation team wants to identify the target venue. Existing approaches usually employ tethered wire/cable to remotely operated vehicles (ROV). When the cable is damaged, the ROV is out of control or not recoverable. In contrast, by deploying a underwater wireless sensor network, as shown in Figure 6.10, the investigation team can control the ROV remotely.

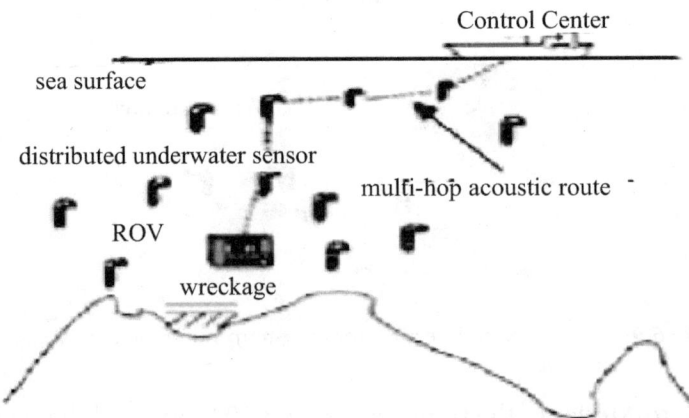

FIGURE 6.10 An illustration of the mobile UWSN architecture for short-term time-critical aquatic exploration applications.

The self-reconfigurable underwater sensor network tolerates more faults than the existing tethered solution. After investigation, the underwater sensors can be recovered by issuing a command to trigger air-bladder devices. In military context, submarine detection is an example of the target short-term time-critical aquatic exploration applications. In the face of state-of-the-art stealth technologies, the acoustic signature of a modern submarine can only be identified within a very short range. Compared to remote sensing technology that has limited accuracy and robustness, the self-configured sensor mesh can identify the enemy's submarine with very high probability since every individual sensor is capable of submarine detection, and more-over, the detection can be reinforced by multiple observations. Figure 6.10 is used to depict this application scenario, with the ROV replaced with the enemy's stealth submarine. The self-reconfigurable wireless sensor network detects the enemy's submarine and notifies the control center via multi-hop acoustic routes. This type of aquatic application demands data rates ranging from very small (e.g., send an alarm that a submarine was detected) to relatively high (e.g., send images, or even live video of the submarine). As it is limited by acoustic physics and coding technology, high data rate networking can only be realized in high frequency acoustic bands in underwater communication. It was demonstrated by empirical implementations that the link bandwidth can reach up to 0.5 Mbps at the distance of 60 meters. Such a high data rate is suitable to deliver even multimedia data. Compared with the first type of UWSN for long-term non-time-critical aquatic monitoring, the UWSN for short-term time-critical aquatic exploration presents the following differences in the protocol design.

- Real-time data transfer is more of concern.

- Energy saving becomes a secondary issue.

- Localization is not a must-do task.

However, reliable, resilient, and secure data transfer is always a desired advanced feature for both types of UWSNs.

6.8 Water Quality Monitoring

Water is a limited resource and essential for agriculture, industry, and the existence of creatures on earth, including human beings. Water quality monitoring is essential to control physical, chemical, and biological characteristics of water. For example, drinking water should not contain any chemical materials that could be harmful to health; water for agricultural irrigation should have low sodium content; and water for industrial uses should be low in certain inorganic chemicals. In addition, water quality monitoring can help with water pollution detection and discharge of toxic chemicals and contamination in water. In most of the river/lake water quality monitoring systems, the following parameters are monitored:

1. Potential of Hydrogen (pH)

pH is a measurement of the concentration of hydrogen ions in the water. A pH sensor measures how acidic or basic the water is, which can directly affect the survival of aquatic organisms. The pH range is from 0 (very acidic) to 14 (very basic), with 7 being neutral. Most water's pH range is from 5.5 to 8.5. Changes in pH can affect how chemicals dissolve in the water. High acidity (such as a pH of less than 4) can be deadly to fish and other aquatic organisms.

2. Dissolved Oxygen (DO)

Dissolved oxygen is the amount of oxygen dissolved in water, measured in milligrams per liter (mg/L). DO measurement tells how much oxygen is available in the water for fish and other aquatic organisms to breathe. The ability of water to hold oxygen in a solution is inversely proportional to the temperature of the water. For example, the cooler the water temperature, the more dissolved oxygen it can hold.

3. Temperature

Temperature measures the warmth or coldness of the water. Temperature is a critical water quality parameter since it directly influences the amount of dissolved oxygen that is available to aquatic organisms. Temperature measurement can also determine the kinds of aquatic organisms that can survive in the water.

4. Conductivity/TDS

Conductivity is the ability of the water to conduct an electrical current, and is an indirect measure of the ion concentration. The more ions are present, the more electricity can be conducted by the water. This measurement is expressed in microsiemens per centimeter (uS/cm). The amount of mineral and salt impurities in the water is called total dissolved solids (TDS). TDS is measured in parts per million. TDS tells how many units of impurities there are for one million units of water. For example, drinking water should be less than 500 ppm, water for agriculture should be less than 1200 ppm.

5. Turbidity

Turbidity is a measure of the clarity of the water. This measurement determines how many particulates are floating around in the water, such as plant debris, sand, silt, and clay, which affect the amount of sunlight reaching aquatic plants. Excess turbidity can reduce reproduction rates of aquatic life when spawning areas and eggs are covered with soil. Turbidity is measured in Nephelometric Turbidity Units (NTU).

A large number of sensor nodes can be deployed to cover a large water monitoring area with enough density. As Figure 6.11 shows, the sensor network consists of one super node and a number of small sensor nodes. Each small sensor node has a low capacity solar panel and two low cost sensors (one temperature sensor and one dissolved oxygen sensor) connected and uses a low power Zigbee radio for data transmission. The super node has a high capacity solar panel and five sensors connected and uses a powerful long-distance 802.11 Ethernet radio for data transmission. The whole network is divided into several clusters based on signal strength. Each cluster has a head node, and the cluster nodes send data to the cluster head node. The cluster head nodes send the gathered data to the super node (sink). All the small sensor nodes use the low power Zigbee radios, and the long distance Ethernet radios are used between the super node and the station at shore.

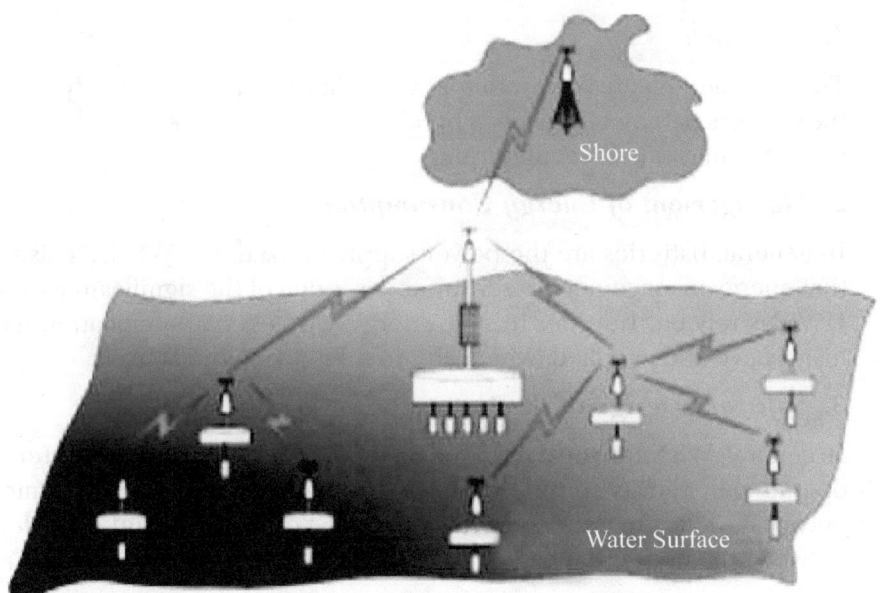

FIGURE 6.11 Water quality monitoring sensor network.

LEACH routing protocol is suitable in the water quality monitoring application. LEACH is one the most popular hierarchical clustering algorithms used in wireless sensor networks. The key idea of LEACH is to form clusters of the sensor nodes based on the received signal strength. LEACH is a distributed protocol and doesn't require global knowledge of the network. Each cluster has a head node, and the head node routes data of the cluster to the sink. This mechanism saves power because only the cluster head nodes do the data transmissions to the sink nodes, and all the nodes in each cluster only need to send the data to the respective cluster head. The cluster head nodes compress the received data from the cluster nodes and send the aggregated packets to the sink node in order to reduce the amount of data sent to the sink node.

Therefore, power consumption is further reduced and the lifetime of the sensor network is increased. The estimated optimal number of the cluster head nodes is 5% of all the sensor nodes. In order to evenly distribute the power dissipation, the cluster heads are rotated over time.

6.9 Limitations of UWSNs

The limitations of UWSNs are given below.

1. Movement

The sea water creates environmental conditions which negatively influence the network parameters, such as breaking up the buoy nodes, and sometimes the WSN may need reconfiguring.

2. Management of Energy Consumption

In general, batteries are the power supply utilized in UWSNs. This means that energy management of sensor nodes is one of the significant issues that UWSNs rely on. In order to save energy, wireless communication mechanisms have been applied, which aim to minimize radio activity.

3. Software Design of the Network

In general, WSNs are heavily based on the Network Embedded System. The operating system is considered to be the core of wireless communication networks. The program code manages the connectivity and data delivery between the nodes, the base station, and the end users.

4. Data Transmission and Security

Communication between UWSN components suffers from a number of issues, such as environmental conditions and network design. For instance, the water environment decreases the radio signal strength of the data transmission and can result in an unstable line of signal between wireless nodes. Additionally, to ensure the confidentiality and integrity of the gathered data, security protocols and techniques must be applied.

6.10 Research Challenges in UWSNs

Major challenges encountered in the design of underwater acoustic networks are as follows:

1. It is necessary to develop less expensive, robust "nano-sensors," for example, sensors based on nanotechnology, which involves development of materials and systems at the atomic, molecular, or macromolecular levels in the dimension range of approximately 1–500 nm.

2. It is necessary to devise periodical cleaning mechanisms against corrosion and fouling, which may impact the lifespan of underwater devices.

3. There is a need for robust, stable sensors on a high range of temperatures since sensor drift of underwater devices may be a concern. To this end, protocols for the calibration of sensors to improve accuracy and precision of sampled data must be developed.

4. There is a need for new integrated sensors for synoptic sampling of physical, chemical, and biological parameters to improve the understanding of processes in marine systems.

5. The underwater channel is impaired because of multi path and fading.

6. Underwater sensors are characterized by high cost because of the extra-protective sheaths needed for sensors, and also a relatively small number of suppliers (i.e., not much economy of scale) are available.

7. Battery power is limited and usually batteries cannot be recharged, as solar energy cannot be exploited.

8. Underwater sensors are more prone to failures because of fouling and corrosion.

Summary

- WSNs based on radio frequency aerial communications (A-WSNs) and Under Water Acoustic Sensor Networks (UW-ASNs) are two broad categories of marine wireless networks depending on the data transmission medium.

- Limitations of UW-ASNs are bandwidth, propagation delay, error rate, connectivity loss, and battery power.

- A-WSN nodes include an RF module, power supply, sensor interface, modules for data, different memory, scientific instruments, and a CPU.

- Three categories of UW-ASNs are static underwater sensor networks, moving underwater sensor networks, and underwater sensor networks with autonomous underwater vehicles.

- A few applications of UWSNs are environmental monitoring, undersea explorations, disaster prevention, assisted navigation, mine reconnaissance, underwater robots, and oceanography.

- Security services include confidentiality, authentication, integrity, and availability.

- Communication methods and node mobility are the distinctions between UWSNs and ground-based WSNs.

- Long-term non-time-critical aquatic monitoring and short-term time-critical aquatic monitoring are two categories of network architecture of UWSNs.

- Water quality monitoring is essential to control physical, chemical, and biological characteristics of water.

- pH, dissolved oxygen, temperature, conductivity, and turbidity of water are monitored.

- Mobility, energy consumption, software design of the network, data transmission, and security are the limitations of UWSNs.

Questions

1. What are the challenges in WSNs for oceanographic monitoring?

2. What are two categories of marine wireless networks depending on the data transmission medium?

3. List the limitations of UW-ASNs.

4. What are the problems of A-WSNs?

5. Draw the A-WSN sensor node for oceanographic monitoring. Explain.

6. Give the diagram of the A-WSN general architecture and explain the blocks.

7. Compare different wireless communication technologies.

8. List oceanographic sensors and surface sensors.

9. Give the configuration of buoys used in A-WSNs.

10. With the help of diagram, what do you mean by underwater sensor networks with AUVs?

11. List the constraints in underwater wireless sensor networks.

12. Draw the schematic of UWSNs.

13. Draw the general architecture of underwater network deployment.

14. List the applications of UWSNs.

15. Draw the sensor node mechanical design.

16. What are the services offered for security?

17. Write about key management in UWSNs security.

18. Explain in detail about the challenges in network protocols for UWSN.

19. Give two categories of UWSNs based on time constraints.

20. In detail write about the UWSNs for long-term non-time-critical aquatic monitoring.

21. Write in detail about the UWSNs for short-term-time critical a quatic exploration.

22. List the parameters momtored for water quality.

23. Draw sensor networks for water quality monitoring.

24. List the limitations of UWSNs.

25. List your own practical research challenges in UWSNs.

Further Reading

1. *Ocean Electronics* by S. R.Vijayalakshmi and S. Muruganand

2. *Ocean Instrumentation, Electronics, and Energy* by S. R.Vijayalakshmi

3. *Underwater Acoustic Sensor Networks* by Yang Xiao

4. *Wireless Sensor Networks: From Theory to Applications* by Ibrahiem M. M. El Emary and S. Ramakrishnan

References

1. *http://www.mdpi.eom/1424-8220/10/7/6948/htm*

2. *https://www.isi.edu/~johnh/PAPERS/Heidemann06a.pdf*

3. *http://www.ijettjournal.org/volume-3/issue-2/IJETT-V3I2P215.pdf*

INTERNET OF THINGS WSNS

This chapter discusses the integration of wireless sensor networks with the Internet of Things (IoT).

7.1 An Introduction to the Internet of Things (IoT)

The Internet of Things (IoTs) can be described as connecting everyday objects like smart phones, Internet TVs, sensors, and actuators to the World Wide Web where the devices are intelligently linked together, enabling new forms of communication between things and people, and between things themselves. Building the IoT has advanced significantly in the last couple of years since it has added a new dimension to the world of information and communication technologies. In 2008, the number of connected devices surpassed connected people, and it has been estimated that by 2020 there will be 50 billion connected devices, which is seven times the world population. Now anyone, from anytime and anywhere, can have connectivity for anything, and it is expected that these connections will extend and create an entirely advanced dynamic network of the IoT. The development of the Internet of Things will revolutionize a number of sectors, from wireless sensors to nanotechnology.

In fact, one of the most important elements in the Internet of Things paradigm is wireless sensor networks (WSNs). WSNs consist of smart sensing nodes with embedded CPUs, low power radios, and sensors which are used to monitor environmental conditions such as temperature, pressure, humidity, vibration, and energy consumption. In short, the purpose of the WSN is to provide sensing services to the users. Since the number of users of the Internet is increasing, it is wise to provide WSN services to this ever-growing community.

The Internet of Things (IoT) concept is based on the pervasive presence around us of a variety of things or objects through which unique addressing schemes are able to interact and cooperate with each other in order to reach common goals. As the name suggests, the purpose of this architecture is to interconnect all kinds of objects over the Internet as shown in Figure 7.1.

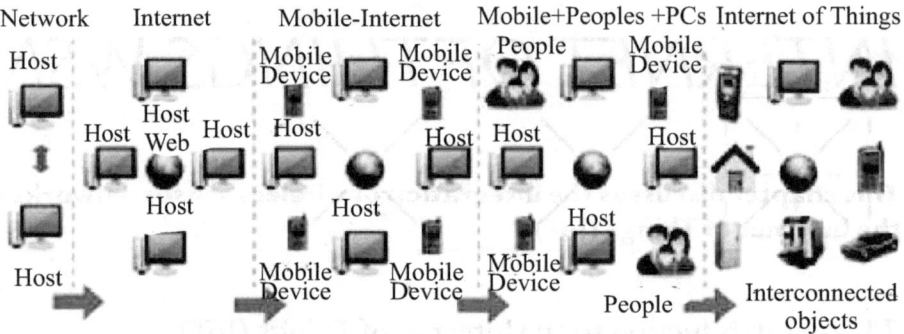

FIGURE 7.1 Evolution of the Internet.

Definition of the IoT

"The IoT allows people and things to be connected Anytime, Anyplace, with Anything and Anyone, ideally using Any path/network and Any service."

FIGURE 7.2 Representation of the first definition of the IoT.

The IoT infrastructure allows combinations of different types of smart items using different but interoperable communication protocols and realizes a dynamic heterogeneous network that can be deployed also in inaccessible or remote spaces (oil platforms, mines, forests, tunnels, pipes, etc.) or in cases of emergencies or hazardous situations (earthquakes, fire, floods, radiation areas, etc.). Giving these objects the possibility to communicate with each other and to elaborate on the information retrieved from the surroundings implies having different areas where a wide range of applications can be deployed. These can be grouped into the following domains: healthcare, personal and social, smart environment (such as at home or in the office), futuristic applications, and transportation and logistics, as represented in Figure 7.3.

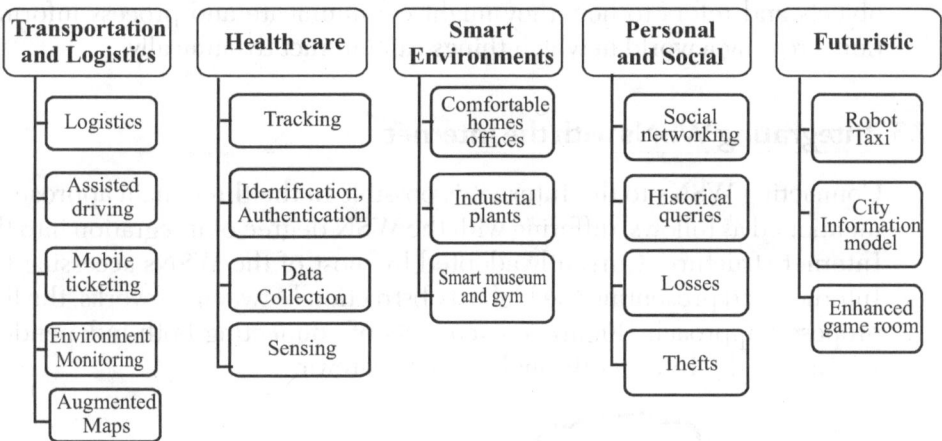

FIGURE 7.3 IoT application areas.

7.2 Context Awareness

Context awareness plays an important role in the IoT to enable services customization according to the immediate situation with minimal human intervention. Acquiring, analyzing, and interpreting relevant context information regarding the user will be a key ingredient to create a whole new range of smart applications.

Context is any information that can be used to characterize the situation of an entity. An entity is a person, place, or object that is considered relevant to the interaction between a user and an application, including the user and applications themselves. Therefore, context awareness is the result gained from utilizing context information, such as the ability to adapt behavior depending on the current situation of the users in context-aware applications.

Ubiquitous Computing

The focus on context-aware computing evolved from desktop applications, web applications, mobile computing, ubiquitous computing, to the IoT over the last decade. This is an era in which computer devices will be embedded in everyday objects, invisible at work in the environment around us; in which intelligent, intuitive interfaces will make computer devices simple to use; and in which communication networks will connect these devices to facilitate anywhere, anytime, always-on communication. Ubiquitous computing is the growing trend toward embedding microprocessors in everyday objects and refers to how they might communicate and process information, creating a world in which things can interact dynamically.

7.3 Integrating WSNs with the Internet

Connecting WSNs to the Internet is possible in the three main approaches mentioned as follows, differing with the WSN degree of integration into the Internet structure. Currently adopted by most of the WSNs accessing the Internet, and presenting the highest abstraction between networks, the first proposed approach (Figure 7.4) consists of connecting both independent WSNs and the Internet through a single gateway.

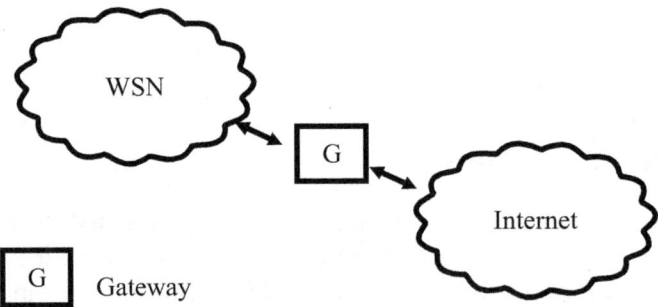

FIGURE 7.4 Independent network.

Showing an increasing integration degree, the second approach (Figure 7.5) forms a hybrid network, still composed of independent networks, where few dual sensor nodes can access the Internet.

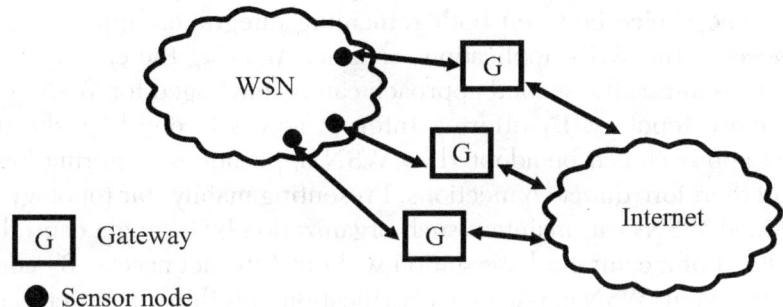

FIGURE 7.5 Hybrid network.

Illustrated by Figure 7.6, the last approach is inspired from current WLAN structure and forms a dense 802.15.4 access point network, where multiple sensor nodes can join the Internet in one hop.

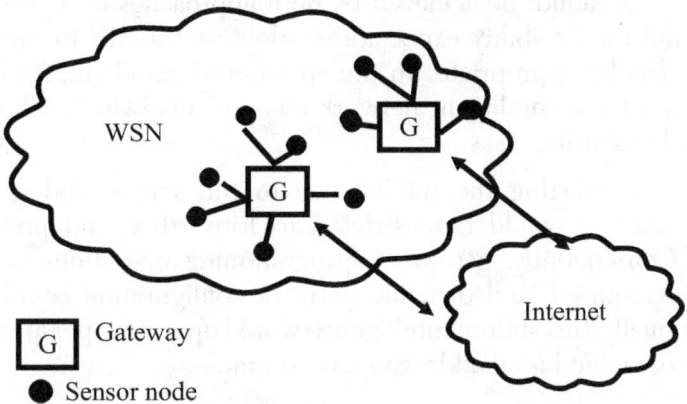

FIGURE 7.6 Access point network.

It is obvious that the first approach presents a single point of failure due to the gateway uniqueness. Gateway dysfunction would break down the connection between the WSN and the Internet. With several gateways and access points, the second and third scenarios do not present such a weakness. To ensure network robustness, they would consequently be preferred, if the application supports this type of network structure.

The choice between both remaining integration approaches is influenced by the WSN application scenario. Allowing the coverage of important distances, the second approach can be envisaged for WSNs organized in mesh topology. By offering Internet access in one hop, the third and last approach can be adopted by WSN applications requiring low latency and therefore direct connections. Presenting mainly star topology, the concerned WSNs can maintain such organization by having a central gateway instead of a common base station without Internet access. By considering the previous WSN application classification, this third approach can be suitable for objects and human beings monitored.

It is important to remark that both the second and third integration approaches support only static network configuration. Indeed, each new device wanting to join the Internet requires time-consuming gateway reprogramming. Therefore, the flexibility wanted by the future Internet of the Things cannot be achieved by both approaches in their current form. To fulfill the flexibility expectation, adopting the "IP to the Field" paradigm may be appropriate. In the considered paradigm, sensor nodes are expected to be intelligent network components, which will no longer be limited to sensing tasks.

By transferring the intelligence to the sensor nodes, the gateway functionalities would be restricted to forwarding and protocol translation. Consequently, gateway reprogramming operations would no longer be required, and dynamic network configuration could be attained. Additionally, this shift of intelligence would open new perspectives, including geographic-based addressing, for example.

7.4 Architecture of the IoT for WSNs

The architecture of the IoT system is divided into three layers (Figure 7.7): the Sensor Layer, the Coordinator Layer, and the Supervision Layer.

The Sensor Layer consists of sensors that interact with the environment. Every sensor was integrated with wireless nodes using an Xbee platform called End Devices. These End Devices form a Mesh network and send the information gathered by the sensors to the Coordinator Layer through the sink node called the base station. Messages are routed from one End Device to another until they reach this base station.

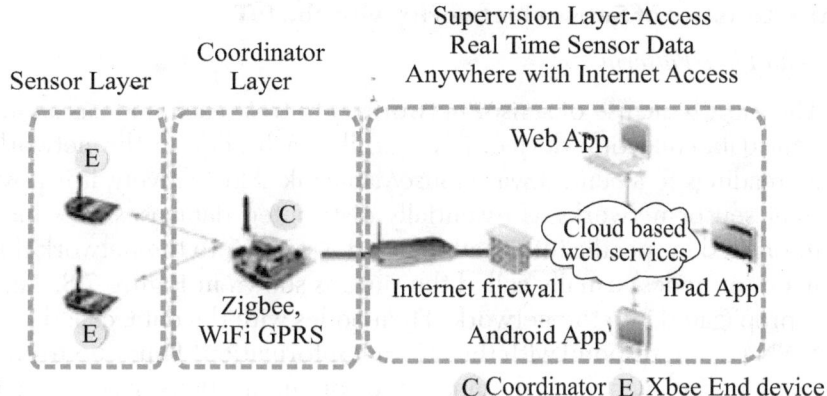

FIGURE 7.7 Architecture for the IoT.

The Coordination Layer is responsible for the management of the data received from the sensor network. It temporarily stores the gathered data into a buffer and sends it to the Supervision layer at predefined intervals. The Base station, which comprises Arduino, an Ethernet shield, and XBee, is connected to the Internet using a cable and is powered using an AC adaptor. It serves as a mobile mini application server between the wireless sensors and the dedicated network and has more advanced computational resources compared to the End Devices found in the Sensor Layer. At the base station, the sink node gathers data from wireless sensors using the Zigbee protocol and sends this data to Cloud-based sensor data platforms.

Finally, the Supervision Layer accommodates the base station with a Web server to connect and publish the sensor data on the Internet. This layer stores the sensor data in a database and also offers a Web interface for the end users to manage the sensor data and generate statistics, thus allowing existing networks to be connected to other applications with minimal changes.

When integrating future large scale Wireless Sensor Networks with the Internet Wireless Sensor Networks (WSN), it is envisioned that WSNs will consist of thousands to millions of tiny sensor nodes, with limited computational and communication capabilities. When networked together, these unattended devices can provide high resolution knowledge about sensed phenomena. Possible applications of these networks range from habitat and ecological sensing, structural monitoring and smart spaces, emergency response, and remote surveillance.

Characteristics of Sensor Networks with the IoT

Data Flow Patterns

The most basic use of sensor networks is to treat each node as an independent data collection device. Periodically, each node in the network sends its readings to a central warehouse/data sink. Alternatively, it is possible to treat sensor networks as essentially distributed databases, in which users interested in specific information insert a query into the network through a node (or nodes) usually called the sink, as shown in Figure 7.8. This query is propagated into the network. Then nodes with the data, called sources in WSN jargon, respond with the relevant information. Thus one-to-many and many-to-one data flows dominate the communications in sensor networks. This can be contrasted with the arbitrary one-to-one addressable flows that are typical of most IP-based networks.

Energy Constraints

The nodes in unattended large-scale sensor networks are likely to be battery powered, with limited recharging capabilities. Under these conditions, the primary network performance metric of interest is the energy efficiency of operation (a related metric is the lifetime of the network measurable in terms of the time when a significant portion of nodes in the network fail due to energy depletion). Typically, communication is significantly more energy expensive than computation.

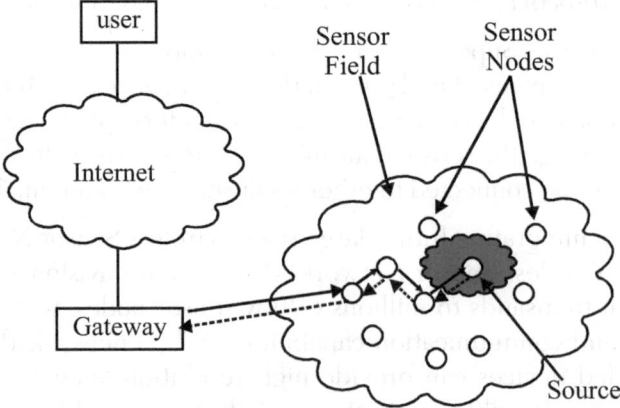

FIGURE 7.8 Communication Architecture using a gateway.

In Figure 7.8 the full arrows represent the dissemination of the query, and the dashed arrows, the data routed back. In this case the gateway is the single point of access to the WSN, and it performs the conversion of the necessary protocols including the IP.

Application-specific networking and data-centric routing

Traditional IP-based networks follow the layering principle, which separates the application level concerns from network layer routing. This is necessary because a multitude of applications are expected to run over a common networking substrate. By contrast, sensor networks are likely to be quite limited in the applications they perform. This calls for cross-layer optimizations and application-specific designs. One design principle that exploits application specificity to significantly reduce communication energy is the use of in-network processing to filter out irrelevant and redundant information. For example, intermediate nodes may be allowed to look at the application-level content of packets in order to aggregate them with information originating from other sources.

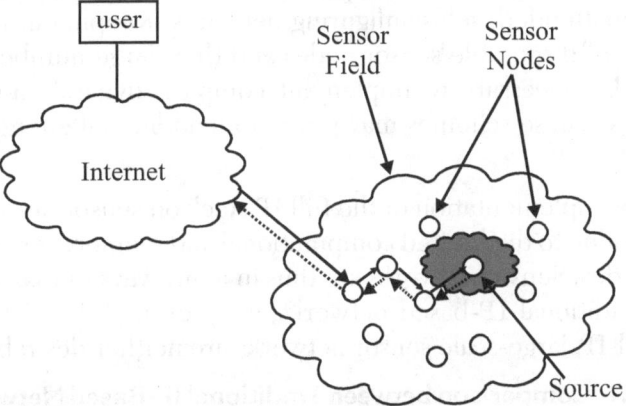

FIGURE 7.9 Communication architecture with direct connection: the difference compared to Figure 7.8 is that in this case every node has an IP address and can be directly accessed from any point in the Internet that has wireless capabilities.

Related to this is the distinction between address-centric and data-centric routing. The Internet was designed around an address-centric ideology, which works when data is usually attached to a specific host. It requires prior knowledge of which host to contact. Almost all transactions (ftp, http, email, etc.) on the Internet have this characteristic, that it is known a priori where the data is located. For this reason, communication on the Internet is usually point-to-point, and this requires the ability to uniquely identify each host through IP addresses.

In sensor networks, however, the query is most likely to be for named data. For example, in a WSN application, the question is unlikely to be: "What is the temperature at sensor number 271?" Rather, the question would be: "Where are the nodes whose temperature exceeds 45 degrees?" The Directed Diffusion protocol has shown that it is possible to do data-centric querying and routing without the use of globally unique IP-like addresses for all nodes in the network.

One advantage of doing without globally unique IDs is that each packet need not carry address information in the header. Many applications for low rate sensing will result in small amounts of data per packet (on the order of a few bytes). IPv6, for example, has 40 bytes of header per packet. Doing without this header addressing information can result in a significant reduction in communication overhead and thus energy.

Another argument for doing without globally unique IDs for all nodes in sensor networks is the complexity of address management in such large-scale, unattended, self-configuring networks. Keeping in mind the limited lifetime of disposable sensor nodes and their large numbers, it would otherwise be necessary to implement complex, dynamic address allocation schemes. These schemes may present an additional energy burden on the network.

The implementation of the full IP stack on sensor networks may not be feasible due to the limited computational and memory resources on component nodes. Sensor networks are thus in many ways fundamentally different from traditional IP-based networks, as given in Table 7.1. For these reasons, all-IP, large-scale sensor networks are neither desirable nor feasible.

Table 7.1 Comparison between Traditional IP-Based Networks and Large-Scale Wireless Sensor Networks

	Traditional IP-Based Networks	Large Scale Wireless Sensor Networks
Networking mode	Application-independent	Application-specific
Routing Paradigm	Address-Centric	Data-centric, Location-centric
Typical Data Flow	Arbitrary, One to one	To/from querying sink, One to-many and many-to-one
Data Rates	High (Mbps)	Low (kbps)
Resource constraints	Bandwidth	Energy (battery-operated nodes), Limited Processing and memory

	Traditional IP-Based Networks	Large Scale Wireless Sensor Networks
Network Lifetime	Long (years–decades)	Short (days–months)
Operation	Attended, administered	Unattended, Self-configuring

7.5 Gateway-Based Integration

Giving an IP address to every sensor node is not the right approach to integrating sensor networks with the Internet. While it is desirable to not have to develop new protocols or perform protocol conversion at gateways, the application-specific property of wireless sensor networks demands this type of solution. Single or multiple independent gateways are called for in homogeneous networks, where all the nodes have the same capability in terms of processing, energy, and communication resources. In addition to gateways, an overlay IP network may be utilized in heterogeneous networks, where some nodes may be more capable than the majority of nodes (for example, when some laptop computers can be part of the network).

Homogeneous Wireless Sensor Networks

The basic solution for integration in the case of a homogeneous wireless sensor network is to use an application-level gateway to interface the sensor network to the Internet. The gateway may be implemented in the form of a web server, for example. In the case of simple sensor networks where nodes are providing information continuously, they can be stored and displayed on a dynamic web page from the gateway node. In the case of more sophisticated sensor networks, the gateway can viewed as a front end to a distributed database. The users accessing the gateway server may issue SQL type queries.

The query optimization is performed through data-centric, in-network processing, and the response is obtained from the network and displayed to the user. One drawback of this approach is that a lot of data has to be routed from and back to the gateway, implying that all the nodes near the gateway will exhaust their energy resources sooner, if they are not rechargeable.

Another possibility is to deploy wireless sensor networks with more than one independent gateway used as points of interface between the network and the Internet. Having several points of access to the network would have two important advantages: eliminating a single point of failure and distributing evenly the energy consumed by the nodes (assuming the queries on the different gateways can be load-balanced).

In homogeneous WSNs, where all the nodes have the same capabilities, the flexibility for other communication architectures is limited.

Heterogeneous Wireless Sensor Networks

Heterogeneous networks allow for the possibility of giving an IP address to the more capable nodes in the network. In general, capable devices could perform more tasks, and hence carry more of the burden in the network. There may also be application-specific reasons why these more capable devices should be addressable from within and without the network.

For example, if the more capable devices are capable of actuation, they may need to be addressed in order to be tasked. In other scenarios, the higher capability nodes may act as addressable cluster heads. In such networks, it may be possible to construct an overlay IP network that sits on top of the underlying wireless sensor network. The technical challenge in this approach is to construct some kind of tunneling mechanism to allow the devices with IP addresses to communicate among themselves in an address-centric manner (Figure 7.10). In general, the IP-addressable nodes in the network may not be adjacent to each other. To create an overlay IP network, then, it will be necessary to create some form of a link abstraction from the multiple hops between nearby IP-addressable nodes. If the intermediate nodes do not have any global identifiers, the link-abstraction will need to be formed in a data-centric manner.

The problem of creating tunnels depends on the characteristics of the wireless sensor network. If the application is more likely to have high IP traffic inside the wireless sensor network, then, multiple paths among the IP-addressable nodes would be preferred, in order to load balance the consumption of energy in the less capable sensor nodes. On the other hand, if the IP traffic is going to be low, then a single route can be enough. Building an overlay network based on a flooded query approach (Directed Diffusion) would be suitable for high traffic, and also building it using a directed-query approach (ACQUIRE) would be suitable for low-traffic conditions.

Directed Diffusion is a good candidate to build up the overlay structure in high-IP-traffic applications. Directed Diffusion is a data-centric communication paradigm that is quite different from the address-centric ideology in traditional networks. The goal of Directed Diffusion is to establish efficient n-way communication between one or more sources and sinks. In basic Directed Diffusion, an interest for named data is first distributed through the network via flooding. The interest description is done by attribute-value pairs. In our case it could be described as:

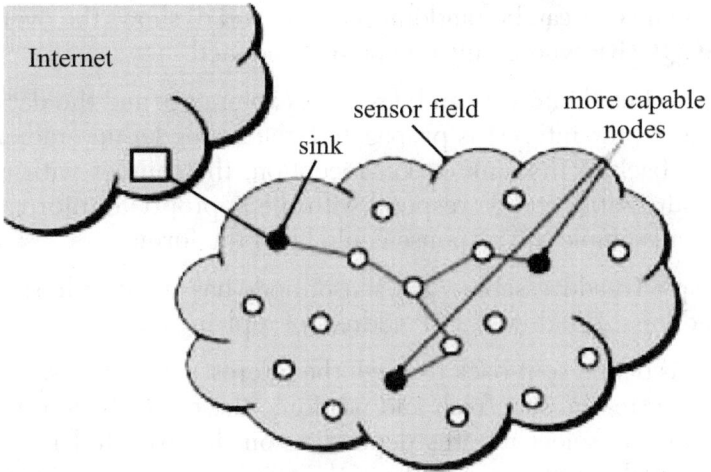

FIGURE 7.10 Heterogeneous Network: the lines show the tunneling communicating the nodes with IP addresses (circles filled dark).

type: IP-addressable // detect nodes that have an IP address
interval: 20ms // send message every 20ms
duration: 200ms // ... a total of 10 messages

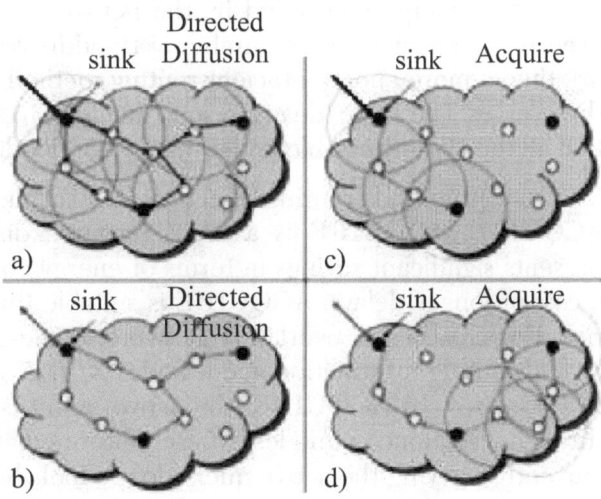

FIGURE 7.11 Heterogeneous WSN with and Overlay IP network.

In Figure 7.11 a) shows the first stage in directed diffusion where the query is flooded to find all the IP-addressable nodes; b) shows that multiple routes are obtained with this mechanism; c) ACQUIRE is used to build up the overlay IP network and observe that the query is sent through a path—one that can be randomly chosen; and d) shows the overlay obtained by ACQUIRE where only one path is obtained.

This initial interest can be seen as exploratory and the data rate should be low. As the interest is propagated, the nodes set up gradients from the source back to the sink. Upon reception, the sources with relevant data (IP-addressable nodes) respond with the appropriate information stream. For our example the response could be of the form:

type: IP-addressable // the sensor node has an IP address
address: IP-address // IP address of replying node

This data is sent back through the interest's gradient path. After reception, the sink must refresh and reinforce the most efficient paths. Finally, the sink can select n-paths depending on the expected IP traffic; higher traffic would imply more paths in the overlay structure. Figure 7.11 a) and b) show the directed diffusion mechanism. Note that once the overlay paths are created, they may be used for arbitrary communication between the IP-addressable nodes, not only between the nodes and the sink.

The main advantage of implementing the overlay structure with Directed Diffusion is that it can provide multiple paths; however, the amount of energy consumed by the network is high. If the IP traffic is expected to be low and the number of IP-addressable nodes is known a priori, then a more energy-efficient routing method can be used to construct the overlay. The basic idea is to send an agent to traverse the network and find all the IP-addressable nodes, instead of flooding.

One proposed routing mechanism that uses agents in WSNs is ACQUIRE. ACQUIRE is a novel resource discovery mechanism that presents significant savings in terms of energy compared with flooding, at a cost of longer delays. ACQUIRE is suitable for one-shot, complex queries. For creating an overlay IP network, a one-shot query could be sent to find routing information about nodes X, Y, Z, which are known to have IP addresses. In ACQUIRE, the active query is forwarded step by step through a sequence of nodes. At each intermediate step the node which is currently carrying the active query does a lookahead of d hops in order to resolve the query partially. Once the resource is found, the required data is sent back. For our purposes, routing information to these nodes must be included in the data that is sent back (e.g., by including the intermediate routing nodes in the data, as in source routing). Figure 7.11 c) and 7.11 d) show the ACQUIRE mechanism.

7.6 The IoT and WSN Design Principles

The Internet of Things (IoT) applications capture data from the physical world, communicate this data to a central server, and run analytics on it. In order to complete the IoT loop, the user has to select sensors, choose a compatible platform to plug in those sensors, write the firmware, select data protocols, establish communication to the server, and ensure security of the transferred data. The IoT creates an intelligent, invisible network fabric that can be sensed, controlled, and programmed. IoT-enabled products employ embedded technology that allows them to communicate, directly or indirectly, with each other or the Internet. To date, the world has deployed about 5 billion "smart" connected things. In a lifetime, one will experience life with a trillion-node network. The industry will only achieve the reality of 50 billion connected devices by simplifying how things connect and communicate today.

A problem arises when the user wants data related to machine utilization, power consumption, the temperature of the system, and excess vibration in machines, and wants to store this data on a computer system, access it on a mobile phone, get instant alerts, and share these reports with other concerned people. The solution to this is the platforms. These platforms comprise hardware and software that connect industrial equipment, machines, and sensors to the Internet in a plug and play model and let users control field devices from a central dashboard. They can be used for a variety of purposes including the following:

1. Connecting industrial protocol equipment to the Internet, customer relationship management, and enterprise resource planning systems.

2. Collecting physical data such as temperature, vibration, stress, and pressure.

3. Establishing solar powered off-grid units to collect sensor data.

4. Controlling units deployed in the field from a central location.

5. Taking decisions based on sensor data.

Given below is a step by step operation to the solution to working with the IoT.

1. Select the sensors required for an application from the catalog of compatible sensors provided.

2. Plug the selected sensors on the small form factor board.

3. Upload the firmware provided.

4. Power the board using a solar panel, lithium-polymer battery, or mains electricity.

5. Connect it to the cloud gateway using either a wired (Ethernet, RS485, etc.) or a wireless (GSM, Zigbee, etc.) solution.

6. Start getting data collected by the sensor on cloud, an Android application, or external cloud services such as web services.

Platforms have a secure digital (SD) card memory onboard. There is an external random access memory (RAM) that increases the power of onboard processing. It also has general-purpose input/output pins to interface with multiple digital and analog devices and sensors. By powering the device on battery or solar energy and transmitting data over the Global System for Mobile Communication (GSM), one can make a true off-the-grid system. One can also set up a wireless (Zigbee) sensor mesh network with solar power in order to collect physical data from regions that have no network coverage or even electricity.

Challenges in implementing the IoT

1. Technical challenges include government regulations with regards to spectrum allocation, security, battery issues, cost, and privacy.

2. Security, standards, and overburdening the network are three requirements that need to be focused on before implementing mass adoption.

3. Good throughput is important for IoT, but there will be trade-offs between data speed and battery life. Multiple input multiple output (MIMO) technology is set to be a key part of these efficiency measures.

4. Existing IoT sensors are not equipped to take advantage of 5G. New devices may need to have multiple antennae for fewer dropouts.

5. An IoT scenario deals with big volumes of data due to the large number of sensors involved. There are three main problems that must be solved: resolution, sensitivity, and reliability. Compressed sensing involves reducing the number of samples collected in an IoT wireless sensor network. Thus, it is possible to create standalone applications that require fewer resources.

6. When an application involves a large number of sensor networks that are sending data on a continuous basis, it becomes important to capture that information effectively and keep as much of it as possible for further analysis. Such analyses are important to understand trends, forecast events, and other functions. In these cases, data size would grow non-linearly in a rather short period of time.

Challenges for WSNs in an IoT

The "IP to the Field" paradigm involves assigning additional responsibilities to sensor nodes in addition to their usual sensing functionality. To highlight and discuss the challenges emerging from such novel responsibility assignment, the three potential tasks that the sensor nodes would have to accomplish are as follows: security, quality of service management, and network configuration.

A. Security

In common WSNs without Internet access, the sensor nodes may already play an important role to ensure data confidentiality, integrity, availability, and authentication depending on the application sensitivity. However, the current identified attack scenarios require a physical presence near the targeted WSN in order to jam, capture, or introduce malicious nodes, for example. By opening WSNs to the Internet, such location proximity would no longer be required, and attackers would be able to threaten WSNs from everywhere. In addition to this novel location diversity, WSNs may have to address new threats like malware introduced by the Internet connection and evolving with attacker creativity. Most current WSNs connected to the Internet are protected by a central and unique powerful gateway ensuring efficient protection. However, a direct reuse of such existing security mechanisms is made impossible by the scarce energy, memory, and computational resources of the sensor nodes. At last, many services on the Internet make use of cryptography with large key lengths such as RSA-1024, which are not currently supported by sensor nodes. Consequently, innovative security mechanisms must be developed according to the resource constraints to protect WSNs from novel attacks originating from the Internet.

B. Quality of Service

With gateways acting only as repeater and protocol translators, sensor nodes are also expected to contribute to quality of service management by optimizing the resource utilization of all heterogeneous devices that are part of the future Internet of Things. Not considered as a weakness, the device heterogeneity opens new perspectives in terms of workload distribution. In fact, resource differences may be exploited to share the current workload between nodes offering available resources. Improving the QoS, such collaborative work is consequently promising for mechanisms requiring a high amount of resources like security mechanisms. Nevertheless, the existing approaches ensuring QoS in the Internet are not applicable in WSNs, as sudden changes in the link characteristics can lead to significant reconfiguration of the WSN topology. It is therefore mandatory to find novel approaches toward ensuring delay and loss guarantees.

C. Configuration

In addition to security and QoS management, sensor nodes can also be required to control the WSN configuration, which includes covering different tasks, such as address administration to ensure scalable network constructions and ensuring self-healing capabilities by detecting and eliminating faulty nodes or managing their own configuration. However, self configuration of participating nodes is not a common feature in the Internet. Instead, the user is expected to install applications and recover the system from crashes. In contrast, the unattended operation of autonomous sensor nodes requires novel means of network configuration and management.

7.7 Big Data and the IoT in WSNS

There are roughly three distinct stages for the Internet of Things (IoT). First, data is collected using sensors. At the next step, this data is analyzed with the help of complex algorithms that were embedded into the IoT device or cloud-based data processing. This is followed by decision making (using analytical engines) and transmission of data to the decision making server. If the information collected so quite large and complex that it becomes difficult to analyze using traditional data processing techniques, we call it Big Data. Results made from this analysis are then transmitted to the actuator system, where the decision is implemented.

The important thing to understand about the IoT is that it is not just about reporting if lights are on a parked car or to switch off a TV. The real magic is in that it can figure out potential situations that are about to happen and then go ahead and implement preventive measures before the problem actually occurs. Machine-to-machine device management service providers look at figuring out which among these enormous data are useful and how to segregate them into consumable chunks that can be sold to different organizations or enterprises. The approach they take is to develop an adaptation for each of these machines that communicates in different languages, and hence shield that chaos from the consumer, using an intelligent gateway that presents a unified, sensible view.

Live digital recording methods, artificial intelligence, and complex analytics software can be used to provide content capturing, management, tracking, and analysis solutions for the department of justice and the judiciary. These products can be used to capture live proceedings, right from the crime scene to police investigations to the court room, in the form of text and high quality audio and video on an enterprise mobility platform. An artificially intelligent computer can use its own algorithms to aggregate the different types of information feeds, correlate them, and present the analyzed data in a form that is easily understood by the viewer. With the technologies like Big Data analytics and enterprise video analytics being used, this method could realize otherwise unnoticeable insights, patterns, and correlations in the huge amount of evidence and data, and hence provide a powerful, predictive, and visual analysis platform. This method can also be used in smart classrooms, where the teacher can assess which student is less focused or misbehaving in class. In medical applications the system can suggest all possible medical conditions of a patient based on his or her symptoms and histories.

Enables Intelligent Manufacturing

Manufacturers use the IoT in agriculture. The plants are farmed by data-driven agricultural technology, powered by sensor arrays working in tandem with cloud computing solutions to crunch data and ensure that the environment remains at optimal levels throughout the growth process. This includes monitoring temperature, humidity, and fertilizer composition as they try to figure out the best growing conditions and ways to control microorganisms. A food and agricultural cloud solution leverages information and communication technology (ICT) to dramatically improve the efficiency of agricultural operations. Manufacturers can continuously analyze agricultural data for highly productive cultivation and facilitate the entire management process, including management, production, and sales, for a more efficient agricultural operation.

Another example is smarter manufacturing with IoT motorcycle plants. Software in the plant keeps records of how different equipment performs, such as the speed of fans in the painting booth. The software then automatically adjusts the machinery when it detects that the fan speed, temperature, or humidity has deviated from acceptable ranges.

Applications incorporating Big Data could be useful for manufacturers who have deployed the IoT in a full fledged manner. To ensure delivery of data collected from factory-wide IoT implementation, manufacturers need networks that can cope with RF challenges in the plant, harsh environmental conditions, and reliability for transmission of alarms and real-time data stream processing. Instead of transmitting huge amounts of unprocessed data over factory floors, the processing can be done in the device itself. This means that the system will now only have to transmit results to the central system, resulting in a lower amount of transmitted data. The programmable SoCs (System on Chips) could be used to crunch this data by leveraging the massive parallel processing power of field programmable gate arrays (FPGAs) with embedded microprocessors. These complete systems on a programmable chip form a sort of reprogrammable CPU architecture. Some examples of vendors with these kind of chips are the Xilinx Zynq, Altera Arria, and ActelSmartFusion families.

Predictive Engine Diagnostics

Big time engine makers like Rolls Royce, BMW, and Mercedes Benz jumped into Big Data and the IoT business. Their engine health monitoring unit combines latest sensor technologies with data collection, management, and analysis techniques, letting them accurately predict engine failures at an early stage. This optimizes engine maintenance and repair schedules, thereby improving safety and providing better consumer service at lower costs.

Taking one step further in the Big Data business model, Rolls Royce developed a snake robot that is equipped with self positioning, reasoning, planning, and adaptation capabilities. The 1.25cm (half-inch) diameter robot named MiRoR (miniaturized robotic systems for holistic in situ repair and maintenance works in restrained and hazardous environments) can take pictures of engine interiors and send it in real time to experts who control it remotely. This will let engine experts quickly find faults in large, complex machines like aircraft.

Mercedes Benz is an example of an automotive manufacturer who has embraced the IoT. The roadside assistance, safety, and security features provided by this popular car manufacturer have been enhanced with the introduction of the mbrace feature in their cars. Their new system now enables remote vehicle controls, performs remote engine diagnostics, and delivers software updates to keep the car running perfectly. These also monitor fleet performance and understand journey management to deliver a variety of solutions that help end users reduce fuel consumption and fleet size while maintaining performance.

Energy Management with Big Data

Several terabytes of data coming from sensors and energy meters can be used for intelligent monitoring of power usage and increasing the efficiency of the whole system. The smart building is a concept of making buildings smarter and energy efficient using the IoT and Big data analyses. An example is the smart, Wi-Fi enabled, sensor driven, programmable, self adapting thermostat. It monitors the user's temperature adjustments and uses sensors (temperature, humidity, and activity sensors) and sophisticated algorithms to learn and identify patterns, which are later used to intelligently control the heating of the home, intuitively. It adjusts the temperature according to the time of the day, weather conditions, and human activities inside the home, thereby providing an effective use of energy. Being Wi-Fi connected, the device follows current weather forecasts and adjusts the room temperature accordingly. It can also be controlled from remote locations using laptops, tablets, and smart phones.

Intelligent solar and wind analytics is another area. It uses Big Data analytics to provide operational intelligence for solar power plants, wind turbines, and other IoT equipment. Data from various sensor components like anemometers (to measure wind speed and direction), pyranometers (to measure solar irradiance on planar surface), pressure sensors, temperature sensors, and humidity sensors can help understand the potential of energy available for conversion. Sensors within the equipment's subsystems, such as generators, rotor systems, gearboxes, solar panels, and inverters, give information on the electrical and mechanical performance of the system.

Interfacing with hundreds of different sensors, supervisory control and data acquisition (SCADA) systems and smart meters pull information collected by them to the cloud and analyze this massive and complex data using the tools for better predictive maintenance and forecast. A web-based platform is made available to the customer, where they can gauge the health of the device monitor and its performance, and view forecasts and predictions of component failure. These factors further increase the efficiency, effectiveness, and hence the revenue of the system.

Life Sciences and Medicine

Big Data analysis in IoT networks is taking the idiom "prevention is better than cure" to a whole new level. Advanced sensors, mobile apps, and extensive databases are helping us reach the goal of personal health improvement.

Brain computer interface (BCI), also known as brain-machine interface (BMI), mind machine interface (MMI), or direct neural interface is a method that conjoins the human brain and machines, and lets you control external devices like bionic modules using thoughts. The system consists of several powerful and precise sensors (typically multi-electrode arrays) that collect brain signals, a decoder that can detect and interpret neural signals related to each activity, and an external module like a computer screen or robotic arm that translates these electronic signals into corresponding actions. This technology is a breakthrough in the field of medicine, as it can help restore functions to patients with disabilities.

Nanotechnology with the IoT

Sensors and actuator units make an IoT device a functional system that is capable of monitoring correct physical parameters and responding correspondingly. These sensors and actuators are prevalent at macro scale in industrial applications and these technologies cannot be scaled for numbers that IoT business roadmaps foresee. Nanotechnology will help build novel sensors and actuators in IoT domains such as ambient monitoring, automobiles, health monitoring of buildings, biomedical applications, and many more. Bulk sensors are replaced with components that are sensitive, rugged, stable, scalable, economical, and have small form factor and low power, making them viable for the IoT. Furthermore, nanotechnology promises to integrate sensors, actuators, processors, radios, and energy harvesting devices to make the Internet of nano-enabled Things in the near future for more potential applications.

Consumers already have connected things like thermostats, energy meters, lighting control systems, music streaming and control systems, remote video streaming boxes, pool systems, and irrigation systems, with more to come. Most of these systems have some connectivity through a

Web site so that a user can manage them through a standard Web browser or a smart phone app, which acts as a personal network operations center.

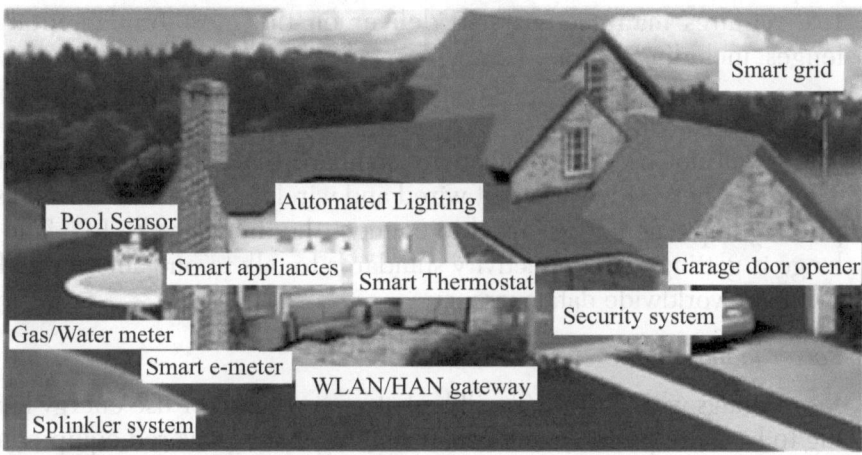

FIGURE 7.12 IoT enabled home with connected devices and appliances working invisibly for consumers.

While both the industrial and consumer scenarios are exciting, deployment is not simplified since they are all disparate vertical systems. The systems may use the exact same protocols and OS underpinnings, but the communications layers are inconsistent. Each also uses open application programming interfaces (APIs) without a horizontal connection, which would lead to easier cross application integration.

Take for example a sprinkler control system. It can have a level of intelligence so it knows when to water based on sensors and Internet weather data under programmable control. However, it does not know anything about motion sensors around a house that might indicate a reason to delay the sprinklers in a zone to avoid drenching the dog or kids. There are no motion sensor inputs on the sprinkler controller, so other motion control vertical integration needs to be used to transfer data to another cloud server. Then the two cloud servers need to be "glued" together somehow. Hopefully, both system integrations allow for some small amount of additional control. However, hope is never a good word in electronic systems. An additional vertical application written in Perl, Python, PHP, or another programming language on a server can program a connection that allows motion to delay the sprinkler zone (or other logic the user may want).

7.8 Challenges in the IoT WSNs

Preparing the lowest layers of technology for the horizontal nature of the IoT requires manufacturers to deliver on the most fundamental challenges, including:

Connectivity

There will not be one connectivity standard that "wins" over the others. There will be a wide variety of wired and wireless standards as well as proprietary implementations used to connect the things in the IoT. The challenge is getting the connectivity standards to talk to one another with one common worldwide data currency.

Power management

More things within the IoT will be battery powered or use energy harvesting to be more portable and self-sustaining. Line-powered equipment will need to be more energy efficient. The challenge is making it easy to add power management to these devices and equipment. Wireless charging will incorporate connectivity with charge management.

Security

With the amount of data being sent within the IoT, security is a must. Built-in hardware security and use of existing connectivity security protocols is essential to secure the IoT. Another challenge is simply educating consumers to use the security that is integrated into their devices.

Complexity

Manufacturers are looking to add connectivity to devices and equipment that has never been connected before to become part of the IoT. Ease of design and development is essential to get more things connected, especially when typical RF programming is complex. Additionally, the average consumer needs to be able to set up and use their devices without a technical background.

Rapid Evolution

The IoT is constantly changing and evolving. More devices are being added every day and the industry is still in its naissance. The challenge facing the industry is the unknown. Unknown devices. Unknown applications. Unknown use cases. Given this, there needs to be flexibility in all facets of development, for example, processors and microcontrollers that range from 16–1500 MHz to address the full spectrum of applications from a microcontroller (MCU) in a small, energy-harvested wireless sensor node to high-performance, multi-core processors for IoT infrastructure. A wide variety of wired and wireless connectivity technologies are needed to meet the various needs of the market. Last, a wide selection of sensors, mixed signals, and power-management technologies are required to provide the user interface to the IoT and energy-friendly designs.

For many engineers, the greatest challenge in designing for the Internet of Things (IoT) is connectivity. Implementing robust and secure access to the Internet or Wide Area Network (WAN) is outside their range of experience. To make design even more difficult, developers need to support access to multiple devices that are limited in their processing capability. Connectivity must also be added in a way that does not adversely impact overall system cost or power efficiency.

The diversity of end points a gateway must support raises design concerns as well. Directly connecting a simple node like a pressure sensor to the Internet can be complex and expensive, especially if the node does not have its own processor. In addition, different types of end equipment support varying interfaces. To collect and aggregate data from a disparate set of nodes requires a means for bridging devices with a range of processing capabilities and interfaces together in a consistent and reliable way.

Gateways offer an elegant means for simplifying the networking of "things." They achieve this by supporting the different ways nodes natively connect, whether this is a varying voltage from a raw sensor, a stream of data over I2C from an encoder, or periodic updates from an appliance via Bluetooth®. Gateways effectively mitigate the great variety and diversity of devices by consolidating data from disparate sources and interfaces and bridging them to the Internet. The result is that individual nodes don't need to bear the complexity or cost of a high-speed Internet interface in order to be connected.

7.9 Simple Versus Embedded Control Gateways

There are several ways to implement an IoT gateway, depending upon the application. Two common approaches are a simple gateway and an embedded control gateway. Both provide consolidated connectivity by aggregating data from multiple end points. In general, a simple gateway organizes and packetizes the data for transport over the Internet. It is also responsible for distributing data back to end points in applications where two-way communication is advantageous or required.

Note that a gateway is different from a router. A router manages similar traffic, and it connects devices that share a common interface. For example, the devices that connect to a home router all use IP. In contrast, because a gateway functions as a bridge, it must be able to route different types of traffic, aggregate data from varying communication interfaces, and convert these streams to a common protocol for access across the WAN. Some devices might use IP natively while others might use PAN-based protocols like Bluetooth, Zigbee, or 6LoWPAN. Nodes that are simple sensors may need to be connected to an ADC to convert their raw analog voltage to a digital value before transport.

An embedded control gateway extends the functionality of a simple gateway by providing processing resources and intelligence for handling local applications. This can take the form of shared processing resources where the gateway performs tasks that would otherwise occur on nodes. For example, an embedded control gateway could evaluate and filter sensor data as well as implement high level management tasks. After evaluating and filtering sensor data, a gateway could determine whether a critical threshold has been passed. If so, it could then trigger an alarm that is passed up through the network to alert an appropriate manager.

Having an intelligent embedded control IoT gateway can reduce the complexity and cost of end points. Depending upon the application, this can result in significant system savings. Consider a security system with an array of sensors to which it connects. Consolidating processing, such as sensor data filtering, in the gateway enables nodes to leverage a shared resource, making each node simpler as well as lower in cost.

The same holds for enabling connectivity. IP is a complex protocol to implement with relatively high overhead for more simple IoT nodes. Instead, simple nodes can connect to a PAN using a wired connection like I2C or a wireless interface like Bluetooth. The gateway also connects to the PAN and then bridges each connection to an IP-based WAN interface like Wi-Fi or Ethernet. In both of these cases, savings include lower processing, memory, and power requirements. Nodes can therefore be less expensive as well as more efficient.

When these savings are spread across a network, they add up quickly. End points that have to house their own intelligence and WAN connectivity require more complex architectures. Using a consolidated or shared architecture, the cost of each end point can be substantially reduced, more than making up for any increase in gateway cost through volume savings. Reducing the complexity of nodes also reduces overall power consumption for applications where nodes have limited battery life or operate on energy harvesting sources.

Distributed intelligence also accelerates the implementation of new applications. Consider smart appliances that use time-of-day information from the utility meter to operate during off peak hours to reduce energy costs. Implementing this intelligence at the node level requires that the washer, dryer, and dishwasher be able to communicate with the utility meter.

When each appliance comes from a different company, the interface to use this feature will likely differ, creating interoperability issues. In addition, to take advantage of this feature, consumers would need to buy new appliances. Enabling intelligence in a gateway addresses both interoperability issues on a local level while minimizing the changes required to connect appliances. Rather than require full intelligence in each appliance, the gateway can provide the base intelligence for all devices. This also has the advantage of consolidating management of new features for consumers; rather than needing to figure out and integrate each new appliance as it enters the home, the consumer only needs to understand how to manage the gateway. An intelligent gateway also better addresses the issues that arise from connecting disparate nodes, compared to users manually connecting each device or appliance to the Internet.

For many applications, an intelligent gateway can eliminate the need for a dedicated onsite management or control end point. For example, with an integrated LCD controller, a gateway can support a user interface so users can directly interact with nodes. Alternatively, an intelligent gateway can provide a web-based user interface, accessible through a PC, tablet, or smart phone, to allow users to easily access additional built-in applications. This enables the gateway to serve as a flexible and dynamically programmable onsite control point. This in turn lowers the cost of installation of new systems as well as enables third parties to introduce new technology and devices with a significantly lower cost of entry.

Finally, a gateway can serve as a fabric between co-located nodes when Internet access is lost or temporarily interrupted. This ensures robust local connectivity without the cloud, thus increasing the reliability of the local network to maintain its intended functions.

There are several ways that an IoT gateway can extend connectivity to nodes, as shown in Figure 7.13. In Figure 7.13(a), nodes connect to the IoT via a gateway. The nodes themselves are not IP-based and thus cannot directly connect to the Internet/WAN. Rather, they use either wired or wireless PAN technology to connect to the gateway with a less expensive and less complex mode of connectivity. The gateway maintains an IoT agent for each node that manages all data to and from the nodes. In this case, application intelligence can also be located in the gateway.

In Figure 7.13(b), nodes connect directly to the Internet using a WAN connection such as Wi-Fi or Ethernet. The gateway serves primarily as a router; in fact, it can be simply a router when nodes have their own IoT agent and autonomously manage themselves. Figure 7.13(c) is similar to 7.13(b), except that nodes connect directly to the Internet using a PAN connection such as 6LoWPAN. In this case, the gateway serves as a translation point between the PAN and WAN.

Today, the highest profile IoT applications are industrial, medical, and security. As this technology matures, it is clear that it will completely change how we live and do business in every industry.

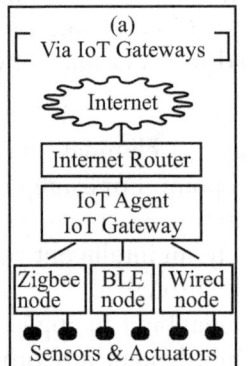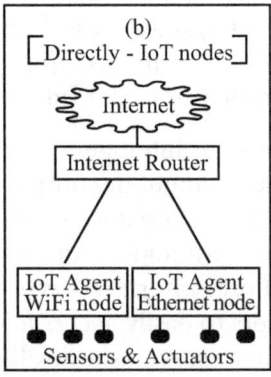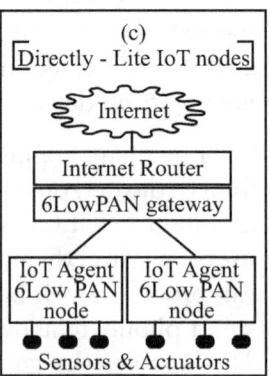

FIGURE 7.13 There are several ways that an IoT gateway can extend connectivity to nodes.

In Figure 7.13 (a) nodes connect to the IoT via a gateway using a less expensive and less complex wired or wireless PAN technology; (b) nodes connect directly to the Internet using a WAN connection such as Wi-Fi or Ethernet; (c) nodes connect indirectly to the Internet using a PAN connection such as 6LoWPAN.

With the wide acceptance of Internet Protocol (IP), it is becoming easier to process data and make meaningful use of information. Lot of companies provide enterprise-level database solutions for data storage and software tools to streamline business processes, such as asset tracking, process control systems, and building management systems as seen in Figure 7.14. Smart phones and tablets provide people with useful and actionable information, such as live parking information or real-time machine health monitoring to inform maintenance schedules. And while there are wireless sensors in place today, there is a hunger for more sensor data to measure and optimize processes that have not been previously measured.

To further enable wide-scale deployment of sensors, IP standard efforts are underway, with the goal of making small wireless sensors as easy to access as web servers. These efforts are the confluence of two driving forces: the proven low power, highly reliable performance of time-synchronized mesh networks, and the ongoing IP standard efforts for seamless integration into the Internet. Together these forces will drive relatively small, low power sensors that communicate reliably and are IP-enabled.

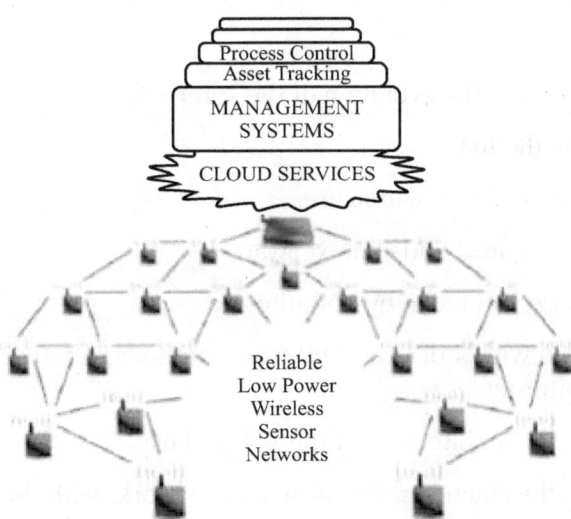

FIGURE 7.14 Making IP-enabled wireless sensors reliable and low power will enable widespread usage.

Summary

- The IoT allows people and things to be connected anytime, anyplace, with anything and anyone, ideally using any network and any service.

- Context is any information that can be used to characterize the situation of a person, place, or object that is considered relevant to the interaction between a user and an application.

- Context awareness is the ability to adapt behavior depending on the current situation of the users in context-aware applications.

- The architecture of the IoT system is divided into three layers: sensor layer, coordinator layer, and supervision layer.

- Traditional IP and WSN networks are different in network mode, routing paradigm, data flow, data rate, resource constraints, network lifetime, and operation.

- Integration of WSNs with the IoT is based on gateways of two types: homogeneous WSNs and heterogeneous WSNs.

- Challenges of WSN IoT parameters are security, quality of service, configuration, connectivity, power management, complexity, and rapid evolution.

Questions

1. Write about the evolution of the Internet.

2. Define the IoT.

3. List the IoT application areas.

4. Define context and context awareness.

5. What do you mean by ubiquitous computing?

6. Explain with a diagram different methods of integrating the WSN to the Internet.

7. Draw the architecture of the IoT and explain.

8. Write the characteristics of sensor networks with the IoT.

9. Compare traditional IP-based networks and large-scale wireless sensor networks.

10. Write two different methods of gateway-based integration in detail.

11. What are the design principles of the IoT and WSNs?

12. List the challenges for WSNs in the IoT.

13. Explain about Big Data and the IoT in WSNs.

14. What do you mean by Big Data?

15. Give the example of an IoT-enabled home structure.

16. Explain about simple versus embedded control gateways.

Further Reading

1. *Learning Internet of Things* by Peter Waher

2. *The Internet of Things* by Samuel Greengard

3. *Internet of Things: A Hands-On Approach* by Arshdeep Bahga and Vijay Madisetti

References

1. *http://www.edn.com/design/analog/4426319/Low-Power-wireless-sensor-networks for-the-Internet-of-Things*

2. *http://www.cs.usc.edu/assets/002/82967.pdf*

3. *http://www.ti.com/ww/en/connect_more/pdf/SWB001.pdf? DCMP=iot&HQS=iot-nslwp*

4. *http://www.ti.com/lit/wp/spmy013/spmy013.pdf*

CHAPTER **8**

WIRELESS MULTIMEDIA SENSOR NETWORKS

This chapter discusses wireless multimedia sensor networks, software, and hardware architecture.

8.1 Introduction to Wireless Multimedia Sensor Networks (WMSNs)

The availability of low-cost hardware such as CMOS cameras and microphones has fostered the development of Wireless Multimedia Sensor Networks (WMSNs), that is, networks of wirelessly interconnected devices that are able to ubiquitously retrieve multimedia content such as video and audio streams, still images, and scalar sensor data from the environment. With rapid improvements and miniaturization in hardware, a single sensor device can be equipped with audio and visual information collection modules. In addition to the ability to retrieve multimedia data, WMSNs will also be able to store, process in real-time, correlate, and fuse multimedia data originating from heterogeneous sources. Wireless multimedia sensor networks will not only enhance existing sensor network applications such as tracking, home automation, and environmental monitoring, but they will also enable several new applications such as:

Multimedia Surveillance Sensor Networks

Wireless video sensor networks will be composed of interconnected, battery-powered miniature video cameras, each packaged with a low-power wireless transceiver that is capable of processing, sending, and receiving data. Video and audio sensors will be used to enhance and complement existing surveillance systems against crime and terrorist attacks. Large-scale networks of video sensors can extend the ability of law enforcement agencies to monitor areas, public events, private properties, and borders.

Storage of Potentially Relevant Activities

Multimedia sensors could infer and record potentially relevant activities (thefts, car accidents, traffic violations), and make video/audio streams or reports available for future query.

Traffic Avoidance, Enforcement, and Control Systems

It will be possible to monitor car traffic in big cities or highways and deploy services that offer traffic routing advice to avoid congestion. In addition, smart parking advice systems based on WMSNs will allow monitoring of available parking spaces and provide drivers with automated parking advice, thus improving mobility in urban areas. Moreover, multimedia sensors may monitor the flow of vehicular traffic on highways and retrieve aggregate information such as average speed and number of cars. Sensors could also detect violations and transmit video streams to law enforcement agencies to identify the violator, or buffer images and streams in case of accidents for subsequent accident scene analysis.

Advanced Health Care Delivery

Telemedicine sensor networks can be integrated with 3G multimedia networks to provide ubiquitous health care services. Patients will carry medical sensors to monitor parameters such as body temperature, blood pressure, pulse oximetry, ECG, and breathing activity. Furthermore, remote medical centers will perform advanced remote monitoring of their patients via video and audio sensors, location sensors, and motion or activity sensors, which can also be embedded in wrist devices.

Automated Assistance for the Elderly and Family Monitors

Multimedia sensor networks can be used to monitor and study the behavior of elderly people as a means to identify the causes of illnesses that affect them such as dementia. Networks of wearable or video and audio sensors can infer emergency situations and immediately connect elderly patients with remote assistance services or with relatives.

Environmental Monitoring

Habitat monitoring uses acoustic and video feeds in which information has to be conveyed in a time-critical fashion. For example, arrays of video sensors are used by oceanographers to determine the evolution of sandbars via image processing techniques.

Person Locator Services

Multimedia content such as video streams and still images, along with advanced signal processing techniques, can be used to locate missing persons, or identify criminals or terrorists.

Industrial Process Control

Multimedia content such as imaging, temperature, or pressure, among others, may be used for time-critical industrial process control. Machine vision is the application of computer vision techniques to industry and manufacturing, where information can be extracted and analyzed by WMSNs to support a manufacturing process such as those used in semiconductor chips, automobiles, food, or pharmaceutical products. For example, in quality control of manufacturing processes, details of final products are automatically inspected to find defects. In addition, machine vision systems can detect the position and orientation of parts of the product to be picked up by a robotic arm. The integration of machine vision systems with WMSNs can simplify and add flexibility to systems for visual inspections and automated actions that require high-speed, high-magnification, and continuous operation.

WMSNs will stretch the horizon of traditional monitoring and surveillance systems by:

Enlarging the View

The Field of View (FoV) of a single fixed camera, or the Field of Regard (FoR) of a single moving pan-tilt-zoom (PTZ) camera is limited. Instead, a distributed system of multiple cameras and sensors enables perception of the environment from multiple disparate viewpoints, and helps overcome occlusion effects.

Enhancing the View

The redundancy introduced by multiple, possibly heterogeneous, overlapped sensors can provide enhanced understanding and monitoring of the environment. Overlapped cameras can provide different views of the same area or target, while the joint operation of cameras and audio or infrared sensors can help disambiguate cluttered situations.

Enabling Multi-Resolution Views

Heterogeneous media streams with different granularity can be acquired from the same point of view to provide a multi-resolution description of the scene and multiple levels of abstraction. For example, static medium-resolution camera views can be enriched by views from a zoom camera that provides a high-resolution view of a region of interest. For example, such feature could be used to recognize people based on their facial characteristics.

Many of the above applications require the sensor network paradigm to be rethought in view of the need for mechanisms to deliver multimedia content with a certain level of quality of service (QoS). There are several main peculiarities that make QoS delivery of multimedia content in sensor networks an even more challenging, and largely unexplored, task:

Resource Constraints

Sensor devices are constrained in terms of battery, memory, processing capability, and achievable data rate. Hence, efficient use of these scarce resources is mandatory.

Variable Channel Capacity

While in wired networks the capacity of each link is assumed to be fixed and predetermined, in multi-hop wireless networks, the attainable capacity of each wireless link depends on the interference level perceived at the receiver. This, in turn, depends on the interaction of several functionalities that are handled by all network devices such as power control, routing, and rate policies. Hence, capacity and delay attainable at each link are location dependent, vary continuously, and may be bursty in nature, thus making QoS provisioning a challenging task.

Cross-Layer Coupling of Functionalities

In multi-hop wireless networks, there is a strict interdependence among functions handled at all layers of the communication stack. Functionalities handled at different layers are inherently and strictly coupled due to the shared nature of the wireless communication channel. Hence, the various functionalities aimed at QoS provisioning should not be treated separately when efficient solutions are sought.

Multimedia In-Network Proces sing

Processing of multimedia content has mostly been approached as a problem isolated from the network-design problem, with a few exceptions such as joint source-channel coding and channel-adaptive streaming. Hence, research that addressed the content delivery aspects has typically not considered the characteristics of the source content and has primarily studied cross-layer interactions among lower layers of the protocol stack. However, the processing and delivery of multimedia content are not independent, and their interaction has a major impact on the levels of QoS that can be delivered. WMSNs will allow performing multimedia in-network processing algorithms on the raw data. Hence, the QoS required at the application level will be delivered by means of a combination of both cross-layer optimization of the communication process, and in-network processing of raw data streams that describe the phenomenon of interest from multiple views, with different media, and on multiple resolutions. Hence, it is necessary to develop application independent and self-organizing architectures to flexibly perform in-network processing of multimedia contents.

8.2 Factors Influencing the Design of Multimedia Sensor Networks

Wireless Multimedia Sensor Networks (WMSNs) will be enabled by the convergence of communication and computation with signal processing and several branches of control theory and embedded computing. This cross-disciplinary research will enable distributed systems of heterogeneous embedded devices that sense, interact, and control the physical environment. There are several factors that mainly influence the design of a WMSN.

Application-Specific QoS Requirements

The wide variety of applications envisaged on WMSNs will have different requirements. In addition to data delivery modes typical of scalar sensor networks, multimedia data include snapshot and streaming multimedia content. Snapshot-type multimedia data contain event-triggered observations obtained in a short time period. Streaming multimedia content is generated over longer time periods and requires sustained information delivery. Hence, a strong foundation is needed in terms of hardware and supporting high-level algorithms to deliver QoS and consider application-specific requirements. These requirements may pertain to multiple domains and can be expressed, among others, in terms of a combination of bounds on energy consumption, delay, reliability, distortion, or network lifetime.

High Bandwidth Demand

Multimedia content, especially video streams, require transmission bandwidth that is orders of magnitude higher than that supported by currently available sensors. For example, the nominal transmission rate of state-of-the-art IEEE 802.15.4 compliant component motes is 250 Kbit/s. Data rates at least one order of magnitude higher may be required for high-end multimedia sensors, with comparable power consumption. Hence, high data rate and low-power consumption transmission techniques need to be leveraged.

Multimedia Source Coding Techniques

Uncompressed raw video streams require excessive bandwidth for a multihop wireless environment. For example, a single monochrome frame in the NTSC-based Quarter Common Intermediate Format (QCIF, 176 • 120) requires around 21 Kbytes, and at 30 frames per second (fps), a video stream requires over 5 Mbit/s. Hence, it is apparent that efficient processing techniques for lossy compression are necessary for multimedia sensor networks. Traditional video coding techniques used for wire line and wireless communications are based on the idea of reducing the bit rate generated by the source encoder by exploiting source statistics. To this aim, encoders rely on intra-frame compression techniques to reduce redundancy within one frame, while they leverage inter-frame compression (also known as predictive encoding or motion estimation) to exploit redundancy among subsequent frames to reduce the amount of data to be transmitted and stored, thus achieving good rate-distortion performance. Since predictive encoding requires complex encoders, powerful processing algorithms, and entails high energy consumption, it may not be suited for low-cost multimedia sensors. However, it has recently been shown that the traditional balance of complex encoder and simple decoder can be reversed within the framework of the so-called distributed source coding, which exploits the source statistics at the decoder, and by shifting the complexity at this end, allows the use of simple encoders. Clearly, such algorithms are very promising for WMSNs and especially for networks of video sensors, where it may not be feasible to use existing video encoders at the source node due to processing and energy constraints.

Multimedia In-Network Processing

WMSNs allow performing multimedia in-network processing algorithms on the raw data extracted from the environment. This requires new architectures for collaborative, distributed, and resource-constrained processing that allow for filtering and extraction of semantically relevant information at the edge of the sensor network. This may increase the system scalability by reducing the transmission of redundant information, merging data originated from multiple views, on different media, and with multiple resolutions. For example, in video security applications, information from uninteresting scenes can be compressed to a simple scalar value or not be transmitted altogether, while in environmental applications, distributed filtering techniques can create a time-elapsed image. Hence, it is necessary to develop application-independent architectures to flexibly perform in-network processing of the multimedia content gathered from the environment.

Power Consumption

Power consumption is a fundamental concern in WMSNs, even more than in traditional wireless sensor networks. Infact, sensors are battery-constrained devices, while multimedia applications produce high volumes of data, which require high transmission rates, and extensive processing. While the energy consumption of traditional sensor nodes is known to be dominated by communication functionalities, this may not necessarily be true in WMSNs. Therefore, protocols, algorithms, and architectures to maximize the network lifetime while providing the Quality of Service (QoS) required by the application are a critical issue.

Flexible Architecture to Support Heterogeneous Applications

WMSN architectures will support several heterogeneous and independent applications with different requirements. It is necessary to develop flexible, hierarchical architectures that can accommodate the requirements of all these applications in the same infrastructure.

Multimedia Coverage

Some multimedia sensors, in particular video sensors, have larger sensing radii and are sensitive to direction of acquisition (directivity). Furthermore, video sensors can capture images only when there is an unobstructed line of sight between the event and the sensor. Hence, coverage models developed for traditional wireless sensor networks are not sufficient for pre deployment planning of a multimedia sensor network.

Integration with Internet (IP) Architecture

It is of fundamental importance for the commercial development of sensor networks to provide services that allow querying the network to retrieve useful information from anywhere and at any time. For this reason, future WMSNs will be remotely accessible from the Internet, and will therefore need to be integrated with the IP architecture. The characteristics of WSNs rule out the possibility of all-IP sensor networks and recommend the use of application-level gateways or overlay IP networks as the best approach for integration between WSNs and the Internet.

Integration with Other Wireless Technologies

Large-scale sensor networks may be created by interconnecting local "islands" of sensors through other wireless technologies. This needs to be achieved without sacrificing the efficiency of the operation within each individual technology.

8.3 Network Architecture of WMSNs

The problem of designing a scalable network architecture is of primary importance. Most proposals for wireless sensor networks are based on a flat, homogenoeus architecture in which every sensor has the same physical capabilities and can only interact with neighboring sensors. Traditionally, the research on algorithms and protocols for sensor networks has focused on scalability, that is, how to design solutions whose applicability would not be limited by the growing size of the network. Flat topologies may not always be suited to handle the amount of traffic generated by multimedia applications including audio and video. Likewise, the processing power required for data processing and communications, and the power required to operate it, may not be available on each node.

Reference Architecture

Figure 8.1 is a reference architecture for WMSNs, where three sensor networks with different characteristics are shown, possibly deployed in different physical locations. The first cloud on the left shows a single-tier network of homogeneous video sensors. A subset of the deployed sensors have higher processing capabilities, and are thus referred to as processing hubs. The union of the processing hubs constitutes a distributed processing architecture. The multimedia content gathered is relayed to a wireless gateway through a multi-hop path. The gateway is interconnected to a storage hub that is in charge of storing multimedia content locally for subsequent retrieval. Clearly, more complex architectures for distributed storage can be implemented when allowed by the environment and the application needs, which may result in energy savings, since by storing it locally, the multimedia content does not need to be wirelessly relayed to remote locations. The wireless gateway is also connected to a central sink, which implements the software front-end for network querying and tasking. The second cloud represents a single-tiered clustered architecture of heterogeneous sensors (only one cluster is depicted). Video, audio, and scalar sensors relay data to a central cluster head, which is also in charge of performing intensive multimedia processing on the data (processing hub). The cluster head relays the gathered content to the wireless gateway and to the storage hub. The last cloud on the right represents a multi-tiered network, with heterogeneous sensors. Each tier is in charge of a subset of the functionalities. Resource-constrained, low-power scalar sensors are in charge of performing simpler tasks, such as detecting scalar physical measurements, while resource-rich, high-power devices are responsible for more complex tasks. Data processing and storage can be performed in a distributed fashion at each different tier.

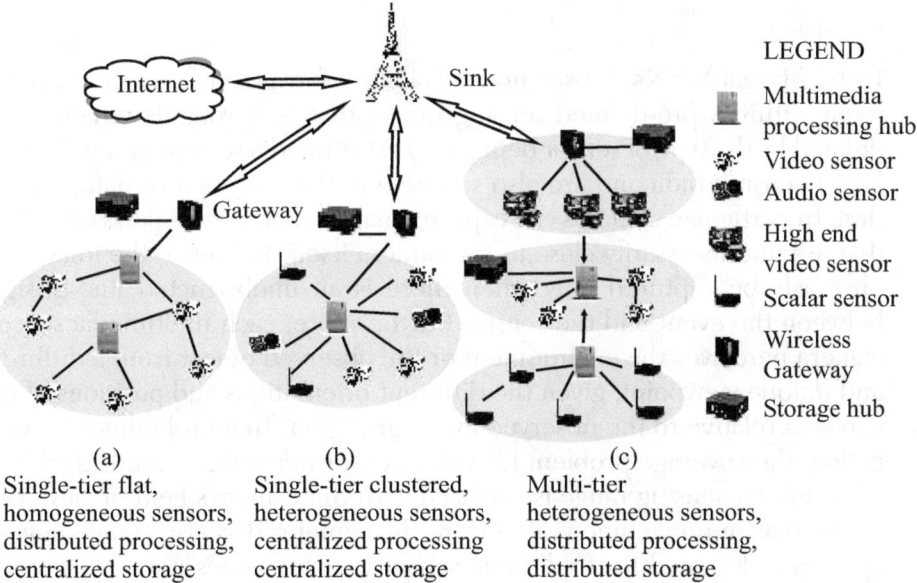

(a)	(b)	(c)
Single-tier flat, homogeneous sensors, distributed processing, centralized storage	Single-tier clustered, heterogeneous sensors, centralized processing centralized storage	Multi-tier heterogeneous sensors, distributed processing, distributed storage

FIGURE 8.1 Reference Architecture of a wireless multimedia sensor network.

Single-Tier vs. Multi-Tier Sensor Deployment

One possible approach for designing a multimedia sensor application is to deploy homogeneous sensors and program each sensor to perform all possible application tasks. Such an approach yields a flat, single-tier network of homogeneous sensor nodes. An alternative, multi-tier approach is to use heterogeneous elements. In this approach, resource-constrained, low-power elements are in charge of performing simpler tasks, such as detecting scalar physical measurements, while resource rich, high-power devices take on more complex tasks. For instance, a surveillance application can rely on low-fidelity cameras or scalar acoustic sensors to perform motion or intrusion detection, while high-fidelity cameras can be woken up on-demand for object recognition and tracking. A multi-tier architecture is advocated for video sensor networks for surveillance applications. The architecture is based on multiple tiers of cameras with different functionalities, with the lower tier consisting of low-resolution imaging sensors, and the higher tier composed of high-end pan-tilt-zoom cameras. Such an architecture offers considerable advantages with respect to a single-tier architecture in terms of scalability, lower cost, better coverage, higher functionality, and better reliability.

Coverage

In traditional WSNs, sensor nodes collect information from the environment within a pre-defined sensing range, that is, a roughly circular area defined by the type of sensor being used. Multimedia sensors generally have larger sensing radii and are also sensitive to the direction of data acquisition. In particular, cameras can capture images of objects or parts of regions that are not necessarily close to the camera itself. However, the image can obviously be captured only when there is an unobstructed line-of-sight between the event and the sensor. Furthermore, each multimedia sensor/camera perceives the environment or the observed object from a different and unique viewpoint, given the different orientations and positions of the cameras relative to the observed event or region. In a preliminary investigation, the coverage problem for video sensor networks is conducted. The concept of a sensing range is replaced with the camera's field of view, that is, the maximum volume visible from the camera. It is also shown how an algorithm designed for traditional sensor networks does not perform well with video sensors in terms of coverage preservation of the monitored area.

Multimedia Sensor Hardware

High-end pan-tilt-zoom cameras and high resolution digital cameras are widely available on the market. However, while such sophisticated devices can find application as high-quality tiers of multimedia sensor networks, one can concentrate on low-cost, low energy consumption imaging and processing devices that will be densely deployed and provide detailed visual information from multiple disparate viewpoints, help overcoming occlusion effects, and thus enable enhanced interaction with the environment.

Low-Resolution Imaging Motes

The recent availability of CMOS imaging sensors that capture and process an optical image within a single integrated chip, thus eliminating the need for many separate chips required by the traditional charged-coupled device (CCD) technology, has enabled the massive deployment of low-cost visual sensors. CMOS image sensors are already in many industrial and consumer sectors, such as cell phones, personal digital assistants (PDAs), and consumer and industrial digital cameras. CMOS image quality is now matching CCD quality in the low and mid range, while CCD is still the technology of choice for high-end image sensors. The CMOS technology allows integrating a lens, an image sensor, and image processing algorithms, including image stabilization and image compression, on the same chip. With respect to CCD, cameras are smaller, lighter, and consume less power. Hence, they constitute a suitable technology to realize imaging sensors to be interfaced with wireless motes. However, existing CMOS imagers are still designed to be interfaced with computationally rich host devices, such as cell phones or PDAs. The design of an integrated mote for wireless image sensor networks is driven by the need to endow motes with adequate processing power and memory size for image sensing applications. It is argued that 32-bit processors are better suited for image processing than their 8-bit counterparts. It is shown that the time needed to perform operations such as 2-D convolution on an 8-bit processor such as the ATMEL ATmega128 clocked at 4 MHz is 16 times higher than with a 32-bit ARM7 device clocked at 48 MHz, while the power consumption of the 32-bit processor is only six times higher. Hence, an 8-bit processor turns out to be slower and more energy-consuming.

Medium-Resolution Imaging Motes

The medium imaging board is a high-performance processing platform designed for sensor, signal processing, control, robotics, and sensor network applications. It is based on Intel's PXA-255 XScale 400 MHz RISC processor, which is the same processor found in many handheld computers including the Compaq IPAQ and the Dell Axim, and has 32 Mbyte of Flash memory, 64 Mbyte of SDRAM, and Bluetooth or IEEE 802.11

cards. Hence, it can work as a wireless gateway and as a computational hub for in-network processing algorithms. When connected with a webcam or other capturing device, it can function as a medium-resolution multimedia sensor, although its energy consumption is still high. Moreover, although efficient software implementations exist, XScale processors do not have hardware support for floating point operations, which may be needed to efficiently perform multimedia processing algorithms.

Energy Harvesting

Techniques for prolonging the lifespan of battery-powered sensors have been the focus in multimedia sensor networks. These techniques include hardware optimizations such as dynamic optimization of voltage and clock rate, wake-up procedures to keep electronics inactive most of the time, and energy-aware protocol development for sensor communications. In addition, energy-harvesting techniques, which extract energy from the environment where the sensor itself lies, offer another important means to prolong the lifetime of sensor devices with the technologies to generate energy from background radio signals, thermoelectric conversion, vibrational excitation, and the human body.

As far as collecting energy from background radio signals is concerned, unfortunately, an electric field of 1 V/m yields only 0.26 1W/cm², as opposed to 100 1W/cm² produced by a crystalline silicon solar cell exposed to bright sunlight. Electric fields of intensity of a few volts per meter are only encountered close to strong transmitters. Another practice, which consists in broadcasting RF energy deliberately to power electronic devices, is severely limited by legal limits set by health and safety concerns.

While thermoelectric conversion may not be suitable for wireless devices, harvesting energy from vibrations in the surrounding environment may provide another useful source of energy. Vibrational magnetic power generators based on moving magnets or coils may yield powers that range from tens of microwatts when based on microelectromechanical system (MEMS) technologies to over a milliwatt for larger devices. Other vibrational microgenerators are based on charged capacitors with moving plates, and depending on their excitation and power conditioning, yield power on the order of 10 μW. It is also reported that recent analysis suggested that 1 cm³ vibrational microgenerators can be expected to yield up to 800 μW/cm³ from machine-induced stimuli, which is orders of magnitude higher than that provided by currently available microgenerators. Hence, this is a promising area of research for small battery-powered devices.

While these techniques may provide an additional source of energy and help prolong the lifetime of sensor devices, they yield power that is several orders of magnitude lower as compared to the power consumption of state-of-the-art multimedia devices. Hence, they may currently be suitable only for very low duty cycle devices.

Collaborative In-Network Processing

It is necessary to develop architectures and algorithms to flexibly perform these functionalities in-network with minimum energy consumption and limited execution time. The objective is usually to avoid transmitting large amounts of raw streams to the sink by processing the data in the network to reduce the communication volume.

Given a source of data (e.g., a video stream), different applications may require diverse information (e.g., raw video stream vs. simple scalar or binary information inferred by processing the video stream). This is referred to as application-specific querying and processing. Hence, it is necessary to develop expressive and efficient querying languages, and to develop distributed filtering and in-network processing architectures, to allow realtime retrieval of useful information.

Similarly, it is necessary to develop architectures that efficiently allow performing data fusion or other complex processing operations in-network. Algorithms for both inter-media and intra-media data aggregation and fusion need to be developed, as simple distributed processing schemes developed for existing scalar sensors are not suitable for computation-intensive processing required by multimedia contents. Multimedia sensor networks may require computation-intensive processing algorithms (e.g., to detect the presence of suspicious activity from a video stream). This may require considerable processing to extract meaningful information and/or to perform compression. A fundamental question to be answered is whether this processing can be done on sensor nodes (i.e., a flat architecture of multi-functional sensors that can perform any task), or if the need for specialized devices for example, computation hubs, arises.

Data Alignment and Image Registration

Data alignment consists of merging information from multiple sources. One of the most widespread data alignment concepts, image registration, is a family of techniques, widely used in areas such as remote sensing, medical imaging, and computer vision, to geometrically align different images (reference and sensed images) of the same scene taken at different times, from different viewpoints, and/or by different sensors:

Different Viewpoints (Multi-View Analysis)

Images of the same scene are acquired from different viewpoints, to gain a larger 2D view or a 3D representation of the scene of interest. Main applications are in remote sensing, computer vision, and 3D shape recovery.

Different Times (Multi-Temporal Analysis)

Images of the same scene are acquired at different times. The aim is to find and evaluate changes in time in the scene of interest. The main applications are in computer vision, security monitoring, and motion tracking.

Different Sensors (Multi-Modal Analysis)

Images of the same scene are acquired by different sensors. The objective is to integrate the information obtained from different source streams to gain more complex and detailed scene representation.

Registration methods usually consist of four steps, that is, feature detection, feature matching, transform model estimation, and image resampling and transformation. In feature detection, distinctive objects such as closed-boundary regions, edges, contours, line intersections, corners, and so on are detected. In feature matching, the correspondence between the features detected in the sensed image and those detected in the reference image is established. In transform model estimation, the type and parameters of the so-called mapping functions, which align the sensed image with the reference image, are estimated. The parameters of the mapping functions are computed by means of the established feature correspondence. In the last step, image resampling and transformation, the sensed image is transformed by means of the mapping functions. These functionalities can clearly be prohibitive for a single sensor. Hence, research is needed on how to perform these functionalities on parallel architectures of sensors to produce single data sets.

8.4 WMSNs as Distributed Computer Vision Systems

Computer vision is a subfield of artificial intelligence, whose purpose is to allow a computer to extract features from a scene, an image, or multidimensional data in general. The objective is to present this information to a human operator or to control some process (e.g., a mobile robot or an autonomous vehicle). The image data that is fed into a computer vision system is often a digital image, a video sequence, a 3D volume from a tomography device, or other multimedia content. Traditional computer vision algorithms require extensive computation, which in turn entails high power consumption. WMSNs enable a new approach to computer vision, where visual observations across the network can be performed by means of distributed computations on multiple, possibly low-end, vision nodes.

8.5 Application Layer

The functionalities handled at the application layer of a WMSN are characterized by high heterogeneity, and encompass traditional communication problems as well as more general system challenges.

The services offered by the application layer include:

(i) providing traffic management and admission control functionalities, that is, preventing applications from establishing data flows when the network resources needed are not available;

(ii) performing source coding according to application requirements and hardware constraints, by leveraging advanced multimedia encoding techniques;

(iii) providing flexible and efficient system software, that is, operating systems and middleware, to export services for higher-layer applications to build upon; and

(iv) providing primitives for applications to leverage collaborative, advanced in-network multimedia proces sing techniques.

Traffic Classes

Admission control has to be based on QoS requirements of the overlying application. WMSNs will need to provide support and differentiated service for several different classes of applications. In particular, they will need to provide differentiated service between real-time and delay-tolerant applications, and loss-tolerant and loss-intolerant applications. Moreover, some applications may require a continuous stream of multimedia data for a prolonged period of time (multimedia streaming), while some other applications may require event triggered observations obtained in a short time period (snapshot multimedia content). The main traffic classes that need to be supported are:

a) Real-Time, Loss-Tolerant Multimedia Streams

This class includes video and audio streams, or multi-level streams composed of video/audio and other scalar data (e.g., temperature readings), as well as metadata associated with the stream, that need to reach a human or automated operator in real-time, that is, within strict delay bounds, and that are however relatively loss tolerant (e.g., video streams can be within a certain level of distortion). Traffic in this class usually has high bandwidth demand.

b) Delay-Tolerant, Loss-Tolerant Multimedia Streams

This class includes multimedia streams that, being intended for storage or subsequent offline processing, do not need to be delivered within strict delay bounds. However, due to the typically high bandwidth demand of multimedia streams and to limited buffers of multimedia sensors, data in this traffic class needs to be transmitted almost in real-time to avoid excessive losses.

c) Real-Time, Loss-Tolerant Data

This class may include monitoring data from densely deployed scalar sensors such as light sensors whose monitored phenomenon is characterized by spatial correlation, or loss-tolerant snapshot multimedia data (e.g., images of a phenomenon taken from several multiple viewpoints at the same time). Hence, sensor data has to be received in a timely fashion, but the application is moderately loss-tolerant. The bandwidth demand is usually between low and moderate.

d) Real-Time, Loss-Intolerant Data

This may include data from time-critical monitoring processes such as distributed control applications. The bandwidth demand varies between low and moderate.

e) Delay-Tolerant, Loss-Intolerant Data

This may include data from critical monitoring processes, with low or moderate bandwidth demand that require some form of offline post processing.

f) Delay-Tolerant, Loss-Tolerant Data

This may include environmental data from scalar sensor networks, or non-time-critical snapshot multimedia content, with low or moderate bandwidth demand.

QoS requirements have recently been considered as application admission criteria for sensor networks. An application admission control algorithm is proposed whose objective is to maximize the network lifetime subject to bandwidth and reliability constraints of the application. An application admission control method determines admissions based on the added energy load and application rewards. While these approaches address application level QoS considerations, they fail to consider multiple QoS requirements (e.g., delay, reliability, and energy consumption) simultaneously, as required in WMSNs. Furthermore, these solutions do not consider the peculiarities of WMSNs, that is, they do not try to base admission control on a tight balancing between communication optimizations and in-network computation. There is a clear need for new criteria and mechanisms to manage the admission of multimedia flows according to the desired application-layer QoS.

Multimedia Encoding Techniques

The captured multimedia content should ideally be represented in such a way as to allow reliable transmission over lossy channels (error-resilient coding), using algorithms that minimize processing power and the amount of information to be transmitted. The main design objectives of a coder for multimedia sensor networks are thus:

a) High Compression Efficiency

Uncompressed raw video streams require high data rates and thus consume excessive bandwidth and energy. It is necessary to achieve a high ratio of compression to effectively limit bandwidth and energy consumption.

b) Low Complexity

Multimedia encoders are embedded in sensor devices. Hence, they need to be low complexity to reduce cost and form factors, and low power to prolong the lifetime of sensor nodes.

c) Error resiliency

The source coder should provide robust and error-resilient coding of source data.

8.6 Transport Layer

In applications involving high-rate data, the transport layer assumes special importance by providing end-to-end reliability and congestion control mechanisms. Particularly in WMSNs, the following additional considerations are in order to accommodate both the unique characteristics of the WSNparadigm and multimedia transport requirements.

Effects of Congestion

In WMSNs, the effect of congestion may be even more pronounced as compared to traditional networks. When a bottleneck sensor is swamped with packets coming from several high-rate multimedia streams, apart from temporary disruption of the application, it may cause rapid depletion of the node's energy. While applications running on traditional wireless networks may only experience performance degradation, the energy loss (due to collisions and retransmissions) can result in network partition. Thus, congestion control algorithms may need to be tuned for immediate response and yet avoid oscillations of data rate along the affected path.

Packet Re-ordering Due to Multi-Path

Multiple paths may exist between a given source-sink pair, and the order of packet delivery is strongly influenced by the characteristics of the route chosen. As an additional challenge, in real-time video/ audio feeds or streaming media, information that cannot be used in the proper sequence becomes redundant, thus stressing on the need for transport layer packet reordering.

TCP/UDPand TCP-Friendly Schemes for WMSNs

For real-time applications like streaming media, the User Datagram Protocol (UDP) is preferred over TCP as timelines s is of greater concern than reliability. However, in WMSNs, it is expected that packets are significantly compressed at the source and redundancy is reduced as far as possible owing to the high transmission overhead in the energy constrained nodes. Under these conditions, the following are important characteristics that may necessitate an approach very different from classical wireless networks.

a) Effect of dropping packets in UDP

Simply dropping packets during congestion conditions, as undertaken in UDP, may introduce discernable disruptions in the order of a fraction of a second. This effect is even more pronounced if the packet dropped contains important original content not captured by inter-frame interpolation, like the Region of Interest (ROI) feature used in JPEG2000 or the I-frame used in the MPEG family.

b) Support for traffic heterogeneity

Multimedia traffic comprising video, audio, and still images exhibits a high level of heterogeneity and may be further classified into periodic or event driven. The UDP header has no provision to allow any description of these traffic classes that may influence congestion control policies. As a contrast to this, the options field in the TCP header can be modified to carry data-specific information.

c) Effect of jitter induced by TCP

A key factor that limits multimedia transport based on TCP, and TCP-like rate control schemes, is the jitter introduced by the congestion control mechanism. This can be, however, mitigated to a large extent by playout buffers at the sink, which is typically assumed to be rich in resources.

d) Overhead of the reliability mechanism in TCP

Blind dropping of packets in UDP containing highly compressed video/audio data may adversely affect the quality of transmission. Yet, at the same time, the reliability mechanism provided by TCP introduces an end-to-end message passing overhead and energy efficiency must also be considered. Distributed TCP Caching (DTC) overcomes these problems by caching TCP segments inside the sensor network and by local retransmission of TCP segments. The nodes closest to the sink are the last-hop forwarders on most of the high-rate data paths and thus run out of energy first. DTC shifts the burden of the energy consumption from nodes close to the sink into the network, apart from reducing network-wide retransmissions.

e) Regulating streaming through multiple TCP connections

The availability of multiple paths between source and sink can be exploited by opening multiple TCP connections for multimedia traffic. Here, the desired streaming rate and the allowed throughput reduction in the presence of bursty traffic, like sending of video data, is communicated to the receiver by the sender. This information is used by the receiver, which then measures the actual throughput and controls the rate within the allowed bounds by using multiple TCP connections and dynamically changing its TCP window size for each connection.

Application-Specific and Non-Standard Protocols

Depending on the application, both reliability and congestion control may be equally important functionalities, or one may be preferred over the other.

a) Reliability

Multimedia streams may consist of images, video, and audio data, each of which merits a different metric for reliability. When an image or video is sent with differentially coded packets, the arrival of the packets with the ROI field or the I-frame respectively should be guaranteed. The application can, however, withstand moderate loss for the other packets containing differential information. Thus, we believe that reliability needs to be enforced on a per-packet basis to best utilize the existing networking resources. If a prior recorded video is being sent to the sink, all the I-frames could be separated and the transport protocol should ensure that each of these reach the sink. Reliable Multi-Segment Transport (RMST) or the Pump Slowly Fetch Quickly (PSFQ) protocol can be used for this purpose as they buffer packets at intermediate nodes, allowing for faster retransmission in case of packet loss. However, there is an overhead of using the limited buffer space at a given sensor node for caching packets destined for other nodes, as well as performing timely storage and flushing operations on the buffer.

b) Congestion control

The high rate of injection of multimedia packets into the network causes resources to be used up quickly. While typical transmission rates for sensor nodes may be about 40 kbit/s, indicative data rates of a constant bit rate voice traffic may be 64 kbit/s. Video traffic, on the other hand, may be bursty and in the order of 500 kbit/s, thus making it clear that congestion must be addressed in WMSNs.

c) Use of multi-path

The use of multiple paths for data transfer in WMSNs is necessary for the following two reasons:

1. A large burst of data (say, resulting from an I-frame) can be split into several smaller bursts, thus not overwhelming the limited buffers at the intermediate sensor nodes.

2. The channel conditions may not permit high data rate for the entire duration of the event being monitored. By allowing multiple flows, the effective data rate at each path gets reduced and the application can be supported.

8.7 Network Layer

The network layer addresses the challenging task of providing variable QoS guarantees depending on whether the stream carries time-independent data like configuration or initialization parameters, time-critical low rate data like presence or absence of the sensed phenomenon, high bandwidth video/audio data, and so forth. Each of the traffic classes has its own QoS requirement which must be accommodated in the network layer.

Addressing and localization

In the case of large WMSNs, it is required that the individual nodes be monitored via the Internet. Such an integration between a randomly deployed sensor network and the established wired network becomes a difficult research challenge. The key problem of global addressing could be solved by the use of IPv6, in which the sensor can concatenate its cluster ID with its own MAC address to create the full IPv6 address. However, the 16-byte address field of IPv6 introduces excessive overhead in each sensor data packet. There are several other schemes that assign unique network-wide IDs or leverage location information to create an address free environment, but they, however, run the risk of incompatibility with the established standards of the Internet. Location information is a key characteristic of any sensor network system. The ability to associate localization information with the raw data sampled from the environment increases the capability of the system and the meaningfulness of the information extracted. Localization techniques for WMSNs are unlikely to differ substantially from those developed for traditional sensor networks. Moreover, WMSNs will most likely leverage the accurate ranging capabilities that come with high bandwidth transmissions.

Routing

Data collected by the sensor nodes needs to be sent to the sink, where useful information can be extracted from it. As an example, multiple routes may be necessary to satisfy the desired data rate at the destination node. Also, different paths exhibiting varying channel conditions may be preferred depending on the type of traffic and its resilience to packet loss. In general, they can be classified into routing based on (i) network conditions that leverage channel and link statistics, (ii) traffic classes that decide paths based on packet priorities, and (iii) specialized protocols for real-time streaming that use spatiotemporal forwarding.

QoS routing based on network conditions

Network conditions include interference seen at intermediate hops, the number of backlogged flows along a path, residual energy of the nodes, among others. A routing decision based on these metrics can avoid paths that may not support high bandwidth applications or introduce retransmission owing to bad channel conditions. The use of image sensors is used to gather topology information that is then leveraged to develop efficient geographic routing schemes. A weighted cost function is constructed that takes into account position with respect to the base station, backlogged packets in the queue, and remaining energy of the nodes to decide the next hop along a route. This approach involves an overhead in which nodes must apprise their neighbors of any changes in the cost function parameters. This work also deals with relative priority levels for event-based (high bandwidth) and periodic (low bandwidth) data. The protocol finds a least-cost, energy-efficient path while considering maximum allowed delays.

QoS routing based on traffic classes

Sensor data may originate from various types of events that have different levels of importance. Consequently, the content and nature of the sensed data also varies. As an example that highlights the need for network-level QoS, consider the task of bandwidth assignment for multimedia mobile medical calls, which include patients' sensing data, voice, pictures, and video data. Unlike the typical source-to-sink multi-hop communication used by classical sensor networks, the proposed architecture uses a 3G cellular system in which individual nodes forward the sensed data to a cellular phone or a specialized information-collecting entity. Different priorities are assigned to video data originating from sensors on ambulances, audio traffic from elderly people, and images returned by sensors placed on the body. In order to achieve this, parameters like hand-off dropping rate (HDR), latency tolerance, and desired amount of wireless effective bandwidth are taken into consideration.

Routing protocols with support for streaming

The SPEED protocol provides three types of real-time communication services, namely, realtime unicast, real-time area-multicast and real-time area-anycast. It uses geographical location for routing, and a key difference with other schemes of this genre is its spatio-temporal character, that is, it takes into account timely delivery of the packets. It is specifically tailored to be a stateless, localized algorithm with minimal control overhead. End-to-end soft real-time communication is achieved by maintaining a desired delivery speed across the sensor network through a combination of feedback control and non-deterministic geographic forwarding. As it works satisfactorily under scarce resource conditions and can provide service differentiation, SPEED takes the first step in addressing the concerns of realtime routing in WMSNs.

8.8 MAC Layer

Owing to the energy constraints of the small, battery-powered sensor nodes, it is desirable that the medium access control (MAC) protocol enable reliable, error-free data transfer with minimum retransmissions while supporting application-specific QoS requirements. Multimedia traffic, namely audio, video, and still images, can be classified as separate service classes and subjected to different policies of buffering, scheduling, and transmission.

It allows the allocation of greater importance to certain parts of the image which can then be coded and transmitted over a better quality link or on a priority basis. Especially relevant to systems for military surveillance or fault monitoring, such application layer features could be leveraged by the MAC by differentially treating the Region of Interest (ROI) packets.

Research efforts to provide MAC layer QoS can be classified mainly as:

1. Channel access policies,

2. Scheduling and buffer management, and

3. Error control.

The main causes of energy loss in sensor networks are attributed to packet collisions and subsequent retransmissions, overhearing packets destined for other nodes, and idle listening, a state in which the transceiver circuits remain active even in the absence of data transfer.

Contention-based protocols

Most contention-based protocols have a single-radio architecture. They alternate between sleep cycles (low-power modes with transceiver switched off) and listen cycles (for channel contention and data transmission). However, we believe that their applicability to multimedia transmission is limited owing to the following reasons:

a) The primary concern in the protocols of this class is saving energy, and this is accomplished at the cost of latency and by allowing throughput degradation. A sophisticated duty cycle calculation based on permissible end-to-end delay needs to be implemented, and coordinating overlapping listening period with neighbors based on this calculation is a difficult research challenge.

b) Coordinating the sleep–awake cycles between neighbors is generally accomplished though schedule exchanges. In case of dynamic duty cycles based on perceived values of instantaneous or time averaged end-to-end latency, the overhead of passing frequent schedules also needs investigation in light of the ongoing high data rate video/audio messaging.

c) Video traffic exhibits an inherent bursty nature and can lead to sudden buffer overflow at the receiver. This problem is further aggravated by the transmission policy adopted in T-MAC.

By choosing to send a burst of data during the listen cycle, T-MAC shows performance improvement over S-MAC, but at the cost of monopolizing a bottleneck node. Such an operation could well lead to strong jitters and result in discontinuous real-time playback.

Contention-free single channel protocols

Time Division Multiple Access (TDMA) is a representative protocol of this class in which the clusterhead (CH) or sink helps in slot assignment, querying particular sensors and maintaining time schedules.

TDMA schemes designed exclusively for sensor networks have a small reservation period (RP) that is generally contention based, followed by a contention-free period that spans the rest of the frame. This RP could occur in each frame or at pre-decided intervals in order to assign slots to active nodes, taking into consideration the QoS requirement of their data streams. The length of the TDMA frames and the frequency of the RP interval are some of the design parameters that can be exploited while designing a multimedia system. For real-time streaming video, packets are time constrained and scheduling policies like Shortest Time to Extinction (STE) or Earliest Due Date (EDD) can be adopted. Both of these are similar in principle as packets are sent in the increasing order of their respective delay tolerance but differ in respect that EDD may still forward a packet that has crossed its allowed delay bound. Based on the allowed packet loss of the multimedia stream, the dependencies between packet dropping rate, arrival rate, and delay tolerance can be used to decide the TDMA frame structure and thus ensure smooth replay of data. This allows greater design choices as against, where the frame lengths and slot duration are considered constant.

As sensor nodes are often limited by their maximum data transmission rate, depending upon their multimedia traffic class, the duration of transmission could be made variable. Thus variable TDMA (V-TDMA) schemes should be preferred when heterogeneous traffic is present in the network. Tools for calculating the minimum worst-case delay in such schemes and algorithms for link scheduling are provided. As real-time streaming media is delay bounded, the link-layer latency introduced in a given flow in order to satisfy data rate requirements of another flow needs to be analyzed well when VTDMA schemes are used.

MIMO technology

The high data rate required by multimedia applications can be addressed by spatial multiplexing in MIMO systems that use a single channel but employ interference cancellation techniques. Recently, virtual MIMO schemes have been proposed for sensor networks, where nodes in close proximity form a cluster. Each sensor functions as a single antenna element, sharing information and thus simulating the operation of a multiple antenna array. A distributed compression scheme for correlated sensor data that specially addresses multimedia requirements is integrated into the MIMO framework. However, a key consideration in MIMO-based systems is the number of sensor transmissions and the required signal energy per transmission. As the complexity is shifted from hardware to sensor coordination, further research is needed at the MAC layer to ensure that the required MIMO parameters like channel state and desired diversity/processing gain are known to both the sender and receiver at an acceptable energy cost.

Contention-Free Multi-Channel Protocols

Along with improving hardware and thus increasing cost, an alternate approach is to efficiently utilize the available bandwidth. By using multiple channels in a spatially overlapped manner, existing bandwidth can be efficiently utilized for supporting multimedia applications.

Scheduling

MAC layer scheduling in the context of WMSNs differs from the traditional networking model in the sense that apart from choosing the queue discipline that accounts for latency bounds, rate/power control and consideration of high channel error conditions need to be incorporated. An optimal solution is a function of all of these factors, appropriately weighted and seamlessly integrated with a suitable channel access policy.

Queues at the MAC layer have been extensively researched, and several schemes with varying levels of complexity exist. Of interest to multimedia applications is the development of schemes that allow a delay bound and thus assure smooth streaming of multimedia content. Depending upon the current residual energy in the network, it is possible to adapt the scheme for greater energy savings, albeit at the cost of a small, bounded increase in worst-case packet latency.

Current network calculus results have been mostly derived for wired networks, and assume static topologies and fixed link capacity, which are clearly unreasonable assumptions in sensor networks. Extending network calculus results to WMSNs is a challenging but promising research thrust, likely to produce important advancements in the ability to provide provable QoS guarantees in multi-hop networks.

Multimedia Packet Size

In wireless networks, the successful reception of a packet depends upon environmental factors that decide the bit error rate (BER) of the link. Packet length clearly has a bearing on reliable link-level communication and may be adjusted according to application delay sensitivity requirements. The Dynamic Packet Size Mechanism (DPSM) scheme for wireless networks follows an additive increase, multiplicative decrease (AIMD) mechanism to decide the packet length, analogous to the congestion control performed by TCP at the transport layer. As an example, if a packet fails the checksum, the sender is intimated and the subsequent packets are sent with a multiplicative decrease in length.

Grouping smaller packets together in order to reduce contention has been explored in Packet Frame Grouping (PFG) and PAcket Concatenation (PAC). Originally devised for 802.11-like protocols, here the header overhead is shared by the frames. In PFG, the individual frames may be addressed to different senders and require per-frame ACKs while PAC requires buffering, as all frames need to have a common destination.

Depending upon the information content of the frame and the channel conditions, variable length forward error-correcting codes (FEC) can be used to reduce the effects of transmission errors at the decoder. The trade-off between the increase of packet length due to the additional parity bits and energy constraints is evaluated, where FEC is shown to perform better than retransmissions.

8.9 Physical Layer

The ultra wide band (UWB) technology has the potential to enable low-power consumption and high data rate communications within tens of meters, characteristics that make it an ideal choice for WMSNs. UWB signals have been used for several decades in the radar community

Although UWB signals, as per the specifications of the Federal Communications Commission (FCC), use the spectrum from 3.1 GHz to 10.6 GHz with appropriate interference limitation, UWB devices can operate using the spectrum occupied by existing radio services without causing interference, thereby permitting scarce spectrum resources to be used more efficiently. Instead of dividing the spectrum into distinct bands that are then allocated to specific services, UWB devices are allowed to operate overlaid and thus interfere with existing services, at a low enough power level that existing services would not experience performance degradation.

There exist two main variants of UWB. The first, known as Time-Hopping Impulse Radio UWB (TH-IR-UWB), is based on sending very short duration pulses (in the order of hundreds of picoseconds) to convey information. Time is divided into frames, each of which is composed of several chips of very short duration. Each sender transmits one pulse in a chip per frame only, and multi-user access is provided by pseudo-random time hopping sequences (THS) that determine in which chip each user should transmit. A different approach, known as Multi-Carrier UWB (MC-UWB), uses multiple simultaneous carriers, and is usually based on Orthogonal Frequency Division Multiplexing (OFDM).

MC-UWB is particularly well suited for avoiding interference because its carrier frequencies can be precisely chosen to avoid narrowband interference to or from narrowband systems. However, implementing a MC-UWB front-end power amplifier can be challenging due to the continuous variations in power over a very wide bandwidth. Moreover, when OFDM is used, high-speed FFT processing is necessary, which requires significant processing power and leads to complex transceivers. TH-IR-UWB signals require fast switching times for the transmitter and receiver and highly precise synchronization. Transient properties become important in the design of the radio and antenna. The high instantaneous power during the brief interval of the pulse helps to overcome interference to UWB systems, but increases the possibility of interference from UWB to narrowband systems. The RF front-end of a TH-IR-UWB system may resemble a digital circuit, thus circumventing many of the problems associated with mixed signal integrated circuits. Simple TH-IR-UWB systems can be very inexpensive to construct.

The TH-IRUWB is particularly helpful and useful for WMSNs for the following reasons:

- It enables high data rate, very low-power wireless communications, on simple-design, low-cost radios (carrier less, baseband communications).

- Its fine delay resolution properties are appropriate for wireless communications in dense multipath environments, by exploiting more resolvable paths.

- Provides large processing gain in presence of interference.

- Provides flexibility, as data rate can be traded or power spectral density and multi-path performance.

- Finding suitable codes for THS is trivial (as opposed to CDMA codes), and no assignment protocol is necessary.

- It naturally allows for integrated MAC/PHY solutions; moreover, interference mitigation techniques allow realizing MAC protocols that do not require mutual temporal exclusion between different transmitters. In other words, simultaneous communications of neighboring devices are feasible without complex receivers as required by CDMA.

- The large instantaneous bandwidth enables fine time resolution for accurate position estimation and for network time distribution (synchronization).

- UWB signals have extremely low-power spectral density, with low probability of intercept/detection (LPI/D), which is particularly appealing for military covert operations.

8.10 Cross-Layer Design

In multi-hop wireless networks there is a strict interdependence among functions handled at all layers of the communication stack. The physical,

MAC, and routing layers together impact the contention for network resources. The physical layer has a direct impact on multiple access of nodes in wireless channels by affecting the interference at the receivers. The MAC layer determines the bandwidth allocated to each transmitter, which naturally affects the performance of the physical layer in terms of successfully detecting the desired signals. On the other hand, as a result of transmission schedules, high packet delays and/or low bandwidth can occur, forcing the routing layer to change its route decisions. Different routing decisions alter the set of links to be scheduled, and thereby influence the performance of the MAC layer. Moreover, specifically for multimedia transmissions, the application layer does not require full insulation from lower layers, but needs instead to perform source coding based on information from the lower layers to maximize the multimedia performance. Existing solutions often do not provide adequate support for multimedia applications since the resource management, adaptation, and protection strategies available in the lower layers of the stack are optimized without explicitly considering the specific characteristics of multimedia applications. Similarly, multimedia compression and streaming algorithms do not consider the mechanisms provided by the lower layers for error protection and resource allocation.

Most of the existing studies decompose the resource allocation problem at different layers, and consider allocation of the resources at each layer separately. In most cases, resource allocation problems are treated either heuristically, or without considering cross-layer interdependencies, or by considering pair-wise interactions between isolated pairs of layers.

The cross-layer transmission of multimedia content over wireless networks is formalized as an optimization problem. Several different approaches for cross-layer design of multimedia communications are discussed, including the bottom-up approach, where the lower layers try to insulate the higher layers from losses and channel capacity variations, and top-down, where the higher layer protocols optimize their parameters at the next lower layer. However, only single-hop networks are considered.

In particular, the improvements of adaptive link layer techniques such as adaptive modulation and packet size optimization, joint allocation of capacity and flows (i.e., MAC and routing), and joint scheduling and rate allocation are discussed. While still maintaining a strict layered architecture, it is shown how these cross-layer optimizations help improve the spectral efficiency at the physical layer and the peak signal-to-noise ratio (PSNR) of the video stream perceived by the user. Clearly, energy constrained multimedia sensors may need to leverage cross-layer interactions one step further. At the same time, optimization metrics in the energy domain need to be considered as well.

Crosslayer design is based on the following principles:

Network Layer QoS Support Enforced by a Crosslayer Controller

The proposed system provides QoS support at the network layer, that is, it provides packet-level service differentiation in terms of throughput, end-to-end packet error rate, and delay. This is achieved by controlling operations and interactions of functionalities at the physical, MAC, and network layers, based on a unified logic that resides on a cross-layer controller that manages resource allocation, adaptation, and protection strategies based on the state of each functional block. The objective of the controller is to optimize some objective function, that is, minimize energy consumption, while guaranteeing QoS requirements to application flows. While all decisions are jointly taken at the controller, implementation of different functionalities is kept separate for ease of design and upgradeability.

UWB Physical/MAC Layer

The communication architecture is based on an integrated TH-IRUWB MAC and physical layer. Similarly to CDMA, TH-IR-UWB allows several transmissions in parallel. Conversely, typical MAC protocols for sensor networks, such as contention-based protocols based on CSMA/CA, require mutual temporal exclusion between neighboring transmitters. This allows devising MAC protocols with minimal coordination. While CDMA usually entails complex transceivers and cumbersome code assignment protocols, this is achievable with simple transceivers in TH-IR-UWB.

Receiver-Centric Scheduling for QoS Traffic

One of the major problems in multi-hop wireless environments is that channel and interference vary with the physical location of devices. For this reason, QoS provisioning should be based on receiver-centric scheduling of packets. This way, the receiver can easily estimate the state of the medium at its side. Thus, it can optimally handle loss recovery and rate adaptation, thereby avoiding feedback overheads and latency, and be responsive to the dynamics of the wireless link using the information obtained locally.

Dynamic Channel Coding

Adaptation to interference at the receiver is achieved through dynamic channel coding, which can be seen as an alternative form of power control, as it modulates the energy per bit according to the interference perceived at the receiver.

Geographical Forwarding

We leverage UWB's positioning capabilities to allow scalable geographical routing. The routing paths are selected by the cross-layer controller by applying an admission control procedure that verifies that each node on the path be able to provide the required service level. The required packet error rate and maximum allowed delay are calculated at each step based on the relative advance of each hop toward the destination.

Hop-by-Hop QoS Contracts

End-to-end QoS requirements are guaranteed by means of local decision. Each single device that participates in the communication process is responsible for locally guaranteeing given performance objectives. The global, end-to-end requirement is enforced by the joint local behaviors of the participating devices.

Multi-Rate Transmission

TH-IR-UWB allows varying the data rate at the physical layer by modifying the pulse repetition period. While this functionality has not been fully explored so far, it is possible to devise adaptive systems that modify the achievable data rate at the physical layer based on the perceived interference and on the required power consumption.

8.11 Wireless Video Sensor Networks

Figures 8.2 and 8.3 describe hardware and software architecture for sensor networks and video networks. The hardware architecture consists of database servers that collect and store sensing and video data by regional groups, backup database servers, control servers, GIS servers, and web servers. The integrated control system comprises the integrated database server to consolidate local databases, the integrated GIS server, and the integrated web server. The local control system also interoperates with the databases that belong to other regions.

In Figure 8.2, sensor networks that adopt an IP-WSN to fulfill a global WSN and techniques to easily interoperate with existing CCTVs may be standardized for better support of integrated monitoring. The software architecture shows software required for sensor networks and camera networks and for the integrated control system and local servers.

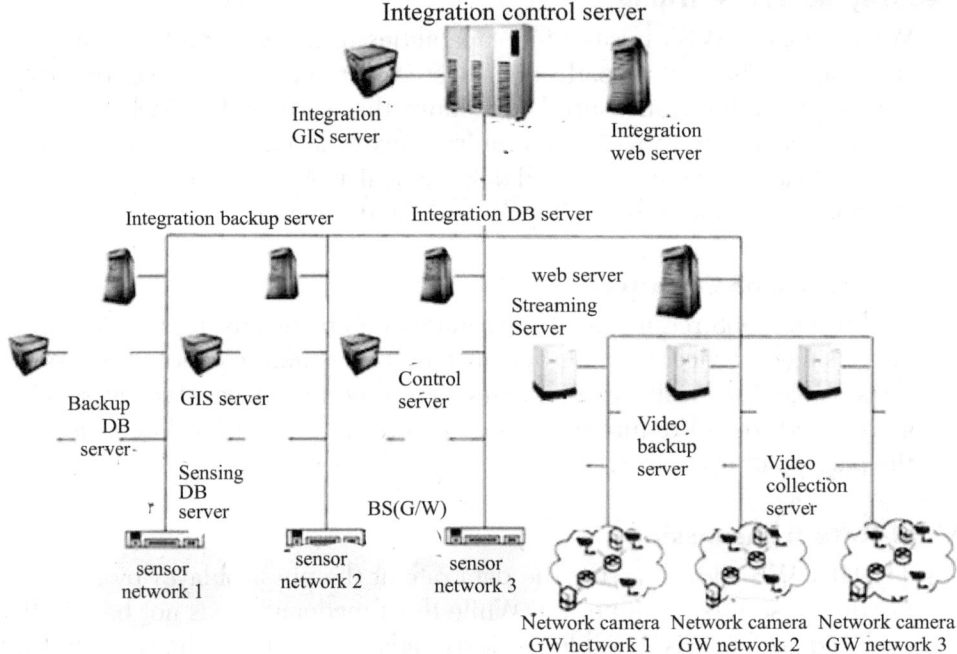

FIGURE 8.2 Hardware architecture for sensor networks and video networks.

Gateways and sensor nodes are necessary to build the sensor networks and gateways, and network cameras are required to form camera networks. Moreover, the integrated control server, streaming servers, web servers, database servers, and GIS servers are necessary for the integrated control system and local servers.

This software architecture of Figure 8.3 shows software required in two parts: the integrated control server and the sensor network including the network camera. The integrated control server mainly uses desktop PCs and workstation PCs, so that there is hardly a limitation on these kinds of software. That is, it is easy to implement, install, and use. Therefore, software blocks mentioned previously are essential components for integrated monitoring. The sensor network, however, is different from the integrated control flow. Because most of its hardware is an embedded device based on ARM, MSP, and Atmel, its software has limitations.

Specifically, software lively is very important for seamless services. Also, additional software may be demanded because a variety of applications utilizing sensing data and video information from network cameras may appear in the future.

Integrated Control System

Integrated Control Sensor			Streaming server	Web server	DB server	GIS server
State display S/W	Control Processing S/W		Image processing S/W	JSP,SHP, PHP	DBMS	GIS engine
Image Processing S/W	Network Management	Device Control S/W	Streaming S/W	Web S/W	DB middleware	Backup agent
Backup agent			Backup agent	Backup agent	Backup agent	OS(Linux)
OS(Linux)			OS(Linux)	OS(Linux)	OS(Linux)	
Network composition S/W	TCP/IP		Network S/W · TCP/IP	TCP/IP	TCP/IP	TCP/IP

Sensor Network / Network Camera

Gateway / sink node		Sensor node	Camera
Sensing data processing S/W			Image processing S/W
OS		Sensor application	Backup agent
Network composition S/W	TCP/IP	Sensor OS	OS(Linux)
RF comm. protocol		RF comm. protocol	TCP/IP

FIGURE 8.3 Software architecture for sensor networks and video networks.

8.12 Three-Tier Architecture of Video Sensor Networks

Hardware Architecture

The hardware architecture is shown in Figure 8.4. It consists of three tiers. Tier 1 has lower-power camera sensors (Cyclop or CMUcam) and a low-power sensor platform (Mote). Tier 2 has webcams (Logitech), a sensor platform (Intel Stargate) and a low-power wakeup circuit (Mote). Tier 3 has a high-performance PZT camera and a mini-ITX embedded PC (Sony).

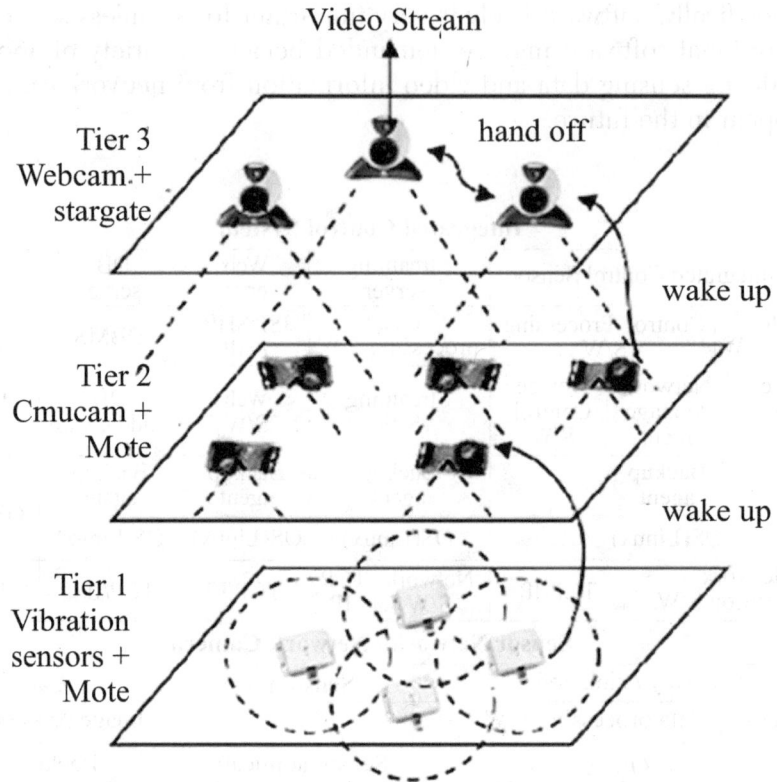

FIGURE 8.4 Three-tier architecture of video sensor networks.

Figure 8.5 shows multi-tier hardware architecture and Figure 8.6 shows software architecture.

Benefits of a multi-tier network:

- Reduces power consumption
- Achieves similar performance of single-tier network
- Low cost
- High coverage
- High reliability
- High functionality

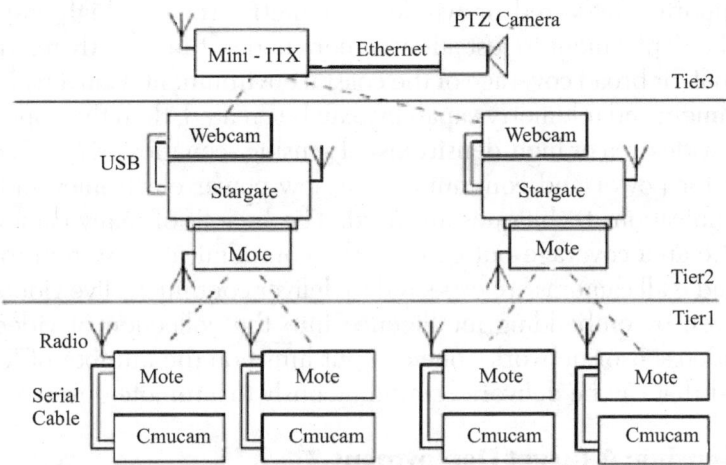

FIGURE 8.5 Multitier hardware architecture.

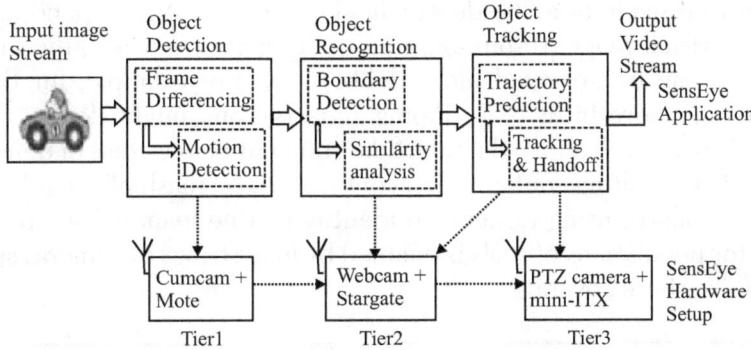

FIGURE 8.6 Multitier software architecture.

8.13 A Wireless Video Sensor Network for Autonomous Coastal Sensing

Video is an important medium for the observation of a variety of phenomena in the physical world. For example, in the coastal setting cameras can be used to monitor sea state including wave height and period, evaluate land erosion, and observe a variety of animal species. One of the severe limitations in deploying video cameras in these settings is their cost; in addition to sensing and processing electronics, they must be enclosed in weatherproof housings, provided with a source of energy, and, if live remote access is desired, must be supported by wired or wireless telemetry. Indeed, high-cost cameras limit the deployment to just a few camera nodes, thus greatly restricting the potential for broad coverage of the coastal environment. Conversely, if a low-cost camera and telemetry capability can be created, then the opportunity to create wide-area or high-density visual sensing scenarios arises. To eliminate wiring for power and communication, low-power electronics and wireless communications techniques are used. The benefit of many cameras yields fantastic area coverage but exposes the communication system to extreme overload if all cameras are expected to deliver continuous live video streams. However, by embedding intelligence into the collection of video sensors (the video sensor network), one can put limits on the number of features of interest that can be delivered simultaneously to a remote observer.

Coastal Sensing: A Target Deployment

A video sensor network (VSN) is best represented schematically as a grid as in Figure 8.7. In this illustration cameras are assumed to be panoramic and thus capable of a 360-degree field of view (FOV). Depending on the camera visual range, resolution, and the targets of interest, the individual cameras can be arranged more deliberately. For example, for the study of nesting shore birds (e.g., Piping Plovers), one might deploy the units densely in a swath above the intertidal zone but below areas of dense beach grass. The resulting configuration continues to be mesh of irregular dimensions. Similarly, using cameras to identify marine mammal standings or to monitor populations of seals is enabled by long strings of cameras spaced to maximize shore coverage.

FIGURE 8.7 Video Sensing Application (gray seals); VSN grid schematic.

A network of these cameras would operate as follows. Each camera operates autonomously, but with similar instructions: wake up periodically and sample the image in the field of view.

Compare the image with prior recordings and determine any changes. If there are none, go back to sleep to conserve energy. Otherwise, perform more advanced video processing to determine if the change in the image represents a target of interest (e.g., a stranded marine mammal). If a target is detected, the camera notifies a gateway to initiate telemetry and/or recording of the target. Neighboring units participate in routing live video from the source camera to the gateway for observation by a human. This is a representative application.

Design of a Video Sensor Node Platform

The general requirements of the video camera unit, or video node (VN Figure 8.8) are listed as follow:

Wireless

The VN must be deployable in remote unattended locations without access to wiring infrastructure. This implies operation on batteries and/or environmental energy harvesting. Data logging must be performed in the system or communicated out of the system to a gateway or to the Internet.

Broad Visual Field of View

To maximize utility, each VN should have a flexible directionality and zoom. Ideally, the field of view is matched to the range of the wireless communication used (e.g., a FOV supporting ranges from 0–100 m corresponding to the RF range of 802. lib).

Variable-Resolution Imagery

The VN should support sub-sampling in both temporal and spatial dimensions to vary the volume of data generated.

Capable of Streaming Video to a Gateway

An important goal is to deliver full video streaming to an end-user once an event or object of interest has been identified.

Infrequent Maintenance

Access to remote locations is inconvenient and potentially costly. The system must operate without need for service for long periods. For example, to study breeding birds can require installation without service for several months. Services to minimize include battery replacement, lens cleaning, data download (if not performed in real time). This requirement puts difficult constraints on the energy used by the camera, communications, and microcontroller (MCU).

Weatherproof and Low Impact

The mechanical enclosure including the energy harvester (solar panel) must be robust to survive in the harsh coastal environment. It must also be non-invasive with respect to studied phenomena and must not cause significant harm to the environment in which it is installed.

Inexpensive

The VNs must be designed for low cost when produced in quantity.

The components of this VN and the motivations for the selection of components are described as follow.

Camera Optics

To achieve a wide field of view we will use a catadioptric configuration using an omni directional mirror fabricated to yield a panoramic image. This approach eliminates the moving parts of most pan-tilt-zoom configurations, reduces corresponding energy use, and reduces directionality challenges.

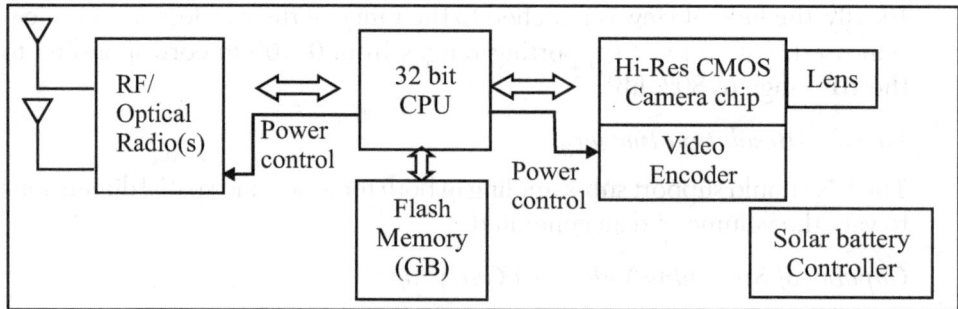

FIGURE 8.8 Major components of our Video Node design.

Image Sensor

Advances in single-chip cameras based on CMOS technology have led to the ubiquity of camera phones. Selection of a 3–5M pixel CMOS chip may provide flexible temporal and spatial sampling, and low active current.

MCU and Data Storage

There are many options here. Specific desirable options are: very low sleep current for periods of inactivity, multimedia instruction sets (for image manipulation), 32-bit architecture, programmable clock rate, and access to large amounts of flash memory.

Communications and Networking

Two modes of communication are anticipated. They are inter-VN communication and Multi-hop communications. The inter-VN communication supports collaborative processing and target tracking, and video streaming is achieved via multihop communications to the gateway. The former is suitable for low power, low-data rate technologies such as 802.15.4, whereas the latter is appropriate for intermittent WTFi (802.1 la,b,g) or other technologies (e.g., optical or IR). Each will be under power control to minimize communication energy costs.

Energy Harvesting and Control

The energy required for video sourcing and streaming is approximately 2500mW for the Axis camera. The energy harvesting unit, a solar panel and a super capacitor bank, is sized to encompass a nominal load profile/duty cycle of 5% over a 24-hour period, and an energy source model that can tolerate up to 30% reduction in incident solar radiation. Thus the design is one of daily energy input to super capacitors with conventional disposable batteries to bridge periods of cloud cover.

Software Control

Software control includes image capture, processing, analysis, data routing, and energy management. In order to support duty cycling of each VN, robust image change detection algorithms will be developed on the target platform. Particular attention will be applied to low-complexity algorithms with minimal memory requirements (the so-called "fast and lean" algorithms) because the available computing power and on-board memory are severely limited. Since VNs are stationary, the algorithms will be based on the concept of background subtraction. The background will be statistically modeled over time and each new image will be tested against this statistical model. For cameras with overlapping fields of view, inter-node collaboration will be permitted by means of message passing (e.g., communicating partial change detection results). If a change is detected, a video stream from the node will be transmitted to the gateway using standard video compression techniques (e.g., MJPEG,MPEG-4,H-264).

Summary

- Wireless multimedia sensor network applications are multimedia surveillance sensor networks, storage of potentially relevant activities, traffic avoidance, advanced health care delivery, automated assistance for the elderly monitors, environmental monitoring, person locator services, and industrial process control.

- Factors influencing the design of WMSNs are application-specific QoS requirements, high bandwidth demand, multimedia source coding techniques, multimedia in-network processing, power consumption, flexible architecture to support heterogeneous applications, multimedia coverage, and integration with IP architecture and with other wireless technologies.

- Computer vision is a subfield of AI, whose purpose is to allow computer to extract features from an image or multidimensional data.

- Application layer services are traffic management, source code to application requirements, efficient system software, and advanced in-network multimedia processing techniques.

- The transport layer is for congestion avoidance and packet reordering due to multipath.

- The network layer is for addressing and routing.

- The MAC layer is for channel access policies, scheduling, buffer management, and error control.

- Ultra Wide Band technology is used in WMSNs.

Questions

1. What do you mean by wireless multimedia sensor networks? List their applications.

2. What are the factors influencing the design of multimedia sensor networks?

3. Draw the reference network architecture of a WMSN. Explain each block in detail.

4. Write about WMSNs as distributed computer vision systems.

5. Write the functions of the application layer in detail.

6. List the services provided by the transport layer.

7. What are challenges of the network layer in WMSNs?

8. Explain about the design of the MAC layer for WMSNs.

9. Write a note on the physical layer for WMSNs.

10. What do you mean by cross-layer design in the case of WMSNs?

11. With a neat diagram, write about the hardware architecture for Wireless Video Sensor Networks.

12. What are the software architecture requirements for sensor networks and video networks?

13. Draw the hardware and software architecture for multi-tier networks. Explain them.

14. What are the benefits of multi-tier networks?

15. Give the applications of WVSNs for autonomous coastal sensing.

16. Give the design requirements of a video sensor node.

Further Reading

1. *Wireless Multimedia Sensor Networks on Reconfigurable Hardware: Information Reduction Techniques* by Li-minn Ang and Kah Phooi Seng

2. *Wireless Sensor Multimedia Networks: Architectures, Protocols, and Applications* by Mohamed Mostafa A. Azim and Xiaohong Jiang

References

http://bwn.ece.gatech.edu/surveys/multimedia.pdf

CHAPTER 9

MOBILE AD HOC NETWORKS

This chapter discusses wireless ad hoc sensor networks, mobile ad hoc networks, and vehicular ad hoc networks.

9.1 Wireless Ad Hoc Sensor Networks

Wireless communication enables information transfer among a network of disconnected and often mobile users. Popular wireless networks such as mobile phone networks and wireless LANs are traditionally infrastructure-based, that is base stations, access points, and servers are deployed before the network can be used. In contrast, ad hoc networks are dynamically formed among a group of wireless users and require no existing infrastructure or pre-configuration. This is shown in Figure 9.1.

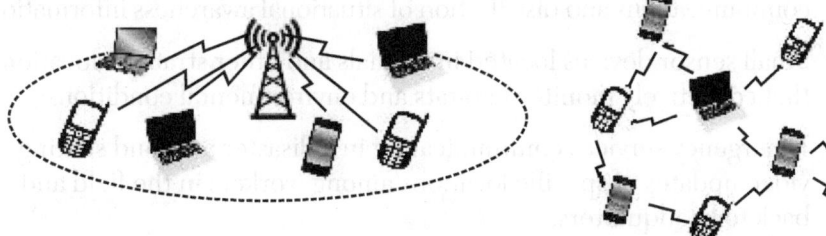

Infrastructure-based wireless network   Ad hoc Network

FIGURE 9.1 Comparison of infrastructure wireless and ad hoc network.

The initial development of ad hoc networks was primarily driven by military applications, where rapid network formation and survivability are key requirements. Relying on a system centralized around base stations is simply not an option because the base stations must first be deployed in the correct location (almost impossible in a hostile environment) and the network is subject to failure if one or several base stations are destroyed. On the other hand, a distributed network architecture, with all nodes having equal responsibility and using broadcast radio, is ideally suited to military requirements. To overcome the limited radio transmission ranges (i.e., not all nodes are within the range of every other node), nodes are equipped with the ability to forward information on behalf of others, that is, multi-hop communications. Combined with packet switching technology (hence the term *packet radio networks,* which is often used interchangeably with ad hoc networks; other synonyms and related terms include: *mobile ad hoc networks, wireless ad hoc networks, self-organizing networks, multi-hop wireless networks* and *mesh networks*) and suitable medium access control protocols, multi-hop communication provides the basis for resilient, large-scale military ad hoc networks.

A wireless ad hoc sensor network consists of a number of sensors spread across a geographical area. Each sensor has wireless communication capability and some level of intelligence for signal processing and networking of the data. The dynamic and self-organizing nature of ad hoc networks makes them particularly useful in situations where rapid network deployments are required or it is prohibitively costly to deploy and manage network infrastructure. Some example applications include:

- Attendees in a conference room sharing documents and other information via their laptops and handheld computers;

- Armed forces creating a tactical network in unfamiliar territory for communications and distribution of situational awareness information;

- Small sensor devices located in animals and other strategic locations that collectively monitor habitats and environmental conditions;

- Emergency services communicating in a disaster area and sharing video updates of specific locations among workers in the field and back to headquarters;

- Military sensor networks to detect and gain as much information as possible about enemy movements, explosions, and other phenomena of interest;

- Sensor networks to detect and characterize Chemical, Biological, Radiological, Nuclear, and Explosive (CBRNE) attacks and material;

- Sensor networks to detect and monitor environmental changes in plains, forests, oceans, etc.;
- Wireless traffic sensor networks to monitor vehicle traffic on highways or in congested parts of a city;
- Wireless surveillance sensor networks for providing security in shopping malls, parking garages, and other facilities; and
- Wireless parking lot sensor networks to determine which spots are occupied and which are free.

The previous list suggests that wireless ad hoc sensor networks offer certain capabilities and enhancements in operational efficiency in civilian applications as well as assist in the national effort to increase alertness to potential terrorist threats.

Wireless ad hoc sensor networks are classified into two ways. They are whether or not the nodes are individually addressable, and whether the data in the network is aggregated. The sensor nodes in a parking lot network should be individually addressable, so that one can determine the locations of all the free spaces. This application shows that it may be necessary to broadcast a message to all the nodes in the network. If one wants to determine the temperature in a corner of a room, then addressability may not be so important. Any node in the given region can respond. The ability of the sensor network to aggregate the data collected can greatly reduce the number of messages that need to be transmitted across the network.

The basic goals of a wireless ad hoc sensor network generally depend upon the application, but the following tasks are common to many networks:

1. *Determine the value of some parameter at a given location:* In an environmental network, one might want to know the temperature, atmospheric pressure, amount of sunlight, and the relative humidity at a number of locations. This example shows that a given sensor node may be connected to different types of sensors, each with a different sampling rate and range of allowed values.

2. *Detect the occurrence of events of interest and estimate parameters of the detected event or events:* In the traffic sensor network, one would like to detect a vehicle moving through an intersection and estimate the speed and direction of the vehicle.

3. *Classify a detected object:* Is a vehicle in a traffic sensor network a car, a mini-van, a light truck, a bus, etc.?

4. *Track an object:* In a military sensor network, track an enemy tank as it moves through the geographic area covered by the network.

In these four tasks, an important requirement of the sensor network is that the required data be disseminated to the proper end users. In some cases, there are fairly strict time requirements on this communication. For example, the detection of an intruder in a surveillance network should be immediately communicated to the police so that action can be taken.

Wireless ad hoc sensor network requirements include the following:

1. *Large number of (mostly stationary) sensors:* Aside from the deployment of sensors on the ocean surface or the use of mobile, unmanned, robotic sensors in military operations, most nodes in a smart sensor network are stationary. Networks of 10,000 or even 100,000 nodes are envisioned, so scalability is a major issue.

2. *Low energy use:* Since in many applications the sensor nodes will be placed in a remote area, service of a node may not be possible. In this case, the lifetime of a node may be determined by the battery life, thereby requiring the minimization of energy expenditure.

3. *Network self-organization:* Given the large number of nodes and their potential placement in hostile locations, it is essential that the network be able to self-organize; manual configuration is not feasible. Moreover, nodes may fail (either from lack of energy or from physical destruction), and new nodes may join the network. Therefore, the network must be able to periodically reconfigure itself so that it can continue to function. Individual nodes may become disconnected from the rest of the network, but a high degree of connectivity must be maintained.

4. *Collaborative signal processing:* To improve the detection/estimation performance, it is often quite useful to fuse data from multiple sensors. This data fusion requires the transmission of data and control messages, and so it may put constraints on the network architecture.

5. Querying ability: A user may want to query an individual node or a group of nodes for information collected in the region. Depending on the amount of data fusion performed, it may not be feasible to transmit a large amount of the data across the network. Instead, various local sink nodes will collect the data from a given area and create summary messages. A query may be directed to the sink node nearest to the desired location.

In the wireless ad hoc sensor networks, each node may be equipped with a variety of sensors, such as acoustic, seismic, infrared, still/motion video camera, and so forth. These nodes may be organized in clusters such that a locally occurring event can be detected by most, if not all, of the nodes in a cluster. Each node may have sufficient processing power to make a decision, and it will be able to broadcast this decision to the other nodes in the cluster. One node may act as the cluster master, and it may also contain a longer range radio using a protocol such as IEEE 802.11 or Bluetooth.

9.2 Mobile Ad Hoc Networks (MANET)

An Ad hoc network is a collection of mobile nodes which forms a temporary network without the aid of centralized administration or standard support devices regularly available as conventional networks. These nodes generally have a limited transmission range, so each node seeks the assistance of its neighboring nodes in forwarding packets, and hence the nodes in an ad hoc network can act as both routers and hosts. Thus, a node may forward packets between other nodes as well as run user applications. By nature these types of networks are suitable for situations where either no fixed infrastructure exists or deploying a network is not possible. The mobile ad hoc networks have found many applications in various fields like military, emergency, conferencing, and sensor networks. Each of these application areas has its specific requirements for routing protocols. The mobile ad hoc network has the following features:

- Autonomous terminal

- Distributed operation

- Multi-hop routing

- Dynamic network topology

- Fluctuating link capacity

- Lightweight terminals

Autonomous Terminal

In MANET, each mobile terminal is an autonomous node, which may function as both a host and a router. In other words, besides the basic processing ability as a host, the mobile nodes can also perform switching functions as a router. So, usually endpoints and switches are indistinguishable in MANET.

Distributed Operation

Since there is no background network for the central control of the network operations, the control and management of the network is distributed among the terminals. The nodes involved in a MANET should collaborate among themselves, and each node acts as a relay as needed to implement functions like security and routing.

Multi-Hop Routing

Basic types of ad hoc routing algorithms can be single-hop and multi-hop, based on different link layer attributes and routing protocols. Single-hop MANET is simpler than multi-hop in terms of structure and implementation, with the lesser cost of functionality and applicability. When delivering data packets from a source to its destination out of the direct wireless transmission range, the packets should be forwarded via one or more intermediate nodes.

Dynamic Network Topology

Since the nodes are mobile, the network topology may change rapidly and unpredictably and the connectivity among the terminals may vary with time. MANET should adapt to the traffic and propagation conditions as well as the mobility patterns of the mobile network nodes. The mobile nodes in the network dynamically establish routing among themselves as they move about, forming their own network on the fly. Moreover, a user in the MANET may not only operate within the ad hoc network, but may require access to a public fixed network (e.g., Internet).

Fluctuating Link Capacity

The nature of high bit-error rates of wireless connections might be more profound in a MANET. One end-to-end path can be shared by several sessions. The channel over which the terminals communicate is subjected to noise, fading, and interference, and has less bandwidth than a wired network. In some scenarios, the path between any pair of users can traverse multiple wireless links, and the link themselves can be heterogeneous.

Lightweight Terminals

In most of the cases, the MANET nodes are mobile devices with less CPU processing capability, small memory size, and low power storage. Such devices need optimized algorithms and mechanisms that implement the computing and communicating functions.

Table 9.1 compares WSN and MANET.

Table 9.1 Comparison between WSN and MANET

Factors/Issues	WSNs	MANETs
Interaction	Focus on interaction with the environment	Close to humans, e.g., laptops, PDAs, mobile radio terminals.
Nodes deployed	Very large	Not many
Population of nodes	Densely populated	Sparsely populated
Failure rate	High	Low
Communication	Broadcast	Point-to-point
Communication range	Short	Long
Metrics	Efficiency, resolution, latency, scalability, robustness	Receipt rate, dissemination, speed, redundancy
Power	Limited	Not an issue
Bandwidth deficient	Sometimes	Yes
Identification	Not unique	Unique ID by its MAC address
Memory	Limited	High
Fault tolerance	Needed only if nodes exhaust available energy or are moved	Needed as mobility increases
Data redundancy	Sometimes	No
Routing protocols	Flooding, Gossiping, flat Routing, Hierarchical, Location based	Pro-active, Reactive, Hybrid
Topology	Dynamic	Dynamic
Standards	Zigbee, IEEE 802.15.4, ISA 100, IEEE 1451	IEEE802.11

Issues to be considered when deploying MANET

The following are some of the main routing issues to be considered when deploying MANETs:

- Unpredictability of environment
- Unreliability of wireless medium
- Resource-constrained nodes
- Dynamic topology

- Transmission errors
- Node failures
- Link failures
- Route breakages
- Congested nodes or links

Unpredictability of environment

Ad hoc networks may be deployed in unknown terrains, hazardous conditions, and even hostile environments where tampering or the actual destruction of a node may be imminent. Depending on the environment, node failures may occur frequently.

Unreliability of wireless medium

Communication through the wireless medium is unreliable and subject to errors. Also, due to varying environmental conditions such as high levels of electro-magnetic interference (EMI) or inclement weather, the quality of the wireless link may be unpredictable.

Resource-constrained nodes

Nodes in a MANET are typically battery powered as well as limited in storage and processing capabilities. Moreover, they may be situated in areas where it is not possible to recharge and thus have limited lifespans. Because of these limitations, they must have algorithms which are energy efficient as well as operating with limited processing and memory resources. The available bandwidth of the wireless medium may also be limited because nodes may not be able to sacrifice the energy consumed by operating at full link speed.

Dynamic topology

The topology in an ad hoc network may change constantly due to the mobility of nodes. As nodes move in and out of range of each other, some links break while new links between nodes are created.

As a result of these issues, MANETs are prone to numerous types of faults including the following:

Transmission errors

The unreliability of the wireless medium and the unpredictability of the environment may lead to transmitted packets being garbled and thus with errors.

Node failures

Nodes may fail at any time due to different types of hazardous conditions in the environment. They may also drop out of the network either voluntarily or when their energy supply is depleted.

Link failures

Node failures as well as changing environmental conditions (e.g., increased levels of EMI) may cause links between nodes to break. Link failures cause the source node to discover new routes through other links.

Route Breakages

When the network topology changes due to node/link failures and/or node/link additions to the network, routes become out-of-date and thus incorrect. Depending upon the network transport protocol, packets forwarded through stale routes may either eventually be dropped or be delayed.

Congested Nodes or Links

Due to the topology of the network and the nature of the routing protocol, certain nodes or links may become overutilized, that is, congested. This will lead to either larger delays or packet loss.

9.3 Classification of Routing Protocols for MANETs

Ad hoc routing protocols can be categorized as table-driven or source initiated. Table-driven or proactive routing protocols find routes to all possible destinations ahead of time. The routes are recorded in the nodes' routing tables and are updated within the predefined intervals. Proactive routing protocols are faster in decision making, but cause problems if the topology of the network continually changes. These protocols require every node to maintain one or more tables to store updated routing information from every node to all other nodes. Source-initiated, or reactive, routing protocols are on-demand procedures and create routes only when requested to do so by source nodes.

A route request initiates a route-discovery process in the network and is completed once a route is discovered. If it exists, at the time of request, a route is maintained by a route-maintenance procedure until either the destination node becomes irrelevant to the source or the route is no longer needed. Control overhead of packets is smaller than that of proactive protocols.

Table Driven/Proactive

Destination sequenced distance vector [DSDV]

The DSDV is a table-driven-based routing algorithm. Each DSDV node maintains two routing tables. They are a table for forwarding packets and a table for advertising incremental updates. The nodes will maintain a routing table that consists of a sequence number. The routing table is periodically exchanged so that every node will have the latest information. DSDV is suitable for small networks. The algorithm works as follows: a node or a mobile device will make an update in its routing table and send the information to its neighbor upon receiving the updated information and make an update in its own routing table. The sequence number received is compared with the present sequence number; if the new sequence number is greater, then the new one will be used. A link failure in one of the nodes will change the metric value to infinity and broadcast the message.

Cluster head gateway switch router [CGSR]

CGSR is also a table driven routing protocol. In this algorithm the mobile devices will be grouped to form a cluster. The grouping is based on the range, and each cluster is controlled by the cluster head. All the mobile devices will maintain two tables, the cluster member table and the routing table. The cluster member table will have the information about the cluster head for each destination. The routing table will have routing information. In this protocol the packet cannot be directly sent to the destination; instead, cluster heads are used for routing. CGSR routing involves cluster routing, where a node finds the best route over cluster heads from the cluster member table.

Wireless routing protocol [WRP]

WRP is also based on a table driven approach. This protocol makes use of four tables:

1. Distance table: Which contains information like destination, next hop, distance

2. Routing table: Which contains routing information

3. Link cost table: Which contains cost information to each neighbor

4. Message retransmission list table: This table provides the sequence number of the message, a retransmission counter, acknowledgements, and list of updates sent in an update message

Whenever there is a change in the network, an update will be made, which will be broadcasted to other nodes. Other nodes upon receiving the updated information will make an update in their table. If there is no update in the network, a hello message should be sent.

Source Initiated/Reactive Protocol

Dynamic source routing [DSR]

DSR is a source-initiated or on-demand routing protocol in which the source finds an unexpired route to the destination to send the packet. It is used in the network where mobile nodes move with moderate speed. Overhead is significantly reduced, since nodes do not exchange routing table information. It has two phases:

1. Route discovery

2. Route maintenance

The source which wants to send the information to the destination will create a **route request** message by adding its own identification number and broadcasts it in the network. The intermediate nodes will continue the broadcast but will add their own identification numbers. When the destination is reached, a **route reply** message is generated which will be sent back to the source. The source can receive multiple route replies indicating the presence of multiple paths. The source will pick up one of the paths and will use it for transmission. If there is a link failure, one of the nodes will detect it and will create a **route error message** which will be sent back to the source; in this case the path has to be **reestablished** for further transmission.

Associated based routing [ABR]

ABR is an efficient on-demand or source-initiated routing protocol. In ABR, the destination node decides the best route, using node associativity. ABR is suitable for small networks, as it provides fast route discovery and creates the shortest paths through associativity. Each node keeps track of associativity information by sending messages periodically. If the associativity value is more, it means node mobility is less. If the associativity value is less, it means node mobility is more. In ABR, the source which wants to send the packet to the destination will create a query packet and broadcast in the network. Query packet generation is required for discovering the route. The broadcast continues as long as destination is reached. Once the destination is reached, it creates the reply packet and sends back to the source. The query packet will have the following information:

1. Source ID

2. Destination ID

3. All intermediate node IDs

4. Sequence number

5. CRC

6. Time to live [TTL]

A node sends an updated packet to the neighbors and waits for the reply; if an update is received back, then the associative tick will be incremented higher and it means the mobile device is still a part of the network; otherwise it might not be.

Ad hoc on demand distance vector [AODV]

It is a source-initiated routing protocol in mobile ad hoc networks. The algorithm consists of two phases:

1. Route discovery phase

2. Route maintenance phase

In the route discovery phase the path from source to destination is identified by broadcasting a route request packet [RREQ]. When the intermediate node receives the RREQ, it will create a backward pointer and continue the broadcast. When the route request packet reaches the destination, a route reply would be generated [RREP]. The route reply will have information about the path that can be chosen for the packet transmission. The route request packet can have following information:

1. Source ID

2. Destination ID

3. Sequence number

4. Backward pointer information

5. CRC

6. Time to live [TTL]

In the previous network the RREQ will be broadcasted by source node 1 to its neighbor, and neighbors will check whether the RREQ is already processed. If it is already processed, the packet will be discarded. If it is not processed a backward pointer is created and the broadcast continues. When the packet reaches the destination, a route reply is created [RREP] in the previous network. The first RREP sent to the source can have the path information as 1-2-4-6-8. When the source receives this information, it will be stored in the routing table. Meanwhile, the destination can create one more RREP, which can have the information as 1-3-7-8. The destination will send this RREP to the source and will also ask the source to discard the old path as the new path has the minimum number of hops.

In the route-maintainence phase, the nodes in the network periodically exchange hello messages to inform that they are still a part of the network and the path is valid. Whenever there is a link failure detected, a route error packet [RERR] will be sent to the source, indicating that the path is no longer valid.

Temporary ordered routing algorithm [TORA]

It is also a source-initiated routing algorithm, which creates multiple routes for any source/ destination pair. The advantage of multiple routes is that route discovery is not required for every alteration in the network topology. TORA consists of three phases:

1. Route creation/discovery

2. Route maintenance

3. Route erasure

TORA uses three types of packets. They are query packets for route creation, and update packets for both creation and maintenance. The route will be discovered from the source to destination only when a request is made for the transmission. In this algorithm the source will generate a query packet which will be broadcasted in the network. This continues as long as anode that is directly connected to the destination is identified. When the destination is identified, an update packet will be generated and sent back to the source. The update packet will have the path information if there is more than one update packet received by the source. It means there are multiple paths to the destination, and the source has to choose the best path available.

9.4 Security in Ad Hoc Networks

The following are the security threats in ad hoc networks:

1. *Limited computational capabilities:* The nodes in the mobile ad hoc network are modular, independent, and will have limited computational capability. It becomes a source of vulnerability when they handle public key cryptography.

2. *Limited power supply:* Since nodes have limited power supply, an attacker can exhaust batteries by giving excessive computations to be carried out.

3. *Challenging key management:* The key management becomes extremely difficult as the mobile devices will be under movement.

Types of Attack in an Ad Hoc Network

The attack can be classified into two types. They are Passive and Active. **In a passive attack**, the normal operation of a routing protocol is not interrupted. The attacker just tries to gather the information. **In an active attack**, the attacker can insert some arbitrary packets and therefore might affect the normal operation of the network. An attack can also be one of the following types:

1. *Pin attack:* With the pin attack, an unauthorized node pretends to have shortest path to the destination. The attacker can listen to the path setup phase and become part of the network.

2. *Location disclosure attack:* By knowing the locations of intermediate nodes, the attacker can find out the location of the target node.

3. *Routing table overflow:* The attacker can create some routes whose destination does not exist. It will have a major impact on proactive based-routing.

4. *Energy exhaustion attack:* The attacker tries to forward unwanted packets or send unwanted requests, which can conserve the battery of the nodes.

Criteria for a Secure Routing Protocol

The attack in an ad hoc network can be prevented by using a secure routing protocol. It should have the following properties:

1. *Authenticity:* When a routing table is updated, it must verify whether updates were provided by an authenticated node.

2. *Integrity of information:* When a routing table is updated, the information must be verified whether it is modified or not.

3. *In order updates:* Sequence numbers or some mechanism must be used to maintain updates in order.

4. *Maximum update time:* Updates in routing tables must be done as quickly as possible.

5. *Authorization:* Only authorized nodes must be able to send updated packets.

Security Criteria for Mobile Ad Hoc and Sensor Networks

The following are the widely used criteria to evaluate if the mobile ad hoc network is secure:

1. *Availability*

The term availability means that a node should maintain its ability to provide all the designed services regardless of its security state. This security criterion is challenged mainly during denial-of-service attacks. In this, all the nodes in the network can be the attack target, and thus, some selfish nodes make some of the network services unavailable, such as the routing protocol or the key management service.

2. *Integrity*

Integrity guarantees the identity of the messages when they are transmitted. Integrity can be compromised mainly in two ways:

- Malicious altering

- Accidental altering

A message can be removed, replayed, or revised by an adversary with a malicious goal, which is regarded as malicious altering; on the contrary, if the message is lost or its content is changed due to some benign failures, which may be transmission errors in communication or hardware errors such as a hard disk failure, then it is categorized as accidental altering.

3. *Confidentiality*

Confidentiality means that certain information is only accessible to those who have been authorized to access it. In other words, in order to maintain the confidentiality of some confidential information, we need to keep it secret from all entities that do not have the privilege to access them.

4. *Authenticity*

Authenticity is essentially assurance that participants in communication are genuine and not impersonators. It is necessary for the communication participants to prove their identities as what they have claimed using some techniques so as to ensure authenticity. If there is not such an authentication mechanism, the adversary could impersonate a benign node and thus get access to confidential resources, or even propagate some fake messages to disturb the normal network operations.

5. *No repudiation*

Non-repudiation ensures that the sender and the receiver of a message cannot deny that they have ever sent or received such a message. This is useful especially when one needs to determine if a node with some abnormal behavior is compromised or not. If a node recognizes that the message it has received is erroneous, it can then use the incorrect message as evidence to notify other nodes that the node sending out the improper message may have been compromised.

6. *Authorization*

Authorization is a process in which an entity is issued a credential which specifies the privileges and permissions it has, and it cannot be falsified by the certificate authority. Authorization is generally used to assign different access rights to different levels of users. For instance, one needs to ensure that network management function is only accessible by the network administrator. Therefore, there should be an authorization process before the network administrator accesses the network management functions.

7. *Anonymity*

Anonymity means that all the information that can be used to identify the owner or the current user of the node should default to be kept private and not be distributed by the node itself or the system software. This criterion is closely related to privacy preserving, in which we should try to protect the privacy of the nodes from arbitrary disclosure to any other entities.

Distributed Management Services in Mobile Ad Hoc and Sensor Networks

Mobile ad hoc networks need an effective distributed management solution which can handle various tasks in emergency and rescue operations. This solution should be able to tackle some of the other components of a management system, such as resource management, privacy management, and key management. The dynamic nature of a mobile ad hoc network makes middleware services unsuitable for synchronous communication because they are too vulnerable to communication disruptions. For example, devices might suddenly be out of reach or turned off. The classic alternative to synchronous solutions is a distributed event notification system, as shown in Figure 9.2. Among several technical challenges of mobile ad hoc networks, the sharing of information among various computing devices can be regarded as one of the important issues. Information access and sharing is difficult in mobile ad hoc networks because of their dynamic nature, scarce resources, and heterogeneous user devices.

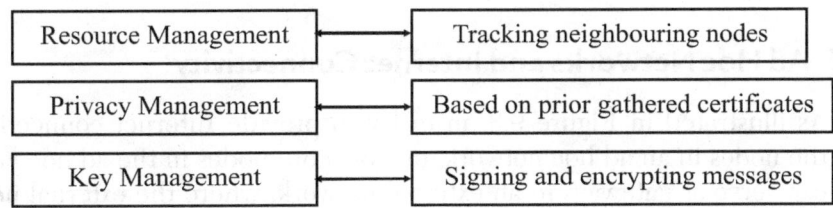

FIGURE 9.2 Distributed event management services.

Network-Wide Broadcasting, Handling Data Loss in Mobile Ad Hoc Networks

Mobile ad hoc networks offer a unique art of network formation where mobile devices can communicate with each other without a pre-existing infrastructure. Ad hoc networks have been considered to be the foundation for new technologies. One important issue is to try to reduce the packet or the data loss during an active transmission. In mobile ad hoc networks, mobile link transmission errors, mobility, and network congestion are some of the major causes of data loss. Data loss due to transmission errors is mainly affected by the physical condition of the channel and the region where networks are deployed. These losses can't be reduced with the improvement in ad hoc routing protocols.

Each mobile device in an ad hoc network has to rely on others for forwarding data packets to other nodes in the network. Routing protocols of a mobile ad hoc network are another way to transmit data from one device to another. NWB (Network-wide Broadcast) is considered to be one of the routing or data exchange related operations, and it is used to discover routes for both unicast (one-to-one) and multi-cast (one-to-many) data exchange operations. NWB can also be defined as a process through which one mobile device sends a packet to all other devices in the network. NWB provides important control and route establishment functionality to different protocols of mobile ad hoc networks. It is especially important for paging, alarming, location updates, route discoveries, or even routing in highly mobile ad hoc environments. Network-wide broadcasting is normally achieved via flooding. In a flooding or broadcasting task, a source mobile device floods or broadcasts the same message to all the devices in the network. Some of the desirable properties of a scalable flooding scheme are reliability, power, and bandwidth efficiency, which can be measured by savings in rebroadcasts.

9.5 Ad Hoc Networks and Internet Connectivity

As illustrated in Figure 9.3, in order to provide Internet connectivity to the nodes in an ad hoc network, one or more nodes in the ad hoc network can serve as gateways to an external network, where the external network can be an infrastructure network such as a LAN, the Internet, or a cellular network, or even an infrastructure-less network such as another ad hoc network. As a solution, an integration of Mobile IP and ad hoc networks is implemented, such that the Mobile IP enables nodes to move between different gateways while maintaining the connectivity, and ad hoc routing protocols provide connectivity among the nodes within the ad hoc network. In other words the Mobile IP provides macro mobility, and ad hoc routing protocols provide micro mobility. Micro mobility, also called Intra-Domain mobility, is the movement of a mobile node within its own network, while macro mobility, also called Inter-Domain mobility, is the movement of mobile node between different networks.

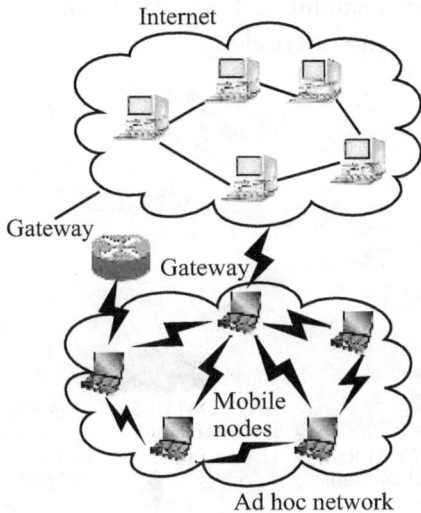

FIGURE 9.3 Internet connectivity in ad hoc networks.

9.6 Mobile Ad Hoc Networking for the Military

Driven by technologies such as data networking, GPS, real-time video feeds from Unmanned Aerial Vehicles (UAVs), and satellite intelligence, today's modern military has access to a plethora of real-time data. However, getting this information to the war fighter at the "edge of the network" is still problematic. Getting real-time voice, data, and streaming video to the soldier at the edge is no easy task. Soldiers are mobile and need high-performance, high-bandwidth networks that move with them to deliver the information they need.

A portion of a military IP network can be based on fixed wired infrastructure, utilizing satellites and networking equipment in operations and command centers. But it is not practical or even possible to create a fixed, wired network infrastructure on a battlefield; the only practical way to provide a networking infrastructure is to create a mobile wireless network. Since most soldiers are typically in or near some sort of fighting vehicles, an effective way to create a mobile wireless network is to make use of such fighting vehicles to carry the infrastructure necessary to build and maintain these networks on the move, as shown in Figure 9.4.

Mobile wireless networks are built using a variety of IP which enable radios and specialized embedded network routers. Network infrastructure radios are called backhaul radios, which can communicate with other backhaul radios or with satellites. Each backhaul radio is connected to an IP router to create a network node.

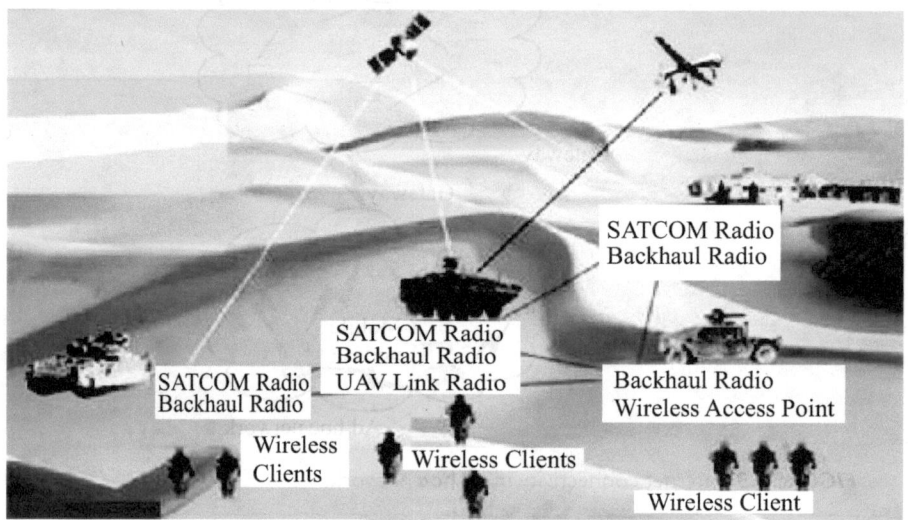

FIGURE 9.4 A view of a battlefield illustrating a mobile, wireless network capable of communicating voice, data, and video traffic. The networking infrastructure is located in vehicles equipped with backhaul radios and embedded routers.

At the "edge of the network" are a variety of clients. Dismounted soldiers carry some, such as handheld radios, man-pack radios, laptops, cameras, and PDAs; multiple clients can connect wirelessly to the same IP network node. There are also a variety of clients that reside inside vehicles. Often an Ethernet switch is interfaced to the router in a vehicle, providing a vehicle local area network (LAN) for clients such as radios, laptops, battlefield display systems, and mission-control computers. This enables the same IP-networking system in a vehicle to support both the internal vehicle and external vehicle communications simultaneously. For example, externally mounted cameras will stream video feeds to the dashboard utilizing the vehicle LAN while in-vehicle and dismounted soldiers communicate with each other.

With fixed wired networks, the network nodes are fixed, and the only components that move are clients that are not used to route other traffic, such as cellular and Wi-Fi-enabled devices. On the battlefield, not only are the clients mobile, but so are the basic building blocks of a mobile wireless network, for example, the radios and specialized embedded routers. This creates a fluid and ever-changing network with dynamic nodes and frequent routing table changes. These are referred to as Mobile Ad hoc NETworks (MA-NET).

Ad hoc networks have numerous distinguishing characteristics when compared with conventional networking solutions, as shown in Table 9.2. Ad hoc networks deliver a compelling advantage wherever highly mobile soldiers, unsupported by fixed infrastructure, need to share IP-based information. They offer superior information sharing at all levels, enabling improved situational awareness, a clearer understanding of the leader's intent, and the ability for remote users to self synchronize.

Table 9.2 Advantages of Deployed Ad Hoc Networks

Self-forming	Nodes that come within radio range of each other can establish a network association without any preconfiguration or manual intervention.
Self-healing	Nodes can join or leave rapidly without affecting operation of the remaining nodes.
No infrastructure	In an ad hoc network, mobile nodes form their own network and essentially become their own infrastructure.
Peer-to-peer	Traditional networks typically support end systems operating in client-server mode. In an ad hoc network, mobile nodes can communicate and exchange information without prior arrangement and without reliance on centralized resources.
Predominantly wireless	Historically, networks have been mostly wired and enhanced or extended through wireless access. The ad hoc environment is essentially wireless, but can be extended to support wired resources.
Highly dynamic	Mobile nodes are in continuous motion, and ad hoc networking topologies are constantly changing

The fact that they are self forming and self healing facilitates deployment and minimizes the need for manual configuration and intervention. Meanwhile, their multi-hop networking nature extends network coverage and provides redundant paths for increased resilience. With ad hoc networks operators also have the ability to operate with or without connectivity to a centralized network. Such networks are a key enabler for new applications such as vehicle-to-vehicle networking, intelligent transportation systems, sensor networking, telemetry monitoring, and more.

Specialized Embedded Routers

When there is a change to the network infrastructure requiring a change to the way packets are routed, routing tables have to be updated and propagated. Network operators in the fixed networks know the paths available and can engineer the routing changes using costing in the rare cases where neighbors change. There is typically the luxury of having monitoring points send an alarm to a network operations center (NOC) in the event of such network events. There can be months of planning for network or routing changes within known maintenance windows. A traditional network router would not cope with dynamic routing table changes that can occur with nodes participating in a MANET.

Weather, terrain, and mobility make radio-based communications dynamic; therefore, routers must be aware of each radio's condition in order to make effective routing decisions with built-in mechanisms to prevent constant re-routing and human intervention. Mobile networks delivering real-time services, such as video and data, cannot tolerate prolonged network changes due to radio dynamics. To address this challenge, IP routers are deployed with technology to minimize network disruption due to network re-convergence. These routers support features such as radio-aware routing, traffic optimization, firewall/network security, and voice services.

Mobile ad hoc networks for military applications pose hardware and platform challenges because todays networking devices must be optimized from a Size, Weight, and Power (SWaP) perspective and also be made to work reliably in harsh environments. In the peer-to-peer world, anybody or anything that moves can potentially be a wireless networking node. Military ad hoc networking requires a variety of platforms, ranging from vehicle-based to hand-carried or wearable, and all offering equivalent network services.

Whether a router and a Gbit Ethernet switch are deployed in a small two-slot box to provide network connectivity and an in-vehicle LAN, or a router is being integrated into a vehicle's existing mission control computer, space and power are at a premium. And the hardware needs to be able to survive the harsh environment in a vehicle on a battlefield. Size, weight, and power are even more critical for clients carried by dismounted soldiers. When the ruggedized, embedded routers are coupled with today's high-performance, IP-enabled radios, they do much more than just create mobile ad hoc networks. They help ensure that the networks and the data are highly secure, critical applications are prioritized, and bandwidth is optimized. They deliver vital data to dismounted soldiers, such as live streaming video. They allow commanders to get a total, integrated view of the battlespace.

9.7 Vehicular Ad Hoc and Sensor Networks

Vehicular Ad Hoc Networks (VANETs) have grown out of the need to support the growing number of wireless products that can now be used in vehicles. These products include remote keyless entry devices, personal digital assistants (PDAs), laptops, and mobile telephones. As mobile wireless devices and networks become increasingly important, the demand for Vehicle-to-Vehicle (V2V) and Vehicle to-Roadside (VRC) or Vehicle-to-Infrastructure (V2I) Communication will continue to grow. VANETs can be utilized for a broad range of safety and non-safety applications, allow for value-added services such as vehicle safety, automated toll payment, traffic management, enhanced navigation, location-based services such as finding the closest fuel station, restaurant, or travel lodge, and infotainment applications such as providing access to the Internet.

Intelligent transportation systems (ITSs)

In intelligent transportation systems, each vehicle takes on the role of sender, receiver, and router to broadcast information to the vehicular network or transportation agency, which then uses the information to ensure safe, free flow of traffic. For communication to occur between vehicles and RoadSide Units (RSUs), vehicles must be equipped with some sort of radio interface or OnBoard Unit (OBU) that enables short-range wireless ad hoc networks to be formed. Vehicles must also be fitted with hardware that permits detailed position information such as a Global Positioning System (GPS) or a Differential Global Positioning System (DGPS) receiver. Fixed RSUs, which are connected to the backbone network, must be in place to facilitate communication. The number and distribution of roadside units is dependent on the communication protocol that is to be used. For example, some protocols require roadside units to be distributed evenly throughout the whole road network, some require roadside units only at intersections, while others require roadside units only at region borders. Though it is safe to assume that infrastructure exists to some extent and vehicles have access to it intermittently, it is unrealistic to require that vehicles always have wireless access to roadside units.

Figures 9.5, 9.6, and 9.7 depict the possible communication configurations in intelligent transportation systems. These include inter-vehicle, vehicle-to-roadside, and routing-based communications. Inter-vehicle, vehicle-to-roadside, and routing-based communications rely on very accurate and up-to-date information about the surrounding environment which, in turn, requires the use of accurate positioning systems and smart communication protocols for exchanging information. In a network environment in which the communication medium is shared, highly unreliable, and with limited bandwidth, smart communication protocols must guarantee fast and reliable delivery of information to all vehicles in the vicinity. It is worth mentioning that Intra-vehicle communication uses technologies such as IEEE 802.15.1 (Bluetooth), IEEE 802.15.3 (Ultra-wide Band), and IEEE 802.15.4 (Zigbee) that can be used to support wireless communication inside a vehicle.

1. *Inter-vehicle communication*

The inter-vehicle communication configuration (Figure 9.5) uses multi-hop multicast/broadcast to transmit traffic related information over multiple hops to a group of receivers. In intelligent transportation systems, vehicles need only be concerned with activity on the road ahead and not behind (an example of this would be for emergency message dissemination about an imminent collision or dynamic route scheduling). There are two types of message forwarding in inter-vehicle communications: naive broadcasting and intelligent broadcasting. In naive broadcasting, vehicles send broadcast messages periodically and at regular intervals. Upon receipt of the message, the vehicle ignores the message if it has come from a vehicle behind it. If the message comes from a vehicle in front, the receiving vehicle sends its own broadcast message to vehicles behind it. This ensures that all enabled vehicles moving in the forward direction get all broadcast messages. The limitations of the naive broadcasting method is that large numbers of broadcast messages are generated, therefore increasing the risk of message collision, resulting in lower message delivery rates and increased delivery times. Intelligent broadcasting with implicit acknowledgement addresses the problems inherent in naive broadcasting by limiting the number of messages broadcast for a given emergency event. If the event-detecting vehicle receives the same message from behind, it assumes that at least one vehicle in the back has received it and ceases broadcasting. The assumption is that the vehicle in the back will be responsible for moving the message along to the rest of the vehicles. If a vehicle receives a message from more than one source, it will act on the first message only.

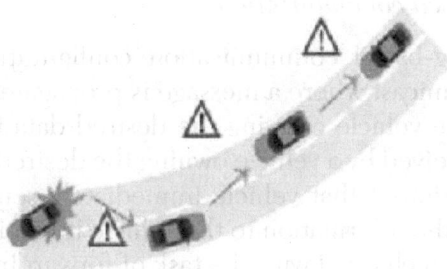

FIGURE 9.5 Inter-vehicle communication.

2. *Vehicle-to-roadside communication*

The vehicle-to-roadside communication configuration (Figure 9.6) represents a single hop broadcast where the roadside unit sends a broadcast message to all equipped vehicles in the vicinity. Vehicle-to-roadside communication configuration provides a high bandwidth link between vehicles and roadside units. The roadside units may be placed every kilometer or less, enabling high data rates to be maintained in heavy traffic. For instance, when broadcasting dynamic speed limits, the roadside unit will determine the appropriate speed limit according to its internal timetable and traffic conditions. The roadside unit will periodically broadcast a message containing the speed limit and will compare any geographic or directional limits with vehicle data to determine if a speed limit warning applies to any of the vehicles in the vicinity. If a vehicle violates the desired speed limit, a broadcast will be delivered to the vehicle in the form of an auditory or visual warning, requesting that the driver reduce his speed.

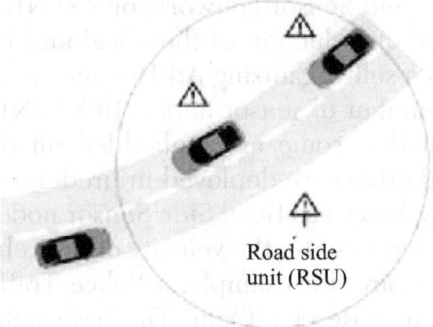

Road side
unit (RSU)

FIGURE 9.6 Vehicle-to-roadside communication.

3. *Routing-based communication*

The routing-based communication configuration (Figure 9.7) is a multi-hop unicast where a message is propagated in a multi-hop fashion until the vehicle carrying the desired data is reached. When the query is received by a vehicle owning the desired piece of information, the application at that vehicle immediately sends a unicast message containing the information to the vehicle it received the request from, which is then charged with the task of forwarding it toward the query source.

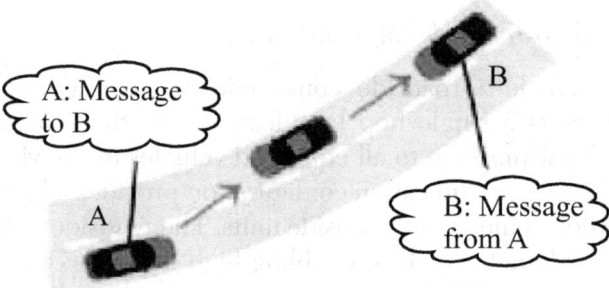

FIGURE 9.7 Routing based communication.

The rapid increase of vehicular traffic and congestion on the highways began hampering the safe and efficient movement of traffic. Consequently, year by year, there is an ascending rate of car accidents and casualties in most countries. Therefore, exploiting the technologies, the vehicular network employing wireless Sensor networks as Vehicular Ad Hoc and Sensor Network, or VASNET for short, is required as a solution of reduction of these sad and reprehensible statistics. VASNET is a self-organizing Ad Hoc and sensor network composed of a large number of sensor nodes. In VASNET there are two kinds of sensor nodes; some are embedded on the vehicles—vehicular nodes—and others are deployed in predetermined distances beside the highway, known as Road Side Sensor nodes (RSS). The vehicular nodes are used to sense the velocity of the vehicle, for instance. Base Stations (BS) are, for example, a Police Traffic Station, a Firefighting Group, or a Rescue Team. The base stations may be stationary or mobile. VASNET provides capability of wireless commumcation between vehicular nodes and stationary nodes, to increase safety and comfort for vehicles on the roads.

Vehicular Ad Hoc Network (VANET)

Vehicular Ad Hoc Networks (VANET), upon implementation, should collect and distribute safety information to massively reduce the number of accidents by warning drivers about the danger before they actually face it. VANET consist of some sensors embedded on the vehicles. The onboard sensors' readings can be displayed to the drivers via monitors to be aware of the vehicle condition or emergency alarms, and also can be broadcast to the other adjacent vehicles. VANET can also be helped by some of the Roadside Units like Cellular Base Stations to distribute the data to the other vehicles, as shown in Figure 9.8. VANET makes extensive use of wireless communication to achieve its aims. VANET is a kind of Mobile Ad Hoc Network (MANET) with some differences, like limitation in power, moving pattern, and mobility.

Limitation in Power: In MANET, power constraint is one of the most important challenges which has over shadowed all other aspects, namely routing and fusion; on the other hand in VANET, a huge battery is carried by the vehicle (i.e., car's battery), so, energy consumption is not a salient issue.

Moving pattern: It is random in the MANET while vehicles tend to move in an organized fashion in VANET.

Mobility: There is high mobility in the VANET in comparison to MANET. However, self-organization and lack of infrastructure are similarities between MANET and VANET. There are some salient challenges in VANET, such as;

1. As mobile nodes (vehicles) are moving with high mobility, quick changes in the VANETs topology are difficult to control.

2. The communication between the vehicles is prone to frequent fragmentation.

3. Rapid change in the link's connectivity cause many paths to be disconnected before they can be utilized.

4. There is no constant density in VANET, as in highways it is high density and in the rural areas, low density.

5. A message can change the topology; for instance, when a driver receives an alarm message, he/she may change his/her direction, which may cause the change in the topology.

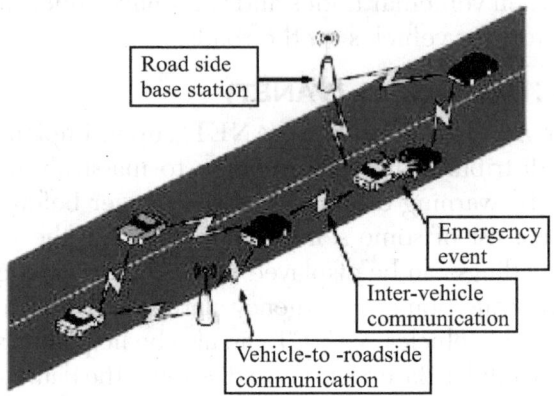

Road side base station

Emergency event

Inter-vehicle communication

Vehicle-to -roadside communication

FIGURE 9.8 VANET communication structure.

9.8 Application of VANET

Crash prevention applications that rely on an infrastructure include road geometry warning to help drivers at steep or curved roads, warning overweight or over height vehicles, highway rail crossing and intersection collision systems to help drivers cross safely, and pedestrian, cyclist, and animal warning systems to inform drivers of possible collisions. These systems become of vital importance at night or under low visibility conditions.

Safety applications which do not rely on an infrastructure include an emergency brake announcement, which is the most important application for crash prevention. The first two cars might not benefit from the emergency brake system, but further cars can avoid the crash. Lane change assistance, road obstacle detection, road departure warning, as well as forward and rear collision warning are all examples of safety V2V applications. Vehicles can also automatically send help requests in case of an accident, which can be vital when no other cars are around.

The system can also help the driver in other ways such as vision enhancement via image processing techniques, and lane keeping assistance and monitoring of onboard systems as well as any cargo or trailers connected to the vehicle. Such systems are generalized as Advanced Driver Assistance Systems (ADAS). The commercial applications of the system cover a wide range of innovative ideas aiding individuals and tourists, such as booking a parking place, downloading tourism information and maps for restaurants and gas stations, navigation and route guidance, payment at toll plazas, Internet access, and connection to home computers. Other devices within the vehicle can also be connected to the On Board Units (OBU) to access any services provided by the network or through the Internet. These applications are not required by the government, but they encourage people to install the system.

Users enter their start position, destination, and time to start their journey, and the server responds with the two best routes. The routes are compiled from a nine-month survey as well as simulations. In its final version the system should be able to collect data from the sensors installed in cars and provide the routes to the OBU. Subscribers to the system get an onboard navigation system that receives information about the weather, road conditions, traffic, and any other related data from road side units and displays it to the user. A number of applications are envisioned for these networks, some of which are:

- Vehicle collision warning
- Security distance warning
- Driver assistance
- Cooperative driving
- Cooperative cruise control
- Dissemination of road information
- Internet access
- Map location
- Automatic parking
- Driverless vehicles

9.9 Routing for VANET

VANETs are a specific class of ad hoc networks; the commonly used ad hoc routing protocols initially implemented for MANETs have been tested and evaluated for use in a VANET environment. Use of these address-based and topology-based routing protocols requires that each of the participating nodes be assigned a unique address. This implies that we need a mechanism that can be used to assign unique addresses to vehicles, but these protocols do not guarantee the avoidance of allocation of duplicate addresses in the network. Thus, existing distributed addressing algorithms used in mobile ad hoc networks are much less suitable in a VANET environment. Specific

VANET-related issues such as network topology, mobility patterns, demographics, density of vehicles at different times of the day, rapid changes in vehicles arriving and leaving the VANET, and the fact that the width of the road is often smaller than the transmission range all make the use of these conventional ad hoc routing protocols inadequate.

1. *Proactive routing protocols*

Proactive routing protocols employ standard distance-vector routing strategies (e.g., Destination-Sequenced Distance-Vector (DSDV) routing) or link-state routing strategies (e.g., Optimized Link State Routing protocol (OLSR) and Topology Broadcast-based on Reverse-Path Forwarding (TBRPF)). They maintain and update information on routing among all nodes of a given network at all times, even if the paths are not currently being used. Route updates are periodically performed regardless of network load, bandwidth constraints, and network size. The main drawback of such approaches is that the maintenance of unused paths may occupy a significant part of the available bandwidth if the topology of the network changes frequently. Since a network between cars is extremely dynamic, proactive routing algorithms are often inefficient.

2. *Reactive routing protocols*

Reactive routing protocols such as Dynamic Source Routing (DSR) and Ad hoc On-demand Distance Vector (AODV) routing implement route determination on a demand or need basis and maintain only the routes that are currently in use, thereby reducing the burden on the network when only a subset of available routes is in use at any time. Communication among vehicles will only use a very limited number of routes, and therefore reactive routing is particularly suitable for this application scenario.

3. *Position-based routing*

Position-based routing protocols require that information about the physical position of the participating nodes be available. This position is made available to the direct neighbors in the form of periodically transmitted beacons. A sender can request the position of a receiver by means of a location service. The routing decision at each node is then based on the destination's position contained in the packet and the position of the forwarding node's neighbors. Consequently, position-based routing does not require the establishment or maintenance of routes.

Examples of position-based routing algorithms include Greedy Perimeter Stateless Routing (GPSR) and Distance Routing Effect Algorithm for Mobility (DREAM). In a position-based routing protocol based on a greedy forwarding mechanism, packets are forwarded through nodes geographically closer to the destination than the previous node. Thus, the position of the next hop will always be closer to the destination node than that of the current hop. The "perimeter routing" mode of GPSR (greedy perimeter stateless routing) that searches for alternate routes that may not be geographically closer is not considered since in a highway scenario the width of the road is often smaller than the range of transmission. Thus, in this scenario there is no way for a route to move away from the destination and still find its way back.

Existing ad hoc networks employ topology-based routing where routes are established over a fixed succession of nodes but which can lead to broken routes and a high overhead to repair these routes. The special conditions and requirements for vehicular communications, including frequent topology changes, short connectivity time, and positioning systems have justified the development of dedicated routing solutions for wireless multi-hop communication based on geographic positions. The use of Global Positioning System (GPS) technology enables forwarding to be decoupled from a node's identity, and therefore the position of the destination node is used rather than a route to it which requires traffic flow via a set of neighbors. Thus position-based routing provides a more scalable and efficient forwarding mechanism appropriate for highly volatile ad hoc networks found in VANETs.

Position-based routing constitutes three core components: beaconing, location service, and forwarding (geographic unicast and geographic broadcast). A position-based routing scheme which employs a unique identifier such as an IP address is used to identify a vehicle along with its current position (GPS coordinate). This scheme only requires that a vehicle knows its own position and that of its one-hop neighbors. Assuming a packet contains the destination position, the router forwards the packet to a node closer to the destination than itself. Given the relatively high speeds of the large number of vehicles involved, this scheme is both adaptive and scalable with respect to network topology.

4. *Beaconing and location service*

Vehicles periodically broadcast short packets with their identifier and current geographic position. Upon receipt of a beacon, a vehicle stores the information in its location table. The requesting vehicle issues a location query message requesting the identification and sequence numbers and hop limit when it needs to know the position of a required vehicle not available in its location table. This message is rebroadcast to nearby vehicles until it reaches the required vehicle or the hop limit is reached. If the request is not a duplicate, the required vehicle answers with a location reply message carrying its current position and timestamp. Upon receipt of the location reply, the originating vehicle updates its location table.

5. *Forwarding*

A geographic unicast transports packets between two nodes via multiple wireless hops. When the requesting node wishes to send a unicast packet, it determines the position of the destination node by looking at the location table. A greedy forwarding algorithm is then used to send the packet to the neighboring vehicle (Figure 9.9), detailing the minimum remaining distance to the destination vehicle, and this process repeats at every vehicle along the forwarding path until the packet reaches its destination. A geographic broadcast distributes data packets by flooding, where vehicles re-broadcast the packets if they are located in the geographic area determined by the packet. The application of advanced broadcasting algorithms help to minimize overhead by reducing the occurrence of broadcast storms. Data and control packet forwarding must be loop-free and toward the destination or target area location. Having packets forwarded across the shortest path toward the destination is not a requirement due to the high network volatility.

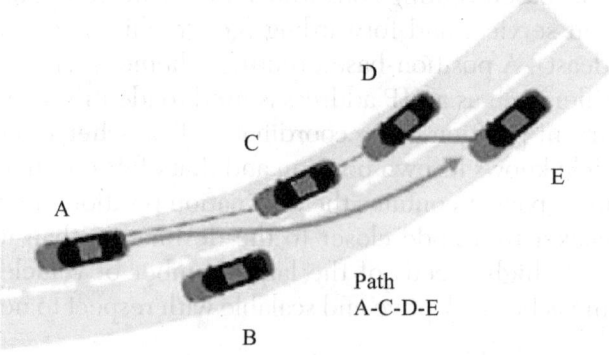

FIGURE 9.9 Cached Greedy Geographic Unicast (CGGC), an example of a greedy unicast transmission based on knowledge of the destination's position.

6. *Protocols for dedicated short-range communication (DSRC)*

Recent research on dedicated short-range communication protocols, namely Coordinated External Peer Communication (CEPEC) and Communications Architecture for Reliable Adaptive Vehicular Ad Hoc Networks (CARAVAN) use mapping and timeslot allocation to reduce the occurrence of denial of service attacks or attacks that burden the limited bandwidth available in vehicular networks.

Communications in a vehicular network are susceptible to denial of service attacks by jamming the communication medium or taxing the limited wireless bandwidth that is available. These attacks are possible due to the DSRC standard specification that a vehicle must wait to transmit until it senses that the channel is idle, allowing a malicious vehicle to constantly transmit noise to prevent transmission from within sensing range of the attacker vehicle.

The Communications Architecture for Reliable Adaptive Vehicular Ad Hoc Networks (CARAVAN) is a solution to these types of communication attacks. CARAVAN uses trusted computing platforms, spread spectrum technology, and a secret pseudorandom spreading code to verify the integrity of the software and hardware of the sending vehicle before allowing the vehicle to transmit messages. CARAVAN includes a new link layer protocol called Adaptive Space Division Multiplexing (ASDM) that allocates timeslots to vehicles to maximize anti-jamming protection. ASDM includes original features that improve on existing Space Division Multiple Access (SDMA) protocols in terms of bandwidth utilization by splitting the roadway into discrete cells that can contain at most one vehicle. A mapping function is then defined that assigns each of the cells a timeslot. No two cells within a predefined range of each other will have the same timeslot. In their approach the physical layer is split into two frequency bands with radio ranges that are selected based on the requirements of the messages carried in each band. Irregularly occurring warning messages place a premium on network connectivity since they are of interest to vehicles far from the message source. These messages are relatively infrequent and therefore require less bandwidth. Periodic messages, on the other hand, are only of interest to vehicles close to the message source, but there are a large number of these messages and they must be generated frequently. The network protocol includes message forwarding rules and a method that leverages the benefits of varying radio ranges to speed delivery of irregular messages.

A cross-layer protocol called Coordinated External Peer Communication (CEPEC) is used for peer to-peer communications in vehicular networks.

The CEPEC protocol coordinates the functions of physical, MAC, and network layers to provide a fair and handoff-free solution for uplink packet delivery from vehicles to roadside units. With CEPEC, the road is logically partitioned into segments of equal length and a relaying head is selected in each segment to perform local packet collecting and aggregate packet relaying. Nodes outside the coverage area of the nearest roadside unit can still get access via a multi-hop route to their roadside unit. Similar to CARAVAN, CEPEC allocates timeslots to vehicles in two steps: first, the roadside unit allocates the timeslots to the segments. Second, intra-segment timeslot allocation occurs where the Segment Head (SH) assigns timeslots to individual vehicles within the segment. Results show that the CEPEC protocol provides higher throughput with guaranteed fairness in multihop data delivery in VANETs when compared with a purely IEEE 802.16-based protocol.

Broadcasting

A geographic broadcast distributes data packets by flooding, where vehicles re-broadcast the packets if they are located in the geographic area determined by the packet. The application of broadcasting algorithms help to minimize overhead by reducing the occurrence of broadcast storms. Data and control packet forwarding must be loop-free and in the direction of the destination or target area location. Having packets forwarded across the shortest path toward the destination is typically found in conventional routing networks and is not a requirement due to the high network volatility.

Several routing efforts have investigated the design of ad hoc routing algorithms suitable for operation in a VANET environment to deal with: a node's mobility, by discovering new routes (reactive routing algorithms), updating existing routing tables (proactive routing algorithms), using geographical location information (position-based routing algorithms), detecting stable vehicle configurations (clusters), using a vehicle's movements to support message transportation, and using broadcasting to support message forwarding.

Vehicles periodically broadcast short packets with their identifiers and current geographic position. Upon receipt of such beacons, a vehicle stores the information in its location table. It is therefore possible to design a Cooperative Collision Avoidance (CCA) system that can assist in collision avoidance by delivering warning messages. When an emergency situation arises, a vehicle that is part of a CCA platoon needs to broadcast a message to all of the vehicles behind it. The vehicles that receive this message selectively forward it based upon the direction from which it came which ensures that all members of the platoon eventually receive this warning.

Mobicasting

The mobicast routing protocol for VANET takes the factor of time into account. The main goal of the mobicast routing protocol is the delivery of information to all nodes that happen to be in a prescribed region of space at a particular point in time. The mobicast protocol is designed to support applications which require spatiotemporary coordination in vehicular ad hoc networks. The spatiotemporary character of a mobicast is to forward a mobicast message to vehicles located in some geographic zone at time t, where the geographic zone is denoted as the Zone Of Relevance (ZOR_t). Vehicles located in ZOR_t at the time t should receive the mobicast message.

Two features are introduced in the mobicast routing protocol for safety and comfort applications, as follows.

To support safety applications, the mobicast routing protocol must disseminate the message on time. Vehicles located in the ZOR_t should receive the mobicast message before time t + 1; therefore, vehicles located in ZOR_t at time t must keep the connectivity to maintain the realtime data communication between all vehicles in ZOR_t. However, the connectivity in ZOR_t is easily lost if any vehicle in ZOR suddenly accelerates or decelerates its velocity, and this leads to a temporary network fragmentation problem. Some vehicles in ZOR_t cannot successfully receive the mobicast messages due to the temporary network fragmentation. To solve this problem, disseminate mobicast messages to all vehicles in ZOR_t via a special geographic zone, known as a Zone Of Forwarding (ZOF_t). This protocol dynamically estimates the accurate ZOF_t to guarantee that the mobicast messages can be successfully disseminated before time t + 1 to all vehicles located in ZOR_t.

In contrast, comfort applications for VANET are usually delay-tolerant. That is, messages initiated from a specific vehicle at time t can be delivered through VANETs to some vehicles within a given constrained delay time λ. For all vehicles located in the zone of relevance at time t (denoted as ZOR_t), the mobicast routing is able to disseminate the data message initiated from a specific vehicle to all vehicles which have ever appeared in ZOR_t at time t. This data dissemination must be done before time t + λ through the multi-hop forwarding and carry-and-forward techniques. The temporary network fragmentation problem is also considered in their protocol design. A low degree of channel utilization should be maintained to reserve the resource for safety applications.

Quality of Service (QoS)

The term Quality of Service (QoS) is used to express the level of performance provided to users. High levels of QoS in traditional networked environments can often be achieved through resource reservation and sufficient infrastructure, however, these cannot be guaranteed in dynamic, ad hoc environments, such as those used in VANETs due to the VANETs' inherent lack of consistent infrastructure and rapidly changing topology. Most QoS routing strategies aim to provide robust routes among nodes and try to minimize the amount of time required to rebuild a broken connection. However, factors such as node velocity, node positioning, the distance between nodes, a reliability of and delay between links can seriously affect the stability of a particular route. QoS can be guaranteed by detecting redundant source nodes and preventing the transmission of duplicate information, thereby restricting redundant broadcasts that limit the application's bandwidth consumption, which improves the latency of messages.

9.10 Security in VANET

The security of VANETs is crucial as their very existence relates to critical, life threatening situations. It is imperative that vital information cannot be inserted or modified by a malicious person. The system must be able to determine the liability of drivers while still maintaining their privacy.

These problems are difficult to solve because of the network size, the speed of the vehicles, their relative geographic position, and the randomness of the connectivity between them. An advantage of vehicular networks over the more common ad hoc networks is that they provide ample computational and power resources. For instance, a typical vehicle in such a network could host several tens or even hundreds of microprocessors. Attackers are classified as having three dimensions: "insider versus outsider," "malicious versus rational," and "active versus passive." The types of attacks against messages can be described as follows: "Bogus Infoimation," "Cheating with Positioning Information," "ID disclosure," "Denial of Service," and "Masquerade." The reliability of a system where information is gathered and shared among entities in a VANET raises concerns about data authenticity. For example, a sender could misrepresent observations to gain advantage (e.g., a vehicle falsely reports that its desired road is jammed with traffic, thereby encouraging others to avoid this route and providing a less congested trip). More malicious reporters could impersonate other vehicles or road-side infrastructure to trigger safety hazards. Vehicles could reduce this threat by creating networks of trust and ignoring, or at least distrusting, information from untrusted senders.

Threats to Availability, Authenticity, and Confidentiality

Attacks can be broadly categorized into three main groups: those that pose a threat to availability, those that pose a threat to authenticity, and those that pose a threat to driver confidentiality. The following sections present threats posed to each of the areas of availability, authenticity, and confidentiality.

Threats to availability

The following threats to the availability of vehicle-to-vehicle and vehicle-to-roadside communication (including routing functionality) have been identified:

Denial of Service Attack: DoS attacks can be carried out by network insiders and outsiders and renders the network unavailable to authentic users by flooding and jamming with likely catastrophic results. Flooding the control channel with high volumes of artificially generated messages, the network's nodes, onboard units, and roadside units cannot sufficiently process the surplus data.

Broadcast Tampering: An inside attacker may inject false safety messages into the network to cause damage, such as causing an accident by suppressing traffic warnings or manipulating the flow of traffic around a chosen route.

Malware: The introduction of malware, such as viruses or worms, into VANETs has the potential to cause serious disruption to its operation. Malware attacks are more likely to be carried out by a rogue insider rather than an outsider and may be introduced into the network when the onboard units and roadside units receive software and firmware updates.

Spamming: The presence of spam messages on VANETs elevates the risk of increased transmission latency. Spamming is made more difficult to control because of the absence of a basic infrastructure and centralized administration.

Black Hole Attack: A black hole is formed when nodes refuse to participate in the network or when an established node drops out. When the node drops out, all routes it participated in are broken, leading to a failure to propagate messages.

Threats to authenticity

Providing authenticity in a vehicular network involves protecting legitimate nodes from inside and/or outside attackers infiltrating the network using a false identity, identifying attacks that suppress, fabricate, alter, or replay legitimate messages, revealing spoofed GPS signals, and impeding the introduction of misinformation into the vehicular network. These include:

Masquerading: Masquerading attacks are easy to perform on VANETs as all that is required for an attacker to join the network is a functioning onboard unit. By posing as legitimate vehicles in the network, outsiders can conduct a variety of attacks such as forming black holes or producing false messages.

Replay Attack: In a replay attack the attacker re-injects previously received packets back into the network, poisoning a node's location table by replaying beacons. VANETs operating in the WAVE framework are protected from replay attacks, but to continue protection an accurate source of time must be maintained, as this is used to keep a cache of recently received messages against which new messages can be compared.

Global Positioning System (GPS) Spoofing: The GPS satellite maintains a location table with the geographic location and identity of all vehicles on the network. An attacker can fool vehicles into thinking that they are in a different location by producing false readings in the GPS positioning system devices. This is possible through the use of a GPS satellite simulator to generate signals that are stronger than those generated by the genuine satellite.

Tunneling: An attacker exploits the momentary loss of positioning information when a vehicle enters a tunnel, and before it receives the authentic positioning information, the attacker injects false data into the onboard unit.

Position Faking: Authentic and accurate reporting of vehicle position information must be ensured. Vehicles are solely responsible for providing their location information, and impersonation must be impossible. Unsecured communication can allow attackers to modify or falsify their own position information to other vehicles, create additional vehicle identifiers (also known as a Sybil Attack), or block vehicles from receiving vital safety messages.

Message Tampering: A threat to authenticity can result from an attacker modifying the messages exchanged in vehicle-to-vehicle or vehicle-to-roadside unit communication in order to falsify transaction application requests or to forge responses.

Message Suppression/Fabrication/Alteration: In this case an attacker either physically disables inter-vehicle communication or modifies the application to prevent it from sending to, or responding from, application beacons.

Key and/or Certificate Replication: Closely related to broadcast tampering is key management and/or certificate replication, where an attacker could undermine the system by duplicating a vehicle's identity across several other vehicles. The objective of such an attack would be to confuse authorities and prevent identification of vehicles in hit-and-run events.

Sybil Attack: Since periodic safety messages are single hop broadcasts, the focus has been mostly on securing the application layer. For example, the IEEE 1609.2 standard does not consider the protection of multi-hop routing. However, when the network operation is not secured, an attacker can potentially partition the network and make delivery of event-driven safety messages impossible.

Threats to confidentiality

Confidentiality of messages exchanged between the nodes of a vehicular network are particularly vulnerable with techniques such as the illegitimate collection of messages through eavesdropping and the gathering of location information available through the transmission of broadcast messages. In the case of eavesdropping, insider and/or outsider attackers can collect information about road users without their knowledge and use the information at a time when the user is unaware of the collection. Location privacy and anonymity are important issues for vehicle users. Location privacy involves protecting users by obscuring the user's exact location in space and time. By concealing a user's request so that it is indistinguishable from other user's requests, a degree of anonymity can be achieved.

Authentication with digital signatures: Authentication with digital signature is a good choice for VANETs because safety messages are normally standalone. Moreover, because of the large number of network members and variable connectivity to authentication servers, a Public Key Infrastructure (PKI) is an excellent method by which to implement authentication where each vehicle would be provided with a public/private key pair. Before sending a safety message, it signs it with its private key and includes the Certification Authority (CA) certificate. By using private keys, a tamper-proof device is needed in each vehicle where secret information will be stored and the outgoing messages will be signed. The large computational burden of verifying a digital signature for every received packet has led to an exploration for alternatives.

9.11 The VSN Architecture for Micro Climate Monitoring

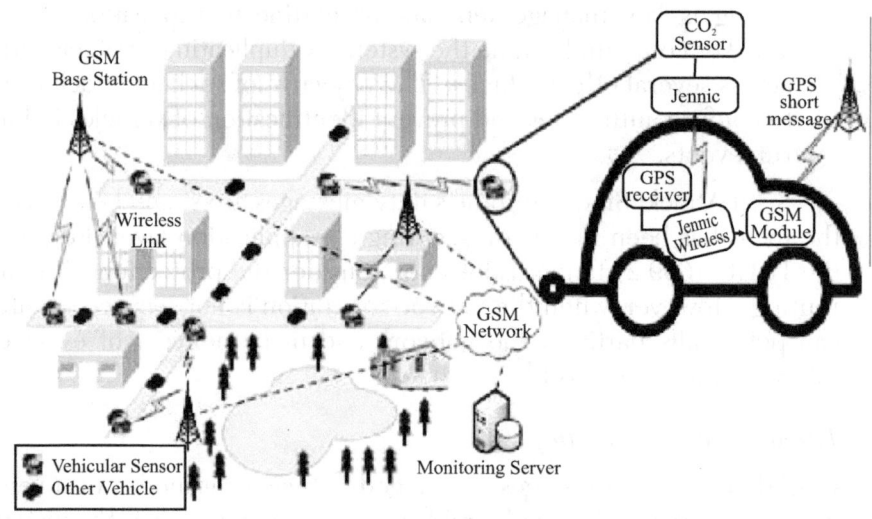

FIGURE 9.10 The VSN architecture for micro climate monitoring.

Figure 9.10 illustrates the proposed VSN architecture for microclimate monitoring. It contains a monitoring server, several vehicular sensors, and GSM networks. Each vehicular sensor is equipped with a CO_2 sensor, a GSM module, and a GPS receiver and periodically reports its sensed CO_2 concentration and its current location to the server through GSM short messages. The monitoring server then calculates the distribution of CO_2 concentration and renders the result on Google Maps. According to the observed distribution and the number of vehicular sensors, the server will ask sensors to adjust their reporting rates. For each vehicular sensor, the intra-vehicle network is a Zigbee network.

VANET simulation models

The environment and topology of VANETs makes it difficult to implement and evaluate them. Outdoor experiments can be used to evaluate VANET protocols and applications, but these can be difficult and expensive to implement because of the high number of vehicles and real-life scenarios involved. It is difficult to perform actual empirical performance measurements because of the inherently distributed, complex environment. To overcome these limitations, simulation tools are used extensively for VANET simulations.

Summary

- A wireless ad hoc sensor network consists of a number of sensors spread across a geographical area.

- Each sensor has wireless communication capability and some level of intelligence for signal processing and networking of the data.

- The requirements of wireless ad hoc sensor networks are a large number of sensors, low energy use, network self-organization, collaborative signal processing, and querying ability.

- Mobile ad hoc networks have features such as autonomous terminals, distributed operation, multi-hop routing, dynamic network topology, fluctuating link capacity, and lightweight terminals.

- The routing issues of MANET are unpredictability of environment, unreliability of wireless medium, resource constrained nodes, and dynamic topology.

- The types of faults in MANETs are transmission errors, node failures, link failures, route breakages, and congested nodes or links.

- Routing protocols are classified into two categories: table driven and source initiated.

- DSDV, CGSR, and WRP are table-driven or proactive-routing protocols.

- DSR, ABR, AODV, and TORA are source-initiated or reactive-routing protocols.

- The attacks are classified into two types, passive and active.

- In a passive attack, the normal operation of routing is not interrupted. The attacker tries to gather the information.

- In an active attack, the attacker inserts some arbitrary packets and therefore might affect the normal operation of the network.

- VANET consists of some sensors embedded on vehicles.

- Applications of VANETs are vehicle collision warning, security distance warning, driver assistance, cooperative driving, cooperative cruise control, dissemination of road information, Internet access, map location, automatic parking, and driverless vehicles.

Questions

1. What are wireless ad hoc sensor networks?

2. What are the classifications of wireless ad hoc sensor networks?

3. What are the requirements of wireless ad hoc sensor networks?

4. List the application of wireless ad hoc sensor networks.

5. List the features of mobile ad hoc networks.

6. What are the routing issues for MANET deployment?

7. Compare WSN and MANET.

8. Explain the classification of routing protocols for MANETs in detail.

9. Write about the security in ad hoc networks.

10. What are the types of attacks in ad hoc networks?

11. List the criteria for a secure routing protocol.

12. In detail write about the security criteria for mobile ad hoc and sensor networks.

13. How can ad hoc networks and the Internet be interfaced?

14. Give the application of mobile ad hoc networking for the military.

15. What are the advantages of deployed ad hoc networks?

16. Explain about the intelligent transportation system.

17. Define VANET.

18. What are the applications of VANET?

19. Explain routing protocols for VANET.

20. Write in detail about the security in VANET.

21. Draw the VSN architecture for micro-climate monitoring. Explain.

Further Reading

1. *Ad Hoc Wireless Networks: Architectures and Protocols* by C. Siva Ram Murthy

2. *Ad Hoc Mobile Wireless Networks: Protocols and Systems* by C. K. Toh

3. *Vehicular Ad Hoc Networks: Standards, Solutions, and Research* by Claudia Campolo, Antonella Molinaro, and Riccardo Scopigno

References

1. *shodhganga.inflibnet.ac.in/bitstream/10603/4106/11/11_chapter%203. pdf*

2. *www.csie.ntpu.edu.tw/~yschen/mypapers/TS-2010-l-2.pdf*

ROUTING AND SECURITY IN WSNS

This chapter discusses the algorithms required for wireless sensor
networks, the network routing protocols, and security in WSNs.

10.1 Algorithms for Wireless Sensor Networks

A wireless sensor network may comprise thousands of sensor nodes. Each
sensor node has a sensing capability as well as a limited energy supply, computing power, memory, and communication ability. Besides military applications, wireless sensor networks may be used to monitor microclimates and
wildlife habitats, the structural integrity of bridges and buildings, building
security, the location of valuable assets (via sensors placed on these valuable
assets), traffic, and so on. However, realizing the full potential of wireless
sensor networks poses myriad research challenges ranging from hardware
and architectural issues, to programming languages and operating systems
for sensor networks, to security concerns, to algorithms for sensor network
deployment, operation, and management. At a high level, the developed
algorithms may be categorized as either centralized or distributed. Because
of the limited memory, computing, and communication capability of sensors, distributed algorithms research has focused on localized distributed
algorithms that require only local (e.g., nearest neighbor) information.

Sensor Deployment and Coverage

In a typical sensor network application, sensors are to be placed (or deployed)
so as to monitor a region or a set of points. In some applications one may
be able to select the sites where sensors are placed while in others (e.g., in
hostile environments) one may simply scatter (e.g., air drop) a sufficiently

large number of sensors over the monitoring region with the expectation that the sensors that survive the air drop will be able to adequately monitor the target region. When site selection is possible, one can use deterministic sensor deployment and when site selection isn't possible, the deployment is nondeterministic. In both cases, it often is desirable that the deployed collection of sensors be able to communicate with one another, either directly or indirectly via multi-hop communication. So, in addition to covering the region or set of points to be sensed, it often requires the deployed collection of sensors to form a connected network. For a given placement of sensors, it is easy to check whether the collection covers the target region or point set and also whether the collection is connected. For the coverage property, it needs to know the sensing range of individual sensors (assume that a sensor can sense events that occur within a distance r, where r is the sensor's sensing range), and for the connected property, it needs to know the communication range, c, of a sensor.

Theorem 1: when the sensor density (i.e., number of sensors per unit area) is finite, $c \geq 2r$ is a necessary and sufficient condition for coverage to imply connectivity.

There is a similar result for the case of k-coverage (each point is covered by at least k sensors) and k-connectivity (the communication graph for the deployed sensors is k connected).

Theorem 2: when $c \geq 2r$, k-coverage of a convex region implies k-connectivity.

Notice that k-coverage with $k > 1$ affords some degree of fault tolerance; one is able to monitor all points so long as no more than $k - 1$ sensors fail. Other variations of the sensor deployment problem also are possible. For example, one may have no need for sensors to communicate with one another. Instead, each sensor communicates directly with a base station that is situated within the communication range of all sensors. In another variant, the sensors are mobile and self deploy. A collection of mobile sensors may be placed into an unknown and potentially hazardous environment. Following this initial placement, the sensors relocate so as to obtain maximum coverage of the unknown environment. They communicate the information they gather to a base station outside of the environment being sensed.

Maximizing Coverage Lifetime

When sensors are deployed in difficult-to-access environments, as is the case in many military applications, a large number of sensors may be air-dropped into the region that is to be sensed. Assume that the sensors that survive the air drop cover all targets that are to be sensed. Since the power supply of a sensor cannot be replenished, a sensor becomes inoperable once it runs out of energy. Define the life of a sensor network to be the earliest time at which the network ceases to cover all targets. The life of a network can be increased if it is possible to put redundant sensors (i.e., sensors not needed to provide coverage of all targets) to sleep and awaken these sleeping sensors when they are needed to restore target coverage. Sleeping sensors are inactive while sensors that are awake are active. Inactive sensors consume far less energy than active ones.

In a decentralized localized protocol, the set of active nodes provides the desired coverage. A sleeping node wakes up when its sleep timer expires and broadcasts a probing signal a distance d (d is called the probing range). If no active sensor is detected in this probing range, the sensor moves into the active state. However, if an active sensor is detected in the probing range, the sensor determines how long to sleep, sets its sleep timer, and goes to sleep. In another distributed localized protocol, sensors may turn themselves on and off. The network operates in rounds, where each round has two phases. They are self-scheduling and sensing. In the self-scheduling phase each sensor decides whether or not to go to sleep. In the sensing phase, the active/awake sensors monitor the region. Sensor s turns itself off in the self-scheduling phase if its neighbors are able to monitor the entire sensing region of s. To make this determination, every sensor broadcasts its location and sensing range. A backoff scheme is proposed to avoid blind spots that would otherwise occur if two sensors turn off simultaneously, each expecting the other to monitor part or all of its sensing region. In this backoff scheme, each active sensor uses a random delay before deciding whether or not it can go to sleep without affecting sensing coverage.

Routing

Traditional routing algorithms for sensor networks are data-centric in nature. Given the unattended and untethered nature of sensor networks, routing must be collaborative as well as energy conserving for individual sensors. In the sensor-centric paradigm, the sensors collaborate to achieve common network-wide goals such as route reliability and path length while minimizing individual costs. The sensor-centric model can be used to define the quality of routing paths in the network (also called path weakness).

Energy conservation is an overriding concern in the development of any routing algorithm for wireless sensor networks. This is because such networks are often located such that it is difficult, if not impossible, to replenish the energy supply of a sensor. Three forms—unicast, broadcast, and multicast—of the routing problem have received significant attention. The overall objective of these algorithms is to either maximize the lifespan (earliest time at which a communication fails) or the capacity of the network (amount of data traffic carried by the network over some fixed period of time).

Assume that the wireless network is represented as a weighted directed graph G that has n vertices/nodes and e edges. Each node of G represents a node of the wireless network. The weight $w(i, j)$ of the directed edge (i, j) is the amount of energy needed by node i to transmit a unit message to node j.

Unicast

In a unicast, one wants to send a message from a source sensor s to a destination sensor t. The five strategies that may be used in the selection of the routing path for this transmission are developed. The first of these is to use a minimum-energy path (i.e., a path in G for which the sum of the edge weights is minimum) from s to t. Such a path may be computed using Dijkstra's shortest path algorithm. However, since in practice messages between several pairs of source-destination sensors need to be routed in succession, using a minimum-energy path for a message may prevent the successful routing of future messages. As an example, consider the graph of Figure 10.1. Suppose that sensors x, b_1, • • •, b_n initially have 10 units of energy each and that u_1, • • •, u_n each have 1 unit. Assume that the first unicast is a unit-length message from x to y. There are exactly two paths from x to y in the sensor network of Figure 10.1. The upper path, which begins at x, goes through each of the u_is, and ends at y, uses $n + 1$ energy units; the lower path uses $2(n + 1)$ energy units. Using the minimum energy path depletes the energy in node u_i, $1 \leq i \leq n$. Following the unicast, sensors u_1, • • •, u_n are unable to forward any messages. So an ensuing request to unicast from u_i to u_j, $i < j$ will fail. On the other hand, had one used the lower path, which is not a minimum energy path, one would not deplete the energy in any sensor and all unit-length unicasts that could be done in the initial network also can be done in the network following the first x to y unicast.

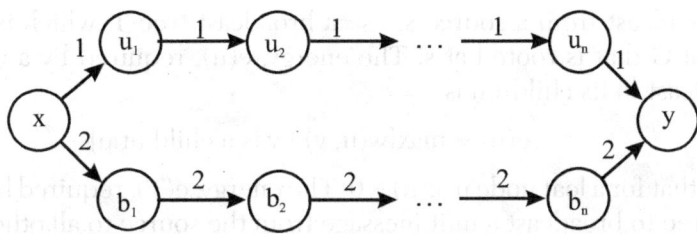

FIGURE 10.1 A sensor network.

The remaining four strategies propose an attempt to overcome the myopic nature of the minimum-energy path strategy, which sacrifices network lifespan and capacity in favor of total remaining energy. Since routing decisions must be made in an online fashion (i.e., if the i_{th} message is to be sent from s_i to t_i, the path for message i must be decided without knowledge of s_j and t_j, $j > i$), we seek an online algorithm with a good competitive ratio. It is easy to see that there can be no online unicast algorithm with constant competitive ratio with respect to network lifespan and capacity. For example, consider the network of Figure 10.1. Assume that the energy in each node is 1 unit. Suppose that the first unicast is from x to y. Without knowledge of the remaining unicasts, one must select either the upper or lower path from x to y. If the upper path is chosen and the source-destination pairs for the remaining unicasts turn out to be (u_1, u_2), (u_2, u_3), ..., (u_{n-1}, u_n), (u_n, y) then the online algorithm routes only the first unicast, whereas an optimal offline algorithm would route all n + 1 unicasts, giving a competitive ratio of n + 1. The same ratio results when the lower path is chosen and the source-destination pairs for the remaining unicasts are (b_1, b_2), (b_2, b_3), ..., (b_{n-1}, b_n), (b_n, y). To maximize lifespan and/or capacity, one needs to achieve some balance between the energy consumed by a route and the minimum residual energy at the nodes along the chosen route.

Multicast and Broadcast

Using an omnidirectional antenna, node i can transmit the same unit message to nodes $j_1, j_2, ..., j_k$, using

$$e_{wireless} = \max\{w(i, j_q) \mid 1 \leq q \leq k\}$$

energy rather than

$$e_{wired} = \sum_{(q=1)}^{K} w(i, j_q)$$

energy. Since, $e_{wireless} \leq e_{wired}$, the reduction in energy needed to broadcast from one node to several others in a wireless network over that needed in a wired network is referred to as the wireless broadcast advantage.

To broadcast from a source s, use a broadcast tree T, which is a spanning tree of G that is rooted at s. The energy, e(u), required by a node of T to broadcast to its children is

$$e(u) = \max\{w(u, v) \mid v \text{ is a child of } u\}$$

Note that for a leaf node u, e(u) = 0. The energy, e(T), required by the broadcast tree to broadcast a unit message from the source to all other nodes is,

$$e(T) = \sum_{(u)}^{K} e(u)$$

The DSA (Dijkstra's shortest paths algorithm) heuristic constructs a shortest path from the source node s to every other vertex in G. The constructed shortest paths are superimposed to obtain a tree T rooted at s. Finally, a sweep is performed over the nodes of T. In this sweep, nodes are examined in ascending order of their index (i.e., in the order 1, 2, 3, ..., n). The transmission energy τ(i) for node i is determined to be

$$\tau(i) = \max\{w(i, j) \mid j \text{ is a child of } i \text{ in } T\}$$

If using τ(i) energy, node i is able to reach any descendents other than its children, then these descendents are promoted in the broadcast tree T and become additional children of i. The MST (minimum spanning tree) uses Prim's algorithm to construct a minimum-cost spanning tree (the cost of a spanning tree is the sum of its edge weights). The constructed spanning tree is restructured by performing a sweep over the nodes to reduce the total energy required by the tree.

The BIP (broadcast incremental power) heuristic begins with a tree T that comprises only the source node s. The remaining nodes are added to T one node at a time. The next node u to add to T is selected so that u is a neighbor of a node in T and e (T U{u}) – e(T) is minimum. Once the broadcast tree is constructed, a sweep is done to restructure the tree so as to reduce the required energy.

Data Collection and Distribution

In the data collection problem, a base station is to collect sensed data from all deployed sensors. The data distribution problem is the inverse problem in which the base station has to send data to the deployed sensors (different sensors receive different data from the base station). In both of these problems, the objective is to complete the task in the smallest amount of time. The data collection and distribution problems are symmetric. Hence, one can derive an optimal data collection algorithm from an optimal data distribution algorithm and vice versa. Therefore, it is necessary to study just one of these two problems explicitly. Here we focus on data distribution.

Let S_1, ..., S_n be n sensors and let S_0 represent the base station. Let p_i be the number of data packets the base station has to send to sensor i, $1 \leq i \leq n$. $p = [p_1, p_2, ..., p_n]$ is the transmission vector.

The distribution of these packets to the sensors is done in a synchronous time-slotted fashion. In each time slot, an S_i may either receive or transmit (but not both) a packet. To facilitate the transmission of the packets, each S_i has an antenna whose range is r. In the unidirectional antenna model, S_i receives a packet only if that packet is sent in its direction from an antenna located at most r away. Because of interference, a transmission from S_i to S_j is successful if the following are true:

1. j is in range, that is $d(i, j) \leq r$, where $d(i, j)$ is the distance between S_i and S_j

2. j is not, itself, transmitting in this time slot

3. There is no interference from other transmissions in the direction of j

Formally, every S_k, $k \neq i$, that is transmitting in this time slot in the direction of S_j is out of range. Here, out of range means $d(k, j) \geq (1 + \delta)r$, where $\delta > 0$ is an interference constant.

In the omni directional antenna model, a packet transmitted by an S_i is received by all S_j (regardless of direction) that are in the antenna's range. The constraints on successful transmission are the same as those for the unidirectional antenna model except that all references to "direction" are dropped.

Sensor Fusion

The reliability of a sensor system is enhanced through the use of redundancy. That is, each point or region is monitored by multiple sensors. A redundant sensor system is faced with the problem of fusing or combining the data reported by each of the sensors monitoring a specified point or region. Suppose that $k > 1$ sensors monitor point p. Let m_i, $1 \leq i \leq k$ be the measurement recorded by sensor i for point p. These k measurements may differ because of inherent differences in the k sensors, the relative location of a sensor with respect to p, as well as because one or more sensors is faulty. Let V be the real value for p. The objective of sensor fusion is to take the k measurements, some of which may be faulty, and determine either the correct measurement V or a range in which the correct measurement lies.

10.2 Routing Protocols

All applications of sensor networks have the requirement of sending the sensed data from multiple points to a common destination called a sink. Resource management is required in sensor nodes regarding transmission power, storage, on-board energy, and processing capacity. There are various routing protocols that have been proposed for routing data in wireless sensor networks due to such problems. The proposed mechanisms of routing consider the architecture and application requirements along with the characteristics of sensor nodes. There are few distinct routing protocols that are based on quality of service awareness or network flow, whereas all other routing protocols can be classified as hierarchical or location based and data-centric.

The routing protocols which are data-centric are based on query and depend on the naming of desired data due to which many redundant transmissions are eliminated. The clustering of nodes in hierarchical routing protocols aims to save the energy by cluster heads that can do some aggregation and reduction of data. The routing protocols that are location based relay data to the desired destination instead of the whole network by utilizing positioning information. In some applications there is requirement of QoS along with the routing functions that are based on network flow modeling.

The other factors which effect routing design are the overhead and data latency. Data latency during network latency is caused by data aggregation and multi-hop relays due to which real-time data is infeasible in these protocols, while in some protocols there are excessive overheads created for the implementation of their algorithm which are not suitable for the networks that energy constrained. So data latency and overhead are the two important factors which affect the designing of routing protocols of WSNs.

Data-Centric Protocols

The sink is used to send queries to certain regions and waits for data from sensors that are located in selected regions in data-centric routing protocols. As queries are used for the requested data, attribute-based naming in order to specify the properties of data is necessary. The first data-centric routing protocol between nodes that considers data negotiation is Sensor protocol for information via negotiation (SPIN) for energy saving and elimination of redundant data.

Flooding and Gossiping

In order to relay data in sensor networks without the need for any routing algorithms and topology maintenance, the two classical methods are flooding and gossiping. A sensor node broadcasts a data packet to all its neighbors, and this process continues until the destination is found, and this technique is known as flooding. In gossiping the packet is not sent to all neighboring nodes but to selected random neighbors which select another random neighbor and in this packet arrives at the destination.

Sensor Protocols for Information via Negotiation

The key feature of SPIN is that meta-data before transmission are exchanged between sensors through a data advertisement mechanism. The new data is advertised by each sensor node to its neighbors, and the interested neighbors which do not have the data send a request message in order to retrieve data. The classic problems of flooding are solved by SPIN's meta-data negotiation.

Directed Diffusion

In this protocol the idea is to diffuse data by using a naming scheme for the data through sensor nodes. To get rid of unnecessary operations of network layer routing in order to save energy is the main idea behind using such a scheme.

Energy-Aware Routing

To increase the lifespan of a network, the method of using set of sub-optimal paths occasionally is introduced. Depending on the energy consumption of the path, these paths are chosen by means of probability functions. The approach is concerned with the main metric of network survivability. This protocol has the following phases:

- Setup phase
- Data communication and route maintenance phase

Rumor Routing

Another variation of Directed Diffusion is rumor routing, which is proposed for contexts in which geographic routing criteria are not applicable.

The query is flooded in the entire network in Directed Diffusion when there is no geographic criterion to diffuse tasks. Thus, the use of flooding is unnecessary in cases where a little amount of data is requested.

Gradient-based Routing

Gradient based routing (GBR) is to maintain a number of hops when the interest is diffused through the network. So minimum numbers of hops are discovered by each hop to the sink that are called the node's height. The gradient is the difference between the node's height and that of its neighbor on that link. With the largest gradient a packet is forwarded on the link.

Hierarchical Protocols

The nodes in hierarchical routing are involved in multi-hop communication within a particular cluster in order to efficiently maintain the energy consumption, and the transmitted messages to the sink are decreased by performing data aggregation and fusion. The formation of a cluster is typically based on a sensor's proximity to the cluster and the energy reserve of sensors. Networking clustering has been pursued in some routing approaches in order to allow the system to cope with additional load and enable to cover a large area of interest without degrading the service. The following are the hierarchical routing protocols:

LEACH (Low Energy Adaptive Clustering Hierarchy)
PEGASIS (Power Efficient Gathering in Sensor Information System) and Hierarchical-PEGASIS
TEEN (Threshold Sensitive Energy Efficient Sensor Network Protocol) and APTEEN. Energy-aware routing for cluster-based sensor networks & Self-organizing protocol

Location-based Protocols

Location information is required for nodes in sensor networks in most of the routing protocols. Energy consumption is estimated by calculating the distance between two particular nodes for which location information is required. As there are no schemes like IP addresses, data is routed in an energy-efficient way by utilizing location information. By using the location of sensors, the query is diffused only in a particular region which is known to be sensed, and a significant number of transmissions will be eliminated. The protocols are designed primarily for MANETs considering the mobility of nodes, whereas they are also applicable to sensor networks in which nodes are fixed or mobility is less. Location-based protocols are as follows:

- Minimum energy communication network (MECN) and small minimum communication energy network (SMECN)

- Geographic Adaptive Fidelity (GAF)

- Geographic and Energy aware routing (GEAR)

Network flow and QoS-aware Protocols

Among the various routing protocols proposed for sensor networks most of them fit in the classification; however, some pursue a somewhat different approach such as QoS and network flow. While setting up the paths in a sensor network, end-to-end delay requirements are considered in QoS-aware protocols. These protocols are:

- Maximum lifespan energy routing

- Maximum lifespan data gathering

- Minimum cost forwarding

- Sequential assignment routing (SAR)

- Energy-aware QoS routing protocol

- SPEED (A Stateless Protocol for Realtime Communication in Sensor Networks)

AODV Routing Protocol

There are two types of routing protocols, which are reactive and proactive. In reactive routing protocols the routes are created only when source wants to send data to a destination, whereas proactive routing protocols are table driven. Because a reactive routing protocol Ad hoc On-demand Distance Vector uses traditional routing tables, one entry per destination and sequence numbers are used to determine whether routing information is up-to-date and to prevent routing loops. The maintenance of time-based states is an important feature of AODV, which means that a routing entry which is not recently used is expired. The neighbors are notified in case of route breakage. The discovery of the route from source to destination is based on query and reply cycles, and intermediate nodes store the route information in the form of route table entries along the route. Control messages used for the discovery and breakage of the route are as follows:

- Route Request Message (RREQ)
- Route Reply Message (RREP)
- Route Error Message (RERR)
- HELLO Messages

Route Request (RREQ)

A route request packet is flooded through the network when a route is not available for the destination from the source. The parameters contained in the route request packet are presented in the following table:

Table 10.1 Route Request Parameters

Source Address	Request ID	Source Sequence Number	Destination Address	Destination Sequence Number	Hop count

A RREQ is identified by the pair source address and request ID; each time when the source node sends a new RREQ, the request ID is incremented. After receiving a request message, each node checks the request ID and source address pair. The new RREQ is discarded if there is already an RREQ packet with same pair of parameters. A node that has no route entry for the destination rebroadcasts the RREQ with an incremented hop count parameter. A route reply (RREP) message is generated and sent back to the source if a node has a route with a sequence number greater than or equal to that of RREQ.

Route Reply (RREP)

On having a valid route to the destination or if the node is the destination, an RREP message is sent to the source by the node. The following parameters are contained in the route reply message:

Table 10.2 Route Reply Parameters

Source Address	Destination Address	Destination Sequence Number	Hop Count	Life Time

Route Error Message (RERR)

The neighborhood nodes are monitored. When a route that is active is lost, the neighborhood nodes are notified by route error message (RERR) on both sides of the link.

Hello Messages

The HELLO messages are broadcast in order to know neighborhood nodes. The neighborhood nodes are directly communicated. In AODV, HELLO messages are broadcast in order to inform the neighbors about the activation of the link. These messages are not broadcast because of short time to live (TTL) with a value equal to one.

Discovery of Route

When a source node does not have routing information about destination, the process of the discovery of the route starts for a node with which the source wants to communicate. The process is initiated by the broadcasting of an RREQ, as shown in Figure 10.2. On receiving the RREP message, the route is established. If multiple RREP messages with different routes are received, then routing information is updated with an RREP message of a greater sequence number.

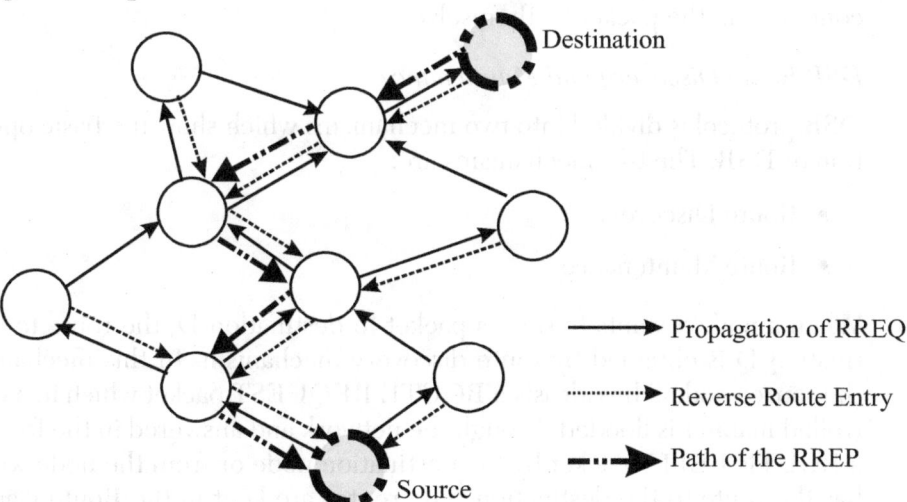

FIGURE 10.2 Discovery of route.

Setup of Reverse Path

The reverse path to the node is noted by each node during the transmission of RREQ messages. The RREP message travels along this path after the destination node is found. The addresses of the neighbors from which the RREQ packets are received are recorded by each node.

Setup of Forward Path

The reverse path is used to send RREP messages back to the source, but a forward path is set up during transmission of the RREP message. This forward path can be called a reverse to the reverse path. The data transmission is started as soon as this forward path is set up. The locally buffered data packets waiting for transmission are transmitted in a FIFO-queue.

DSR Routing Protocol

The Dynamic Source Routing (DSR) protocol is specifically designed for multi-hop ad hoc networks. The difference in DSR and other routing protocols is that it uses source routing supplied by the packet's originator to determine the packet's path through the network instead of independent hop-by-hop routing decisions made by each node. The packet in source routing which is going to be routed through the network carries the complete ordered list of nodes in its header through which the packet will pass. Fresh routing information is not needed to be maintained in intermediate nodes in the design of source routing, since all the routing decisions are contained in the packet by themselves.

DSR Route Discovery and Maintenance

DSR protocol is divided into two mechanisms which show the basic operation of DSR. The two mechanisms are:

- Route Discovery
- Route Maintenance

When a node S wants to send a packet to destination D, the route to destination D is obtained by route discovery mechanism. In this mechanism the source node S broadcasts a ROUTE REQUEST packet which in a controlled manner is flooded through the network and answered in the form of a ROUTE REPLY packet by the destination node or from the node which has the route to the destination. The routes are kept in the Route Cache, which to the same destination can store multiple routes. The nodes check their route cache for a route that could answer the request before repropagation of ROUTE REQUEST. The nodes do not expend effort on obtaining or maintaining the routes that are not currently used for communication; that is, the route discovery is initiated only on-demand.

The other mechanism is the route maintenance by which source node S detects if the topology of the network has changed so that it can no longer use its route to the destination. If the two nodes that were listed as neighbors on the route moved out of the range of each other and the link becomes broken, the source node S is notified with a ROUTE ERROR packet. The source node S can use any other known routes to the destination D, or the process of route discovery is invoked again to find a new route to the destination. This is shown in Figure 10.3.

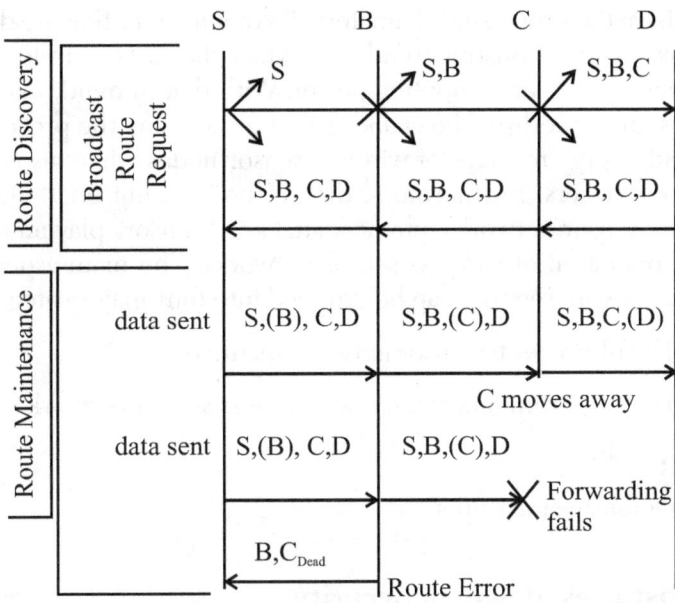

FIGURE 10.3 DSR Route discovery and maintenance.

10.3 Security in Wireless Sensor Networks

As wireless sensor networks continue to grow, so does the need for effective security mechanisms. Because sensor networks may interact with sensitive data and/or operate in hostile unattended environments, it is imperative that these security concerns be addressed from the beginning of the system design. However, due to inherent resource and computing constraints, security in sensor networks poses different challenges from traditional network/computer security. There is currently enormous research potential in the field of wireless sensor network security. Thus, familiarity with the current research in this field will benefit researchers greatly.

Wireless sensor networks are quickly gaining popularity due to the fact that they are potentially low-cost solutions to a variety of real-world challenges. Their low cost provides a means to deploy large sensor arrays in a variety of conditions capable of performing both military and civilian tasks. But sensor networks also introduce severe resource constraints due to their lack of data storage and power. Both of these represent major obstacles to the implementation of traditional computer security techniques in a wireless sensor network. The unreliable communication channel and unattended operation make the security defenses even harder. Indeed, wireless sensors often have the processing characteristics of machines that are decades old (or longer), and the industrial trend is to reduce the cost of wireless sensors while maintaining similar computing power. With that in mind, many researchers have begun to address the challenges of maximizing the processing capabilities and energy reserves of wireless sensor nodes while also securing them against attackers. Furthermore, due to the inherent unattended feature of wireless sensor networks, physical attacks to sensors play an important role in the operation of wireless sensor networks. The main aspects of wireless sensor network security can be grouped into four major categories:

- The obstacles to sensor network security

- The requirements of a secure wireless sensor network

- Attacks

- Defensive measures

10.4 Obstacles of Sensor Security

A wireless sensor network is a special network which has many constraints compared to a traditional computer network. Due to these constraints it is difficult to directly employ the existing security approaches to the area of wireless sensor networks. Therefore, to develop useful security mechanisms while borrowing the ideas from the current security techniques, it is necessary to know and understand these constraints first.

Very Limited Resources

All security approaches require a certain amount of resources for implementation, including data memory, code space, and energy to power the sensor. However, currently these resources are very limited in a tiny wireless sensor.

Limited Memory and Storage Space

A sensor is a tiny device with only a small amount of memory and storage space for the code. In order to build an effective security mechanism, it is necessary to limit the code size of the security algorithm. For example, one common sensor type (TelosB) has an 16-bit, 8 MHz RISC CPU with only 10K RAM, 48K program memory, and 1024K flash storage. With such a limitation, the software built for the sensor must also be quite small. The total code space of TinyOS, the operating system for wireless sensors, is approximately 4K, and the core scheduler occupies only 178 bytes. Therefore, the code size for the all security related code must also be small.

Power Limitation

Energy is the biggest constraint to wireless sensor capabilities. Assume that once sensor nodes are deployed in a sensor network, they cannot be easily replaced (high operating cost) or recharged (high cost of sensors). Therefore, the battery charge taken with them to the field must be conserved to extend the life of the individual sensor node and the entire sensor network. When implementing a cryptographic function or protocol within a sensor node, the energy impact of the added security code must be considered. When adding security to a sensor node, one also has to be interested in the impact that security has on the lifespan of a sensor (i.e., its battery life). The extra power consumed by sensor nodes due to security is related to the processing required for security functions (e.g., encryption, decryption, signing data, verifying signatures), the energy required to transmit the security-related data or overhead (e.g., initialization vectors needed for encryption/decryption), and the energy required to store security parameters in a secure manner (e.g., cryptographic key storage).

Unreliable Communication

Certainly, unreliable communication is another threat to sensor security. The security of the network relies heavily on a defined protocol, which in turn depends on communication.

Unreliable Transfer

Normally the packet-based routing of the sensor network is connectionless and thus inherently unreliable. Packets may get damaged due to channel errors or dropped at highly congested nodes. The result is lost or missing packets. Furthermore, the unreliable wireless communication channel also results in damaged packets. Higher channel error rate also forces the software developer to devote resources to error handling. More importantly, if the protocol lacks the appropriate error handling it is possible to lose critical security packets. This may include, for example, a cryptographic key.

Conflicts

Even if the channel is reliable, the communication may still be unreliable. This is due to the broadcast nature of the wireless sensor network. If packets meet in the middle of transfer, conflicts will occur and the transfer itself will fail. In a crowded (high density) sensor network, this can be a major problem.

Latency

The multi-hop routing, network congestion and node processing can lead to greater latency in the network, thus making it difficult to achieve synchronization among sensor nodes. The synchronization issues can be critical to sensor security, where the security mechanism relies on critical event reports and cryptographic key distribution.

Unattended Operation

Depending on the function of the particular sensor network, the sensor nodes may be left unattended for long periods of time.

Exposure to Physical Attacks

The sensor may be deployed in an environment open to adversaries, bad weather, and so on. The likelihood that a sensor suffers a physical attack in such an environment is therefore much higher than typical PCs, which are located in a secure place and mainly face attacks from a network.

Managed Remotely

Remote management of a sensor network makes it virtually impossible to detect physical tampering (i.e., through tamperproof seals) and physical maintenance issues (e.g., battery replacement). Perhaps the most extreme example of this is a sensor node used for remote reconnaissance missions behind enemy lines. In such a case, the node may not have any physical contact with friendly forces once deployed.

No Central Management Point

A sensor network should be a distributed network without a central management point. This will increase the vitality of the sensor network. However, if designed incorrectly, it will make the network organization difficult, inefficient, and fragile.

10.5 Security Requirements

A sensor network is a special type of network. It shares some commonalities with a typical computer network, but also poses unique requirements of its own.

Data Confidentiality

Data confidentiality is the most important issue in network security. Every network with any security focus will typically address this problem first. In sensor networks, the confidentiality relates to the following:

- A sensor network should not leak sensor readings to its neighbors. Especially in a military application, the data stored in the sensor node may be highly sensitive.

- In many applications nodes communicate highly sensitive data, e.g., key distribution, therefore it is extremely important to build a secure channel in a wireless sensor network.

- Public sensor information, such as sensor identities and public keys, should also be encrypted to some extent to protect against traffic analysis attacks. The standard approach for keeping sensitive data secret is to encrypt the data with a secret key that only intended receivers possess, thus achieving confidentiality.

Data Integrity

With the implementation of confidentiality, an adversary may be unable to steal information. However, this doesn't mean the data is safe. The adversary can change the data, so as to send the sensor network into disarray. For example, a malicious node may add some fragments or manipulate the data within a packet. This new packet can then be sent to the original receiver. Data loss or damage can even occur without the presence of a malicious node due to the harsh communication environment. Thus, data integrity ensures that any received data has not been altered in transit.

Data Freshness

Even if confidentiality and data integrity are assured, it also needs to ensure the freshness of each message. Informally, data freshness suggests that the data is recent, and it ensures that no old messages have been replayed. This requirement is especially important when there are shared-key strategies employed in the design. Typically shared keys need to be changed over time. However, it takes time for new shared keys to be propagated to the entire network. In this case, it is easy for the adversary to use a replay attack. Also, it is easy to disrupt the normal work of the sensor, if the sensor is unaware of the new key change time. To solve this problem a time-related counter can be added into the packet to ensure data freshness.

Availability

Adjusting the traditional encryption algorithms to fit within the wireless sensor network is not free, and will introduce some extra costs. Some approaches choose to modify the code to reuse as much code as possible. Some approaches try to make use of additional communication to achieve the same goal. Some approaches force strict limitations on the data access, or propose an unsuitable scheme (such as a central point scheme) in order to simplify the algorithm. But all these approaches weaken the availability of a sensor and sensor network for the following reasons:

- Additional computation consumes additional energy. If no more energy exists, the data will no longer be available.

- Additional communication also consumes more energy. As communication increases so too does the chance of incurring a communication conflict.

- A single point failure will be introduced if using the central point scheme. This greatly threatens the availability of the network. The requirement of security not only affects the operation of the network, but also is highly important in maintaining the availability of the whole network.

Self-organization

A wireless sensor network is a typically an ad hoc network, which requires every sensor node be independent and flexible enough to be self-organizing and self-healing according to different situations. There is no fixed infrastructure available for the purpose of network management in a sensor network. This inherent feature brings a great challenge to wireless sensor network security as well. For example, the dynamics of the whole network inhibit the idea of pre-installation of a shared key between the base station and all sensors. In the context of applying public-key cryptography techniques in sensor networks, an efficient mechanism for public-key distribution is necessary as well. In the same way that distributed sensor networks must self-organize to support multi-hop routing, they must also self-organize to conduct key management and build trust relations among sensors. If self-organization is lacking in a sensor network, the damage resulting from an attack or even the hazardous environment may be devastating.

Time Synchronization

Most sensor network applications rely on some form of time synchronization. In order to conserve power, an individual sensor's radio may be turned off for periods of time. Furthermore, sensors may wish to compute the end-to-end delay of a packet as it travels between two pair-wise sensors. A more collaborative sensor network may require group synchronization for tracking applications, and so on.

Secure Localization

Often, the utility of a sensor network will rely on its ability to accurately and automatically locate each sensor in the network. A sensor network designed to locate faults will need accurate location information in order to pinpoint the location of a fault. Unfortunately, an attacker can easily manipulate non-secured location information by reporting false signal strengths, replaying signals, and so forth. A device's position is accurately computed from a series of known reference points. In this system, authenticated ranging and distance bounding are used to ensure accurate location of a node. Because of distance bounding, an attacking node can only increase its claimed distance from a reference point. However, to ensure location consistency, an attacking node would also have to prove that its distance from another reference point is shorter. Since it cannot do this, a node manipulating the localization protocol can be found.

The SeRLoc (Secure Range-Independent Localization) algorithm uses locators that transmit beacon information. It is assumed that the locators are trusted and cannot be compromised. Furthermore, each locator is assumed to know its own location. A sensor computes its location by listening for the beacon information sent by each locator. The beacons include the locator's location. Using all of the beacons that a sensor node detects, a node computes an approximate location based on the coordinates of the locators. Using a majority vote scheme, the sensor then computes an overlapping antenna region. The final computed location is the "center of gravity" of the overlapping antenna region. All beacons transmitted by the locators are encrypted with a shared global symmetric key that is pre-loaded to the sensor prior to deployment. Each sensor also shares a unique symmetric key with each locator. This key is also pre-loaded on each sensor.

Authentication

An adversary is not just limited to modifying the data packet. It can change the whole packet stream by injecting additional packets. So, the receiver needs to ensure that the data used in any decision-making process originates from the correct source. On the other hand, when constructing the sensor network, authentication is necessary for many administrative tasks (e.g., network reprogramming or controlling sensor node duty cycle). As discussed previously, message authentication is important for many applications in sensor networks. Informally, data authentication allows a receiver to verify that the data really is sent by the claimed sender. In the case of two-party communication, data authentication can be achieved through a purely symmetric mechanism: the sender and the receiver share a secret key to compute the message authentication code (MAC) of all communicated data.

The basic idea of the μTESLA, another system, is to achieve asymmetric cryptography by delaying the disclosure of the symmetric keys. In this case a sender will broadcast a message generated with a secret key. After a certain period of time, the sender will disclose the secret key. The receiver is responsible for buffering the packet until the secret key has been disclosed. After disclosure the receiver can authenticate the packet, provided that the packet was received before the key was disclosed. One limitation of μTESLA is that some initial information must be unicast to each sensor node before authentication of broadcast messages can begin. An enhancement to the μTESLA system is that it uses broadcasting of the key chain commitments rather than μTESLA's unicasting technique. They present a series of schemes starting with a simple pre-determination of key chains and finally settling on a multi-level key chain technique. The multi-level key chain scheme uses pre-determination and broadcasting to achieve a scalable key distribution technique that is designed to be resistant to denial of service attacks, including jamming.

10.6 Attacks

Sensor networks are particularly vulnerable to several key types of attacks. Attacks can be performed in a variety of ways, most notably as denial of service attacks, but also through traffic analysis, privacy violation, physical attacks, and so on. Denial of service attacks on wireless sensor networks can range from simply jamming the sensor's communication channel to more sophisticated attacks designed to violate the 802.11 MAC protocol or any other layer of the wireless sensor network. Due to the potential asymmetry in power and computational constraints, guarding against a well orchestrated denial of service attack on a wireless sensor network can be nearly impossible. A more powerful node can easily jam a sensor node and effectively prevent the sensor network from performing its intended duty. Attacks on wireless sensor networks are not limited to simply denial of service attacks, but rather encompass a variety of techniques including node takeovers, attacks on the routing protocols, and attacks on a node's physical security.

Definition

A denial of service attack defined as "any event that diminishes or eliminates a network's capacity to perform its expected function." Unfortunately, wireless sensor networks cannot afford the computational overhead necessary in implementing many of the typical defensive strategies. What makes the prospect of denial of service attacks even more alarming is the projected use of sensor networks in highly critical and sensitive applications. For example, a sensor network designed to alert building occupants in the event of a fire could be highly susceptible to a denial of service attack. Even worse, such an attack could result in the deaths of building occupants due to the non-operational fire detection network. Other possible uses for wireless sensors include the monitoring of traffic flows, which may include the control of traffic lights, and so forth. A denial of service attack on such a sensor network could prove very costly, especially on major roads.

Types of Denial of Service attacks

A standard attack on wireless sensor networks is simply to jam a node or set of nodes. Jamming, in this case, is simply the transmission of a radio signal that interferes with the radio frequencies being used by the sensor network. The jamming of a network can come in two forms: constant jamming, and intermittent jamming. Constant jamming involves the complete jamming of the entire network. No messages are able to be sent or received. If the jamming is only intermittent, then nodes are able to exchange messages periodically, but not consistently. This too can have a detrimental impact on the sensor network as the messages being exchanged between nodes may be time sensitive.

Attacks can also be made on the link layer itself. One possibility is that an attacker may simply intentionally violate the communication protocol, for example, Zigbee or IEEE 801.11b (Wi-Fi) protocol, and continually transmit messages in an attempt to generate collisions. Such collisions would require the retransmission of any packet affected by the collision. Using this technique it would be possible for an attacker to simply deplete a sensor node's power supply by forcing too many retransmissions. At the routing layer, a node may take advantage of a multi-hop network by simply refusing to route messages. This could be done intermittently or constantly with the net result being that any neighbor who routes through the malicious node will be unable to exchange messages with, at least, part of the network. The transport layer is also susceptible to attack, as in the case of flooding. Flooding can be as simple as sending many connection requests to a susceptible node. In this case, resources must be allocated to handle the connection request. Eventually a node's resources will be exhausted, thus rendering the node useless.

The Sybil Attack

The Sybil Attack is described as it relates to wireless sensor networks. Simply put, the Sybil attack is defined as a "malicious device illegitimately taking on multiple identities." It was originally described as an attack able to defeat the redundancy mechanisms of distributed data storage systems in peer-to-peer networks. In addition to defeating distributed data storage systems, the Sybil attack is also effective against routing algorithms, data aggregation, voting, fair resource allocation, and foiling misbehaviour detection. Regardless of the target (voting, routing, aggregation), the Sybil algorithm functions similarly. All of the techniques involve utilizing multiple identities. For instance, in a sensor network voting scheme, the Sybil attack might utilize multiple identities to generate additional "votes." Similarly, to attack the routing protocol, the Sybil attack would rely on a malicious node taking on the identity of multiple nodes, and thus routing multiple paths through a single malicious node.

Traffic Analysis Attacks

Wireless sensor networks are typically composed of many low-power sensors communicating with a few relatively robust and powerful base stations. It is not unusual, therefore, for data to be gathered by the individual nodes where it is ultimately routed to the base station. Often, for an adversary to effectively render the network useless, the attacker can simply disable the base station.

To make matters worse, two attacks can identify the base station in a network (with high probability) without even understanding the contents of the packets (if the packets are themselves encrypted). A rate monitoring attack simply makes use of the idea that nodes closest to the base station tend to forward more packets than those farther away from the base station. An attacker need only monitor which nodes are sending packets and follow those nodes that are sending the most packets. In a time correlation attack, an adversary simply generates events and monitors to whom a node sends its packets. To generate an event, the adversary could simply generate a physical event that would be monitored by the sensor(s) in the area (turning on a light, for instance).

Node Replication Attacks

Conceptually, a node replication attack is quite simple: an attacker seeks to add a node to an existing sensor network by copying (replicating) the node ID of an existing sensor node. A node replicated in this fashion can severely disrupt a sensor network's performance: packets can be corrupted or even misrouted. This can result in a disconnected network, false sensor readings, and so forth. If an attacker can gain physical access to the entire network, he can copy cryptographic keys to the replicated sensor and can also insert the replicated node into strategic points in the network. By inserting the replicated nodes at specific network points, the attacker could easily manipulate a specific segment of the network, perhaps by disconnecting it altogether.

Attacks Against Privacy

Sensor network technology promises a vast increase in automatic data collection capabilities through efficient deployment of tiny sensor devices. While these technologies offer great benefits to users, they also exhibit significant potential for abuse. Particularly relevant concerns are privacy problems, since sensor networks provide increased data collection capabilities. Adversaries can use even seemingly innocuous data to derive sensitive information if they know how to correlate multiple sensor inputs. For example, in the "panda-hunter problem," the hunter can infer the position of pandas by monitoring the traffic. The main privacy problem, however, is not that sensor networks enable the collection of information. In fact, much information from sensor networks could probably be collected through direct site surveillance. Rather, sensor networks aggravate the privacy problem because they make large volumes of information easily available through remote access. Hence, adversaries need not be physically present to maintain surveillance. They can gather information in a low-risk, anonymous manner. Remote access also allows a single adversary to monitor multiple sites simultaneously. Some of the more common attacks against sensor privacy are:

Monitor and Eavesdropping

This is the most obvious attack to privacy. By listening to the data, the adversary could easily discover the communication contents. When the traffic conveys the control information about the sensor network configuration, which contains potentially more detailed information than accessible through the location server, the eavesdropping can act effectively against the privacy protection.

Traffic Analysis

Traffic analysis typically combines with monitoring and eavesdropping. An increase in the number of transmitted packets between certain nodes could signal that a specific sensor has registered activity. Through the analysis on the traffic, some sensors with special roles or activities can be effectively identified.

Camouflage

Adversaries can insert their node or compromise the nodes to hide in the sensor network. After that these nodes can masquerade as a normal node to attract the packets, then misroute the packets, for example, forward the packets to the nodes conducting the privacy analysis. It is worth noting that the current understanding of privacy in wireless sensor networks is immature, and more research is needed.

Physical Attacks

Sensor networks typically operate in hostile outdoor environments. In such environments, the small form factor of the sensors, coupled with the unattended and distributed nature of their deployment, make them highly susceptible to physical attacks, that is, threats due to physical node destruction. Unlike many other attacks mentioned previously, physical attacks destroy sensors permanently, so the losses are irreversible. For instance, attackers can extract cryptographic secrets, tamper with the associated circuitry, modify programming in the sensors, or replace them with malicious sensors under the control of the attacker.

10.7 Defensive Measures

The measures for satisfying security requirements, and protecting the sensor network from attacks, are as follows:

- key establishment in wireless sensor networks, which lays the foundation for the security in a wireless sensor network, followed by

- defending against DoS attacks,

- secure broadcasting and multicasting,

- defending against attacks on routing protocols,

- defending against attacks on sensor privacy,

- intrusion detection,

- secure data aggregation,

- defending against physical attacks, and

- trust management.

Key Establishment

One security aspect that receives a great deal of attention in wireless sensor networks is the area of key management. Wireless sensor networks are unique (among other embedded wireless networks) in this aspect due to their size, mobility, and computational/power constraints. This makes secure key management an absolute necessity in most wireless sensor network designs. Encryption and key management/establishment are so crucial to the defense of a wireless sensor network, with nearly all aspects of wireless sensor network defenses relying on solid encryption.

Key establishment is done using one of many public-key protocols. Key exchange techniques use asymmetric cryptography, also called public key cryptography. In this case, it is necessary to maintain two mathematically related keys, one of which is made public while the other is kept private. This allows data to be encrypted with the public key and decrypted only with the private key. The problem with asymmetric cryptography, in a wireless sensor network, is that it is typically too computationally intensive for the individual nodes in a sensor network. Symmetric cryptography is therefore the typical choice for applications that cannot afford the computational complexity of asymmetric cryptography. Symmetric schemes utilize a single shared key known only between the two communicating hosts. This shared key is used for both encrypting and decrypting data. The traditional example of symmetric cryptography is DES (Data Encryption Standard). The use of DES, however, is quite limited due to the fact that it can be broken relatively easily. In light of the shortcomings of DES, other symmetric cryptography systems have been proposed including 3DES (Triple DES), RC5, AES, and so on.

One major shortcoming of symmetric cryptography is the key exchange problem. Simply put, the key exchange problem derives from the fact that two communicating hosts must somehow know the shared key before they can communicate securely. So the problem that arises is how to ensure that the shared key is indeed shared between the two hosts who wish to communicate and no other rogue hosts who may wish to eavesdrop. How to distribute a shared key securely to communicating hosts is a non-trivial problem since pre-distributing the keys is not always feasible.

Key Establishment and Associated Protocols

Random key pre-distribution schemes have several variants. A key pre-distribution scheme is used which relies on probabilistic key sharing among nodes within the sensor network. Their system works by distributing a key ring to each participating node in the sensor network before deployment. Each key ring should consist of a number of randomly chosen keys from a much larger pool of keys generated offline. Using this technique, it is not necessary that each pair of nodes share a key. However, any two nodes that do share a key may use the shared key to establish a direct link to one another.

The LEAP (Light weight Extensible Authentication Protocol) protocol takes an approach that utilizes multiple keying mechanisms. Their observation is that no single security requirement accurately suits all types of communication in a wireless sensor network. Therefore, four different keys are used, depending on whom the sensor node is communicating with. Sensors are preloaded with an initial key from which further keys can be established. As a security precaution, the initial key can be deleted after its use in order to ensure that a compromised sensor cannot add additional compromised nodes to the network.

In PIKE (Peer Intermediaries for Key Establishment in Sensor Network) a mechanism is used for establishing a key between two sensor nodes that is based on the common trust of a third node somewhere within the sensor network. The nodes and their shared keys are spread over the network such that for any two nodes A and B, there is a node C that shares a key with both A and B. Therefore, the key establishment protocol between A and B can be securely routed through C. A hybrid key establishment scheme makes use of the difference in computational and energy constraints between a sensor node and the base station. They posit that an individual sensor node possesses far less computational power and energy than a base station. In light of this, they propose placing the major cryptographic burden on the base station where the resources tend to be greater. On the sensor side, symmetric-key operations are used in place of their asymmetric alternatives.

The sensor and the base station authenticate based on elliptic curve cryptography. Elliptic curve cryptography is often used in sensors due to the fact that relatively small key lengths are required to achieve a given level of security. Use certificates to establish the legitimacy of a public key. The certificates are based on an elliptic curve implicit certificate scheme. Such certificates are useful to ensure both that the key belongs to a device and that the device is a legitimate member of the sensor network. Each node obtains a certificate before joining the network using an out-of-band interface.

Public Key Cryptography

Two of the major techniques used to implement public-key cryptosystems are RSA and elliptic curve cryptography (ECC). Both RSA and elliptic curve cryptography are possible using 8-bit CPUs with ECC, demonstrating a performance advantage over RSA. Another advantage is that ECC's 160 bit keys result in shorter messages during transmission compared the 1024 bit RSA keys. In particular, the point multiplication operations in ECC are an order of magnitude faster than private-key operations within RSA, and are comparable (though somewhat slower) to the RSA public-key operation.

Defending against DoS Attacks

Table 10.3 Sensor Network Layers and DOS Attacks/Defenses

Network Layer	Attacks	Defenses
Physical	Jamming	Spread-spectrum, Priority messages, lower duty cycle, region mapping, mode change
	Tampering	Tamper-proof, hiding

Network Layer	Attacks	Defenses
Link	Collision	Error correcting code
	Exhaustion	Rate limitation
	Unfairness	Small frames
Network and routing	Neglect and greed	Redundancy, probing
	Homing	Encryption
	Misdirection	Egress filtering, authorization monitoring
	Black holes	Authorization, monitoring, redundancy
Transport	Flooding	Client puzzles
	Desynchronization	Authentication

In Table 10.3 the most common layers of a typical wireless sensor network are summarized along with their attacks and defenses. Since denial of service attacks are so common, effective defenses must be available to combat them. One strategy in defending against the classic jamming attack is to identify the jammed part of the sensor network and effectively route around the unavailable portion. A two phase approach is described where the nodes along the perimeter of the jammed region report their status to their neighbors who then collaboratively define the jammed region and simply route around it.

To handle jamming at the MAC layer, nodes might utilize a MAC admission control that is rate limiting. This would allow the network to ignore those requests designed to exhaust the power reserves of a node. This, however, is not foolproof, as the network must be able to handle any legitimately large traffic volumes. Overcoming rogue sensors that intentionally misroute messages can be done at the cost of redundancy. In this case, a sending node can send the message along multiple paths in an effort to increase the likelihood that the message will ultimately arrive at its destination. This has the advantage of effectively dealing with nodes that may not be malicious, but rather may have simply failed, as it does not rely on a single node to route its messages. To overcome the transport layer flooding denial of service attack, suggest using client puzzles as an effort to discern a node's commitment to making the connection by utilizing some of their own resources. A server should force a client to commit its own resources first. Further, a server should always force a client to commit more resources up front than the server. This strategy would likely be effective as long as the client has computational resources comparable to those of the server.

Secure Broadcasting and Multicasting

The research community of wireless sensor networks has progressively reached a consensus that the major communication pattern of wireless sensor networks is broadcasting and multicasting, for example, 1-to-N, N-to-1, and M-to-N, instead of the traditional point-to-point communication on the Internet.

Traditional Broadcasting and Multicasting

Traditionally, multicasting and broadcasting techniques have been used to reduce the communication and management overhead of sending a single message to multiple receivers. In order to ensure that only certain users receive the multicast or broadcast, encryption techniques must be employed. In both a wired and wireless network this is done using cryptography. The problem then is one of key management. To handle this, several key management schemes have been devised: centralized group key management protocols, decentralized management protocols, and distributed management protocols.

In the case of the centralized group key management protocols, a central authority is used to maintain the group. Decentralized management protocols, however, divide the task of group management among multiple nodes. Each node that is responsible for part of the group management is responsible for a certain subset of the nodes in the network. In the last case, distributed key management protocols, there is no single key management authority. Therefore, the entire group of nodes are responsible for key management.

In order to efficiently distribute keys, one well-known technique is to use a logical key tree. Such a technique falls into the centralized group key management protocols. This technique has been extended to wireless sensor networks. While centralized solutions are often not ideal, in the case of wireless sensor networks a centralized solution offers some utility. Such a technique allows a more powerful base station to offload some of the computations from the less powerful sensor nodes.

Secure Multicasting

A directed diffusion-based multicast technique described for use in wireless sensor networks that also takes advantage of a logical key hierarchy. In a standard logical key hierarchy, a central key distribution center is responsible for disbursing the keys throughout the network. The key distribution center, therefore, is the root of the key hierarchy while individual nodes make up the leaves. The internal nodes of the key hierarchy contain keys that are used in the re-keying process. Directed diffusion is a data-centric, energy efficient dissemination technique that has been designed for use in wireless sensor networks. In directed diffusion, a query is transformed into an interest (due to the data-centric nature of the network). The interest is then diffused throughout the network and the network begins collecting data based on that interest.

The dissemination technique also sets up certain gradients designed to draw events toward the interest. Data collected as a result of the interest can then be sent back along the reverse path of the interest propagation. Using the above mentioned directed diffusion technique, the logical key hierarchy is enhanced to create a directed diffusion-based logical key hierarchy. The logical key hierarchy technique provides mechanisms for nodes joining and leaving groups where the key hierarchy is used to effectively re-key all nodes within the leaving node's hierarchy. The directed diffusion is also used in node joining and leaving. When a node declares an intent to join, for example, a join "interest" is generated which travels down the gradient of "interest about interest to join." When a node joins, a key set is generated for the new node based on keys within the key hierarchy. In this case, nodes are grouped based on locality and attach to a security tree. However, they assume that nodes within the mobile network are somewhat more powerful than a traditional sensor in a wireless sensor network.

Secure Broadcasting

A tree-based key distribution scheme is described and suggests a routing-aware-based tree where the leaf nodes are assigned keys based on all relay nodes above them. It takes advantage of routing information being more energy efficient than routing schemes that arbitrarily arrange nodes into the routing tree. Some schemes cases instead use geographic location information (e.g., GPS) rather than routing information. In this case, however, nodes (with the help of the geographic location system) are grouped into clusters with the observation that nodes within a cluster will be able to reach one another with a single broadcast. Using the cluster information, a key hierarchy is constructed.

Defending against Attacks on Routing Protocols

Most current research has focused primarily on providing the most energy-efficient routing. There is a great need for both secure and energy-efficient routing protocols in wireless sensor networks, as attacks such as the sinkhole, wormhole, and Sybil attacks demonstrate. As wireless sensor networks continue to grow in size and utility, routing security must not be an after thought, but rather they must be included as part of the overall sensor network design.

Because wireless sensors are designed to have widely distributed power and computationally constrained networks, efficient routing protocols must be used in order to maximize the battery life of each node. There are a variety of routing protocols in use in wireless sensor networks, so it is not possible to provide a single security protocol that will be able to secure each type of routing protocol.

In general, packet routing algorithms are used to exchange messages with sensor nodes that are outside of a particular radio range. Sensors packets within radio range can be transmitted using a single hop. In such single-hop networks security is still a concern, but it is more accurately addressed through secure broadcasting and multicasting.

The first packet routing algorithm is based on node identifiers similar to traditional routing. In this case, each sensor is identified by an address, and routing to/from the sensor is based on the address. This is generally considered inefficient in sensor networks, where nodes are expected to be addressed by their location, rather than their identifier. A data-centric network is one in which data are stored by name in the sensor network. Data with the same name are stored at the same node. In fact, data need not be stored anywhere near the sensor responsible for generating the data. When searching the network, searches are therefore based on the data's general name, rather than the identity responsible for holding the data.

Techniques for Securing the Routing Protocol

An intrusion-tolerant routing protocol is designed to limit the scope of an intruder's destruction and route despite network intrusion without having to identify the intruder. It should be noted that an intruder need not be an actual intrusion on the sensor network, but might simply be a node that is malfunctioning for no particularly malicious reason. Identifying an actual intruder versus a malfunctioning node can be extremely difficult, and for this reason make no distinction between the two. The first technique described to mitigate the damage done by a potential intruder is to simply employ the use of redundancy. In this case, multiple identical messages are routed between a source and destination. A message is sent once along several distinct paths with the hope that at least one will arrive at the destination. To discern which, if any, of the messages arriving at the destination are authentic, an authentication scheme can be employed to confirm the message's integrity.

It also makes use of an assumed asymmetry between base stations and wireless sensor nodes. They assume that the base stations are somewhat less resource constrained than the individual sensor node. For this reason, it is suggested that the base station be used to compute routing tables on behalf of the individual sensor nodes. This is done in three phases. In the first phase, the base station broadcasts a request message to each neighbor which is then propagated throughout the network. In the second phase, the base station collects local connectivity information from each node. Finally, the base station computes a series of forwarding tables for each node. The forwarding tables will include the redundancy information used for the redundant message transmission.

There are several possible attacks that can be made on the routing protocol during each of the three stages. In the first phase, a node might spoof the base station by sending a spurious request message. A malicious node might also include a fake path(s) when forwarding the request message to its neighbors. It may not even forward the request message at all. To counter this, use a scheme similar to μTESLA where one-way key chains are used to identify a message originating from the base station.

The base station is to broadcast an encrypted message to all of its neighbors. Only those neighbors who are trusted will possess the shared key necessary to decrypt the message. The trusted neighbor(s) then adds its location (for the return trip), encrypts the new message with its own shared key and forwards the message to its neighbor closest to the destination. Once the message reaches the destination, the recipient is able to authenticate the source (base station) using the MAC that will correspond to the base station. To acknowledge or reply to the message, the destination node can simply forward a return message along the same trusted path from which the first message was received.

A wormhole attack is one in which a malicious node eavesdrops on a packet or series of packets, tunnels them through the sensor network to another malicious node, and then replays the packets. This can be done to misrepresent the distance between the two colluding nodes. It can also be used to more generally disrupt the routing protocol by misleading the neighbor discovery process. Often additional hardware, such as a directional antenna, is used to defend against wormhole attacks. This, however, can be cost-prohibitive when it comes to large-scale network deployment. Instead, a visualization approach is used for identifying wormholes. It first computes distance estimation between all neighbor sensors, including possible existing wormholes. Using multi dimensional scaling, they then compute a virtual layout of the sensor network. A surface smoothing strategy is then used to adjust for round-off errors in the multi dimensional scaling. Finally, the shape of the resulting virtual network is analyzed. If a wormhole exists within the network, the shape of the virtual network will bend and curve toward the offending nodes. Using this strategy the nodes that participate in the wormhole can be identified and removed from the network. If a network does not contain a wormhole, the virtual network will appear flat.

Defending against the Sybil Attack

To defend against the Sybil attack, the network needs some mechanism to validate identity. It describes two methods to validate identities, direct validation and indirect validation. In direct validation a trusted node directly tests whether the joining identity is valid. In indirect validation, another trusted node is allowed to vouch for (or against) the validity of a joining node. It primarily describes direct validation techniques, including a radio resource test. In the radio test, a node assigns each of its neighbors a different channel on which to communicate. The node then randomly chooses a channel and listens. If the node detects a transmission on the channel, it is assumed that the node transmitting on the channel is a physical node. Similarly, if the node does not detect a transmission on the specified channel, the node assumes that the identity assigned to the channel is not a physical identity. Another technique to defend against the Sybil attack is to use random key pre-distribution techniques. The idea behind this technique is that with a limited number of keys on a key ring, a node that randomly generates identities will not possess enough keys to take on multiple identities and thus will be unable to exchange messages on the network, due to the fact that the invalid identity will be unable to encrypt or decrypt messages.

Detecting Node Replication Attacks

Two algorithms, randomized multicast and line-selected multicast, are described. Randomized multicast is an evolution of a node-broadcasting strategy. In this each sensor propagates an authenticated broadcast message throughout the entire sensor network. Any node that receives a conflicting or duplicated claim revokes the conflicting nodes. This strategy will work, but the communication cost is far too expensive. In order to reduce the communication cost, a deterministic multicast could be employed where nodes would share their locations with a set of witness nodes. In this case, witnesses are computed based on a node's ID. In the event that a node has been replicated on the network, two conflicting locations will be forwarded to the same witness who can then revoke the offending nodes. But since a witness is based on a node's ID, it can easily be computed by an attacker who can then compromise the witness nodes. Thus, securely utilizing a deterministic multicast strategy would require too many witnesses and the communication cost would be too high.

The line-selected multicast algorithm seeks to further reduce the communication costs of the randomized multicast algorithm. It is based upon rumor routing. The idea is that a location claim traveling from source s to destination d will also travel through several intermediate nodes. If each of these nodes records the location claim, then the path of the location claim through the network can be thought of as a line segment.

Defending against Attacks on Sensor Privacy

Location information that is too precise can enable the identification of a user, or make the continued tracking of movements feasible. This is a threat to privacy. Anonymity mechanisms depersonalize the data before the data is released, which present an alternative to privacy policy-based access control.

Decentralize sensitive data: The basic idea of this approach is to distribute the sensed location data through a spanning tree, so that no single node holds a complete view of the original data.

Secure communication channel: Using secure communication protocols, such as SPINS, the eavesdropping and active attacks can be prevented.

Change Data Traffic

De-patterning the data transmissions can protect against traffic analysis. For example, inserting some bogus data can intensively change the traffic pattern when needed.

Node mobility: Making the sensor movable can be effective in defending privacy, especially the location. For example, the Cricket system is a location-support system for in-building, mobile, location-dependent applications. It allows applications running on mobile and static nodes to learn their physical location by using listeners that hear and analyze information from beacons spread throughout the building. Thus the location sensors can be placed on the mobile device as opposed to the building infrastructure, and the location information is not disclosed during the position determination process and the data subject can choose the parties to which the information should be transmitted.

Policy-based Approaches

Policy-based approaches are currently a hot approach to address the privacy problem. The access control decisions and authentication are made based on the specifications of the privacy policies. In the concept of private authentication, give a general scheme for building private authentication with work logarithms in the number of tags in RFID (radio frequency identification) applications.

The automotive telematics domain has a policy-based framework for protecting sensor information, where an in-car computer can act as a trusted agent. These concepts enable access control based on criteria such as time of the request, location, speed, and identity of the located object. Architecture for a centralized location server is one that controls access from client applications through a set of validator modules that check XML-encoded application privacy policies. In another method it is pointed out that access control decisions can be governed by either room or user policies. The room policy specifies who is permitted to find out about the people currently in a room, while the user policy states who is allowed to get location information about another user.

Information Flooding

Anti-traffic analysis mechanisms are to prevent an outside attacker from tracking the location of a data source, since that information will release the location of sensed objects. The randomized data routing mechanism and phantom traffic generation mechanism are used to disguise the real data traffic, so that it is difficult for an adversary to track the source of data by analyzing network traffic. Based on flooding-based routing protocols, the following methods have been developed for single-path routing to try to solve the privacy problems in sensor networks.

Baseline Flooding

In the baseline implementation of flooding, every node in the network only forwards a message once, and no node retransmits a message that it has previously transmitted. When a message reaches an intermediate node, the node first checks whether it has received and forwarded that message before. If this is its first time, the node will broadcast the message to all its neighbors. Otherwise, it just discards the message.

Probabilistic Flooding

In probabilistic flooding, only a subset of nodes within the entire network will participate in data forwarding, while the others simply discard the messages they receive. One possible weakness of this approach is that some messages may get lost in the network and as a result affect the overall network connectivity. However, this problem does not appear to be a significant factor.

Flooding with Fake Messages

The previous flooding strategies can only decrease the chances of a privacy violation. An adversary still has a chance to monitor the general traffic and even the individual packets. This observation suggests that one approach to alleviate the risk of source-location privacy breaching is to augment the flooding protocols to introduce more sources that inject fake messages into the network. By doing so, even if the attacker captures the packets, he will have no idea whether the packets are real.

Phantom Flooding

Phantom flooding shares the same insights as probabilistic flooding in that they both attempt to direct messages to different locations of the network so that the adversary cannot receive a steady stream of messages to track the source. Probabilistic flooding is not very effective in achieving this goal because shorter paths are more likely to deliver more messages. Therefore, entice the attacker away from the real source suggested and toward a fake source, called the phantom source. In phantom flooding, every message experiences two phases: (1) a walking phase, in which may be a random walk or a directed walk, and (2) a subsequent flooding meant to deliver the message to the sink. When the source sends out a message, the message is unicast in a random fashion within the first hwalk hops (referred to as random walk phase). After the hwalk hops, the message is flooded using the baseline flooding technique (referred to as flooding phase). Similar mechanisms are also used to disguise an adversary from finding the location of a base station by analyzing network traffic. One key problem for these anti-traffic analysis mechanisms is the energy cost incurred by anonymization.

Intrusion Detection

The area of intrusion detection in wireless sensor networks is important. Many secure routing schemes attempt to identify network intruders, and key establishment techniques are used in part to prevent intruders from overhearing network data. However, resource constraints are not the only reason. As such, it is difficult to define characteristics (or signatures) that are specific to a network intrusion as opposed to the normal network traffic that might occur as the result of normal network operations or malfunctions resulting from environment change.

Traditionally, Intrusion Detection has Focused on Two Major Categories.

- Anomaly based intrusion detection (AID)

- Misuse intrusion detection (MID)

Anomaly based intrusion detection relies on the assumption that intruders will demonstrate abnormal behavior relative to the legitimate nodes. Thus, the object of anomaly based detection is to detect intrusion based on unusual system behavior. Typically this is done by first developing a profile of the system in normal use. Once the profile has been generated it can be used to evaluate the system in the face of intruders. The advantage of using an anomaly based system is that it is able to detect previously unknown attacks based only upon knowing that the system behavior is unusual. This is particularly advantageous in wireless sensor networks where it can be difficult to boil an attack down to a signature.

However, such flexible intrusion detection comes at a cost. The first is that the anomaly based approach is susceptible to false positives. This is due largely to the fact that it can be difficult to define normal system behaviors. To help combat this, new profiles can be taken of the network to ensure that the profile in use is up-to-date. However, this takes time. And further, even with the most up-to-date profile possible, it can still be difficult to discern unusual, but legitimate, behavior from an actual intrusion. Another fault in the anomaly based intrusion detection techniques is that the computational cost of comparing the current system activity to the profile can be quite high. In the case of a wireless sensor network, such added computation can severely impact the longevity of the network.

In systems based on misuse intrusion detection, the system maintains a database of intrusion signatures. Using these signatures, the system can easily detect intrusions on the network. Further, the system is less prone to false positives as the intrusion signatures are narrowly defined. Such narrowly defined signatures, while leading to fewer false positives, also imply that the intrusion detection system will be unable to detect unknown attacks. This problem can be somewhat mitigated by maintaining an up-to-date signature database. However, since it can be difficult to characterize attacks on wireless sensor networks, such databases may be inherently limited and difficult to generate. An advantage, however, is that the misuse intrusion detection system requires less computation in order to identify intruders as the comparison of network events to the available signatures is relatively low cost.

Because both techniques have their strengths and weaknesses, traditional intrusion detection systems use systems that implement both anomaly based intrusion detection and misuse intrusion detection models. This allows such systems to utilize the fast evaluation of the misuse intrusion detection system, but still recognize abnormal system behavior.

Intrusion Detection in Wireless Sensor Networks

Typically a wireless sensor network uses cryptography to secure itself against unauthorized external nodes gaining entry into the network. But cryptography can only protect the network against the external nodes and does little to thwart malicious nodes that already possess one or more keys. An intrusion detection system (IDS) is classified into two categories: host-based and network-based. Intrusion detection schemes can be further classified into those that are signature based, anomaly based, and specification based. A host-based IDS system operates on operating systems audit trails, system call audit trails, logs, and so on. A network-based IDS, on the other hand, operates entirely on packets that have been captured from the network. A signature-based IDS simply monitors the network for specific predetermined signatures that are indicative of an intrusion.

In an anomaly-based scheme, a standard behavior is defined, and any deviation from that behavior triggers the intrusion detection system. Finally, a specification-based scheme defines a set of constraints that are indicative of a program's or protocol's correct operation. There are three architectures for intrusion detection in wireless sensor networks. The first is termed the stand-alone architecture. In this case, as its name implies, each node functions as an independent intrusion detection system and is responsible for detecting attacks directed toward itself. Nodes do not cooperate in any way. The second architecture is the distributed and cooperative architecture. In this case, an intrusion detection agent still resides on each node (as in the case of the stand-alone architecture) and nodes are still responsible for detecting attacks against themselves (local attacks), but also cooperate to share information in order to detect global intrusion attempts. The third technique proposed is called the hierarchical architecture. These architectures are suitable for multi-layered wireless sensor networks. In this case, a multi-layered network is described as one in which the network is divided into clusters with cluster-head nodes responsible for routing within the cluster. The multi-layered network is used primarily for event correlation. An intrusion-detection architecture based on the implementation of a local intrusion detection system (LIDS) at each node is described.

In order to extend each node's "vision" of the network, it suggests that the LIDS existing within the network should collaborate with one another. All LIDS within the network will exchange two types of data, security data and intrusion alerts. The security data is simply used to exchange information with other network hosts. The intrusion alerts, however, are used to inform other LIDS of a locally detected intrusion.

A pictorial representation of the LIDS architecture is depicted in Figure 10.4. MIB (management information base) variables are accessed through SNMP running on the mobile host, where the LIDS components are depicted within the block labeled LIDS. The local MIB is designed to interface with the SNMP agent to provide MIB variable collection from the local LIDS agent or mobile agents. The mobile agents are responsible for

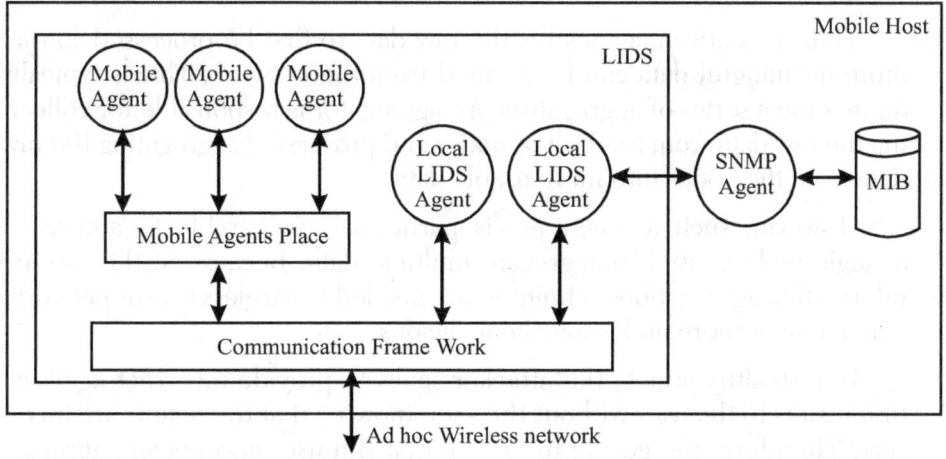

FIGURE 10.4 The LIDS architecture.

both the collection and processing of data from remote hosts, specifically SNMP requests. The agents are capable of migration between individual hosts and are capable of transferring data back to their home LIDS. The local LIDS agent is responsible for detecting and responding to local intrusions as well as responding to events generated by remote nodes.

A method is proposed to use SNMP auditing as the audit source for each LIDS. Rather than simply sending the SNMP messages over an unreliable UDP connection, it is suggested that mobile agents will be responsible for message transporting. When a LIDS detects an intrusion, it should communicate this intrusion to other LIDS on the network. Possible responses include forcing the potential intruder to re-authenticate, or to simply ignore the suspicious node when performing cooperative actions. Although this approach cannot be applied to wireless sensor network directly, it is an interesting idea that explores the local information only, which is the key to any intrusion detection techniques in sensor networks.

Secure Data Aggregation

As wireless sensor networks continue to grow in size, so does the amount of data that the sensor networks are capable of sensing. However, due to the computational constraints placed on individual sensors, a single sensor is typically responsible for only a small part of the overall data. Because of this, a query of the wireless sensor network is likely to return a great deal of raw data, much of which is not of interest to the individual performing the query.

Thus, it is advantageous for the raw data to first be processed so that more meaningful data can be gleaned from the network. This is typically done using a series of aggregators. An aggregator is responsible for collecting the raw data from a subset of nodes and processing/aggregating the raw data from the nodes into more usable data.

However, such a technique is particularly vulnerable to attacks as a single node is used to aggregate multiple data. Because of this, secure information aggregation techniques are needed in wireless sensor networks where one or more nodes may be malicious.

In a stealthy attack, the attacker seeks to provide incorrect aggregation results to the user without the user knowing that the results are incorrect. Therefore, the goal is to ensure that if a user accepts an aggregate value as correct, then there is a high probability that the value is close to the true aggregation value. In the event that the aggregate value has been tampered with, the user should reject the incorrect results with high probability.

The aggregate-commit-prove technique is composed of three phases. In the first stage, aggregate, the aggregator collects data from the sensors and computes the aggregation result according to a specific aggregate function. Each sensor should share a key with the aggregator. This allows the aggregator to verify that the sensor reading is authentic. However, it is possible that a sensor has been compromised and possesses the key, or that the sensor is simply malfunctioning. The aggregate phase does not prevent such malfunctioning.

In the second phase, the commit phase, the aggregator is responsible for committing to the collected data. This commitment ensures that the aggregator actually uses the data collected from the sensors. One way to perform this commitment is to use a Merkle hash-tree construction. Using this technique the aggregator computes a hash of each input value and the internal nodes are computed as the hash of their children concatenated. The commitment is the root value. The hashing is used to ensure that the aggregator cannot change any input values after having hashed them. In the final phase, the aggregator is charged with proving the results to the user. The aggregator first communicates the aggregation result and the commitment. The aggregator then uses an interactive proof to prove the correctness of the results. This generally requires two steps. In the first, the user/home server checks to ensure that the committed data is a good representation of the data values in the sensor network. In the second step, the user/home server decides whether the aggregator is lying. This can be done by checking whether or not the aggregation result is close to the committed result. The interactive proof differs depending on the aggregation function that is being used.

Defending against Physical Attacks

Physical attacks pose a great threat to wireless sensor networks, because of its unattended feature and limited resources. Sensor nodes may be equipped with physical hardware to enhance protection against various attacks. For example, to protect against tampering with the sensors, one defense involves tamper-proofing the node's physical package. Many focus on building tamper-resistant hardware in order to make the actual data and memory contents on the sensor chip inaccessible to attack. Another way is to employ special software and hardware outside the sensor to detect physical tampering. As the price of the hardware itself gets cheaper, tamper-resistant hardware may become more appropriate in a variety of sensor network deployments.

One possible approach to protect the sensors from physical attacks is self-termination. The basic idea is the sensor kills itself, including destroying all data and keys, when it senses a possible attack. This is particularly feasible in the large-scale wireless sensor network which has enough redundancy of information, and the cost of a sensor is much cheaper than the loss of being broken (attacked). The key of this approach is detecting the physical attack.

A simple solution is periodically conducting neighborhood checking in static deployment. For mobile sensor networks, this is still an open problem. Techniques described for extracting protected software and data from smartcard processors include manual micro probing, laser cutting, focused ion-beam manipulation, glitch attacks, and power analysis, most of which are also possible physical attacks on the sensor. Based on an analysis of these attacks, the following are examples of low-cost protection countermeasures that make such attacks considerably more difficult, including:

- Randomized Clock Signal

  Inserting random-time delays between any observable reaction and critical operations that might be subject to an attack.

- Randomized Multithreading

  Designing a multithread processor architecture that schedules the processor by hardware between two or more threads of execution randomly at a per-instruction level.

- Robust Low-frequency Sensor

  Building an intrinsic self-test into the detector. Any attempt to tamper with the sensor should result in the malfunction of the entire processor.

- Destruction of Test Circuitry

  Destroying or disabling the special test circuitry which is for the test engineers, closing the door to micro probing attackers.

- Restricted Program Counter

  Avoid providing a program counter that can run over the entire address space.

- Top-layer Sensor Meshes

  Introducing additional metal layers that form a sensor mesh above the actual circuit and that do not carry any critical signals to be effective annoyances to micro probing attackers.

Directional antennas are introduced to defend against wormhole attacks. *Search-based Physical Attacks* define a search-based physical attack model, where the attacker walks through the sensor network using signal detecting equipment to locate active sensors, and then destroys them. The defense algorithm is executed by individual sensors in two phases: in the first phase, sensors detect the attacker and send out attack notification messages to other sensors; in the second phase, the recipient sensors of the notification message schedule their states to switch.

Trust Management

Trust is an old but important issue in any networked environment, whether social networking or computer networking. Trust can solve some problems beyond the power of the traditional cryptographic security. For example, judging the quality of the sensor nodes and the quality of their services, and providing the corresponding access control, for example, does the data aggregator perform the aggregation correctly? Does the forwarder send out the packet in a timely fashion? These questions are important, but difficult, if not impossible, to answer using existing security mechanisms. The trust issue is emerging as sensor networks thrive. However, it is not easy to build a good trust model within a sensor network given the resource limits.

According to the small-world principle in the context of social networks and peer-to-peer computing, one can employ a path-finder to find paths from a source node to a designated target node efficiently. Based on this observation, a practical approach to compute trust in wireless networks is by viewing individual mobile devices as a node of a delegation graph G and mapping a delegation path from the source node S to the target node T into an edge in the correspondent transitive closure of the graph G, from which the trust value is computed. In this approach, an undirected transitive signature scheme is used within the authenticated transitive graphs. A trust evaluation based security solution is proposed to provide effective security decisions on data protection, secure routing, and other network activities. Logical and computational trust analysis and evaluation are deployed among network nodes. Each node's evaluation of trust on other nodes is based on serious study and inference from trust factors such as experience statistics, data value, intrusion detection results, and references to other nodes, as well as a node owner's preference and policy.

A reputation-based framework for high integrity sensor networks is proposed. This framework employs a beta reputation system for reputation representation, updates, and integration. A mechanism of location-centric isolation of misbehavior and trust routing in sensor networks is also described. In their trust model, the trustworthiness value is derived from the capacity of the cryptography, availability, and packet forwarding. If the trust value is below a specific trust threshold, then this location is considered insecure and is avoided when forwarding packets. The most important issue for building a trust model is adjusting parameters according to environment changes. These suggestions are quite useful for building trust models in the wireless sensor network, given their simplicity and cost savings.

Summary

- When the sensor density (i.e., number of sensors per unit area) is finite, $c > = 2r$ is a necessary and sufficient condition for coverage to imply connectivity.

- For maximum lifespan coverage, the sensor nodes are implemented with sleep and awake algorithms.

- An energy-aware routing algorithm is selected for energy conservation.

- To maximize lifespan, balance between the energy consumed by a route and the minimum residual energy at the nodes along the chosen route is achieved.

- The reliability of a sensor system is enhanced through the use of redundancy.

- Flooding is a way the sensor node broadcasts a data packet to all its neighbors, and this process continues until a destination is found.

- In gossiping, the packet is sent to a selected random neighbor which selects another random neighbor and finally arrives at the destination.

- In reactive routing protocols the routes are created only when source wants to send data to a destination, whereas proactive routing protocols are table driven.

- The main categories of wireless sensor network security are as follows: 1. The obstacles to sensor network security, 2. The requirements of secure wireless sensor networks, 3. Attacks, and 4. Defensive measures.

- Limited resources, memory, and power are a few obstacles in sensor security.

- A denial of service attack is defined as any event that diminishes or eliminates a network's capacity to perform its expected function.

Questions

1. State the theorem for sensor deployment and coverage.

2. How will it possible to maximize coverage lifespan?

3. What do you mean by unicast?

4. Write about multicast and broadcast routing types.

5. Explain about the data collection and distribution at the base station.

6. Write in detail about data-centric protocols.

7. What is flooding?

8. What is gossiping?

9. Write a note on hierarchical protocols.

10. Write in detail about the AODV routing protocol.

11. Write a short note about location-based protocols.

12. Write a method for DSR route discovery and maintenance.

13. List the obstacles in sensor security.

14. List the sensor security requirements.

15. Write the types of attacks in security.

16. Write about the defensive measures.

17. Write about trust management.

18. What do you mean by intrusion detection?

Further Reading

1. *Security for Wireless Sensor Networks* by Donggang Liu and Peng Ning

2. *Security in Wireless Sensor Networks* by George S. Oreku and Tamara Pazynyuk

3. *Security for Wireless Sensor Networks Using Identity-Based Cryptography* by Harsh Kupwade Patil and Stephen A.Szygenda

References

http://www.cise.ufl.edu/~sahni/papers/sensors.pdf

A GUIDE TO SECURING NETWORKS FOR WI-FI (IEEE 802.11 FAMILY) 1.0/ MARCH, 2017

Prepared By
United States Department of Homeland Security (DHS) Cybersecurity Engineering

Table of Contents

1. Introduction

This guide summarizes leading practices and technical guidance for securing networks from wireless threats and for securely implementing wireless access to networks. This document is specifically focused on the wireless technologies commonly referred to as "Wi-Fi" as defined in the Institute of Electrical and Electronics Engineers (IEEE) 802.11 family. This guide does not include commercial mobile networks (e.g., 3GPP, LTE). The recommendations in this guide address wireless threats that are universal to all networks and describe security controls that can work together to mitigate these threats.

Wireless capabilities are widely available, even on networks that are not intentionally providing these services. Wi-Fi signals may infiltrate buildings from commercial providers, adjacent buildings and businesses, and other publicly available services. Authorized and unauthorized Wi-Fi services can be used to gain unauthorized access to networks that are otherwise strongly secured. Due to the pervasive nature of Wi-Fi, it is important to consider the risks associated with these technologies and to examine potential impacts to confidentiality, availability, and integrity when conducting risk and threat analyses. On March 31, 2014, the Federal Communications Commission (FCC) increased the availability of the radio frequency (RF) spectrum for high-speed, high-capacity Wi-Fi in the 5 GHz band in support of the ever-increasing demand for Wi-Fi data connectivity.[1]

In response to the growing number of attacks on networks and the risks associated with the pervasive nature of wireless technologies, a number of wireless security guides have been produced by commercial interests, the Federal Government, and the Department of Defense (DoD). Two of the SANS CIS[2] Critical Security Controls for Effective Cyber Defense v6.0— Boundary Defense (Critical Security Control (CSC) 12) and Wireless Access Control (CSC 15)—are specific to wireless risks and threats.

A major recommendation in the guidance above is to deploy a wireless intrusion detection system (WIDS) and wireless intrusion prevention system (WIPS) on every network, **even when wireless access to that network is not offered**, to detect and automatically disconnect devices using unauthorized wireless services.

CSC 12 and CSC 15 recommend monitoring for communication between networks of different trust levels and specifically calling out WIDS as part

---

[1] Link to FCC announcement: *https://www.fcc.gov/document/fcc-increases-5ghz-spectrum-wi-fi-other-unlicensed-uses*

[2] According to the SANS Institute, the "SANS CIS Critical Security Controls are a recommended set of actions for cyber defense that provide specific and actionable ways to stop today's most pervasive and dangerous attacks." See Appendix A for link to the SANS CIS webpage.

of the technical approach for monitoring communication. DoD Directive 8100.2, Use of Commercial Wireless Devices, Services, and Technologies in the Department of Defense (DoD) Global Information Grid (GIG), includes the DoD policy for addressing Wi-Fi threats to both wireless local area networks (WLANs) as well as wired networks. The directive requires that an active screening capability for wireless devices be implemented on every DoD network. In July 2016, the Office of the Director of National Intelligence issued guidance requiring WIDS capabilities for continuous monitoring.

The significant increase of wireless technology in and around enterprise networks has correspondingly increased the associated risks. These risks include neighboring Wi-Fi networks, hot spots, hotels, mobile hotspot devices such as mobile Wi-Fi (MiFi), and a multitude of mobile devices and smart phones that have Wi-Fi capabilities. The focus on securing enterprise wired networks (through technologies such as firewalls, intrusion prevention systems (IPSs), content filters, and anti-virus and anti-malware detection tools) has made enterprise networks a more difficult target for adversaries. As a result, adversaries are now exploiting less secure end user devices and Wi-Fi networks to penetrate enterprise networks.

In June 2009, Gartner, Inc., a technology research company, performed a study entitled "Next Generation Threats and Vulnerabilities." This study concluded that **Wi-Fi infrastructure attacks had the highest level of severity and the lowest time to invest for the attacker**. While improvements have been made in Wi-Fi technologies since the time of this report that improve the basic security of Wi-Fi systems, users are still a weak link and must have a significant understanding of the technology in order to safeguard against many types of attacks. The automation of connections for ease of use and insecure default configurations can lead users to inadvertently compromise the security of their device or network.

2. Threat Types

By not addressing wireless security, enterprise networks are exposed to the threats listed below. Monitoring for wireless activity and devices enables an enterprise to have better visibility into Wi-Fi use and to identify and mitigate Wi-Fi-related threats. Wi-Fi threats include:

- Hidden or Rogue Access Points (APs) – unauthorized wireless APs attached to the enterprise network may not transmit their service set identifier (SSID) to hide their existence.

- Misconfigured APs – APs with weak or incorrect settings that allow unauthorized devices to connect or expose connection communications to sniffing and replay attacks.

- Banned Devices – devices not allowed on the network by organizational policy (e.g., wireless storage devices).

- Client Mis-association (e.g., department and agency (D/A)) clients connecting to non-D/A networks while at D/A sites) – clients using unsecured and unmonitored networks when secured and monitored network connections are available increases the risk of data loss and system compromise.

- Rogue Clients – unauthorized clients attaching to the network. Rogue clients pose risks of bridging and data loss as well as circumventing established security controls and monitoring efforts.

- Internet Connection Sharing and Bridging Clients – a device that shares its Internet connection or allows connectivity to multiple networks concurrently can be used to bypass network monitoring and security controls and may result in data loss or provide an unsecured network entry point for an attacker.

- Unauthorized Association – an AP-to-AP association that can violate the security perimeter of the network.

- Ad hoc Connections – a peer-to-peer network connection that can violate the security perimeter of the network.

- Honeypot/Evil Twin APs – an AP setup to impersonate authorized APs intercepting network communications and compromising systems that connect to it.

- Denial of Service (DoS) Attacks – an attack that seeks to overwhelm the system causing it to fail or degrade its usability. These attacks are frequently used in conjunction with other attacks (e.g., honeypot) to encourage a wireless client to associate with compromised wireless APs.

3. Threat Remediation

An active WIDS/WIPS enables enterprise networks to create and enforce wireless security policies. WIDS/WIPS provides the ability to centrally monitor and manage enterprise wireless security with respect to the various threats listed above. Alternatively, during an incident related to these threats, an on-site technician would be required to survey the entire enterprise with a laptop or other wireless network detection device in an attempt to locate and identify a rogue AP. Having a WIDS/WIPS capability in place greatly aids in incident remediation.

Successfully identifying and mitigating rogue APs and wireless devices is a challenging and labor-intensive process, as rogue APs are frequently moved and not always powered on. A WIDS/WIPS capability provides immediate automated alerts to the enterprise security operations center (SOC) and can be configured to automatically prevent any clients from attaching to rogue APs. WIDS/WIPS capabilities are also useful for physically locating rogue APs in order to remove them.

4. Recommended Requirements for Enterprise Wireless Networking

Listed below are sample requirements for consideration when securing an enterprise network from wireless threats. These requirements are derived from the sources listed in Appendix A: Authorities and References and should be tailored to specific considerations and applicable compliance requirements. These requirements are currently tailored to guidance applicable to federal Executive Branch D/As.

- Use existing equipment that can be securely configured and is free from known vulnerabilities where possible.

- Meet Federal Information Processing Standards (FIPS) 140-2 compliance for encryption.

- Be compliant to relevant National Institute of Standards and Technology (NIST) 800-53 controls.

- Use the certificates that reside on personal identification verification (PIV) cards for user authentication to comply with Office of Management and Budget (OMB) Homeland Security Presidential Directive 12 (HSPD-12).

- Support an alternative method of certificate authentication where PIV cannot be used.

- Use Extensible Authentication Protocol-Transport Layer Security (EAP-TLS[3]) certificate based methods or better for to secure the entire authentication transaction and communications.

- Minimally use Advanced Encryption Standard (AES) counter mode cipher block chaining message authentication code[4] protocol (CCMP), a form of AES encryption utilized by Wireless Application Protocol (WAP) 2 enterprise networks. More complex encryption technologies supporting the requirement for an enhanced data cryptographic encapsulation mechanism providing confidentiality and the client's capabilities while conforming to FIPS 140-2 may be used as they are developed and approved.

- Allow for enterprise users to operate seamlessly and allow for login scripts and login activities to function normally. Wireless access clients should be able to transition from AP-to-AP with no service disruption while maintaining the security of the connection.

5. Recommended Requirements for WIDS/WIPS

Even wired networks that do not support wireless access should utilize a WIDS/WIPS solution to monitor and detect rogue APs and unauthorized connections. The following list includes specific recommended requirements for WIDS/WIPS sensor networks and should be tailored based on local considerations and applicable compliance requirements. WIDS/WIPS systems should include the following characteristics:

- Rogue client detection capability. The system will reliably detect the presence of a workstation simultaneously broadcasting IP from a second wireless network interface card (NIC).

- Have a rogue WAP detection capability. WAP detection capability should reliably detect the presence of a WAP communicating inside the physical perimeter of the enterprise.

- Have a rogue detection process capability. Rogue client or WAP detection shall occur regardless of authentication or encryption techniques in use by the offending device (e.g., network address

---

[3] RFC 5216

[4] Cipher block chaining message authentication code is abbreviated as CBC-MAC.

translation (NAT), encrypted, and soft WAPs). Rogue detection should combine over-the-air and over-the wire techniques to reliably expose rogue devices.

- Detect and classify mobile Wi-Fi devices such as iPads, iPods, iPhones, Androids, Nooks, and MiFi devices.

- Detect 802.11a/b/g/n/ac devices connected to the wired or wireless network.

- Be able to detect and block multiple WAPs from a single sensor device over multiple wireless channels.

- Be able to enforce a "no Wi-Fi" policy per subnet and across multiple subnets.

- Block multiple simultaneous instances of the following: DoS attacks, ad hoc connections, client mis-associations, media access control (MAC) address spoofing, honeypot WAPs, rogue WAPs, misconfigured WAPs, and unauthorized associations.

- Detect and report additional attacks while blocking the above listed exploits (detection and reporting capabilities will not be affected during prevention).

- Not affect any external (neighboring) Wi-Fi devices. This includes attempting to connect over the air to provide Layer Two fingerprinting; therefore, the use of existing content addressable memory (CAM) tables is not acceptable to fulfill this requirement.

- Provide minimal communications between sensor and server, and a specific minimum allowable Kbps should be identified. The system shall provide automatic classification of clients and WAPs based upon enterprise policy and governance.

- Provide secure communications between each sensor and server to prevent tampering by an attacker.

- Have at least four different levels of permissions allowing WIPS administrators to delegate specific view and admin privileges to other administrators as determined by the D/A.

- Have automated (event triggered) and scheduled reporting.

- Provide customizable reports.

- Segment reporting and administration based on enterprise requirements.

- Produce live packet capture over the air and display directly on analyst workstations.

- Provide event log capture.

- Import site drawings for site planning and location tracking requirements.

- Manually create simple building layouts with auto-scale capability within the application.

- Be able to place sensors and WAPs electronically on building maps to maintain accurate records of sensor placement and future AP locations.

- Meet all applicable federal standards and Federal Acquisition Regulations (FAR)[5] for Federal Government deployments.

6. Recommended Requirements for Wireless Surveys

Many integrators of wireless solutions can perform a predictive or virtual site survey as part of the proposal or estimating process. This approach utilizes a set of building blueprints or floor plans to determine the optimal placement of sensors and APs within the facility. A predictive site survey takes into account the building dimension and structure but cannot account for potential RF sources because no direct examination of the site is conducted. This approach may be sufficient for some enterprises and is significantly less expensive than a more thorough RF site survey.

Alternatively, a wireless site survey, also known as a RF site survey, provides a definitive set of information for developing a wireless deployment and security plan. The survey is a defined set of tasks performed in the facility that documents the wireless characteristics of the physical facilities, coverage areas, and interference sources. This information is essential to understanding the optimal number and placement of WAPs and WIDS/WIPS devices to provide desired coverage and functionality in a facility.

---

[5] Federal standards and Federal Acquisition Regulations (FAR)

Issues that a wireless survey seeks to identify include:

- Multipath Distortion – distortion of RF signals caused by multiple RF reflective paths between the transmitter and receiver.

- RF Coverage Barriers – materials used in construction may not transmit RF signals resulting in unexpected loss of strength and reduced range.

- External and Internal Interference Sources – RF signals used by Wi-Fi are not the only users in that frequency. Identification of interference sources assists in designing a solution that achieves the desired coverage in the most efficient manner.

Before beginning a wireless survey, the following information should be obtained:

- Where in the facility is Wi-Fi access needed?

- Will there be more than one wireless network, such as a work and guest network?

- How many devices and connections will be supported over Wi-Fi?

- What are the data rate needs of these devices over Wi-Fi?

- A facility map or floor plan is essential to overlay the survey results on. This floor plan should be provided to the survey team in a digital file format appropriate to their needs, if possible.

The following list provides specific recommendations for a wireless survey. These recommendations should be tailored based on local considerations and applicable compliance requirements. A survey not intended to serve as a guide for network design and installation, and verification of the wireless communication infrastructure may not require all of the details listed.

The wireless survey should produce the following documents as a product:

- A facilities map(s) showing wireless coverage with the following indicated:

  o Interference sources and strength,

  o Any existing networks' signal strength and coverage contours,

○ External network sources available in the facility with signal strength coverage contours,

○ Identification of areas where multipath distortion may occur,

○ Recommended WAP placement,

○ Recommended WIDS/WIPS placement, and

○ Indication of signal strength coverage contours using recommended placement.

- A report providing details of findings and recommendations including details of risks, threats, and recommended mitigations. The report should include a RF spectrum analysis that will minimally indicate:

  ○ RF interference sources,

  ○ Measurement of signal-to-noise ratio (SNR),

  ○ RF power peaks, and

  ○ Wi-Fi channel interference.

- A detailed list of materials needed to accomplish goals and coverages as identified in the survey maps and reports.

- An estimated labor hours report required to install, test, and validate the installation described in the survey maps and reports.

The survey information enables optimization of AP channels, antenna type, AP transmit power levels, and placement for the proposed wireless network installation.

7. Budget Estimation Guide

Configuration and budget estimation guidance is provided below for the technical solutions outlined in these recommendations. The example information is the product of market research conducted by DHS. This guidance should be used for budgetary purposes only and the final costs will be heavily dependent on the physical characteristics of the facilities being considered. **Accurate cost estimation is best determined by working with your Network Infrastructure Support team and requesting competitive proposals from experienced installers of these solutions**.

The following factors should be accounted for to ensure a comprehensive estimate of the total project costs:

- Site Evaluation – a predictive site survey utilizing the site floor plans with documentation on existing network infrastructure can provide an accurate cost estimation for equipment required to cover the facility. While not as precise as an onsite RF survey, this typically provides sufficient accuracy for budget purposes. If your site is over 50,000 square feet (sq. ft.) or has significant potential RF interference sources (e.g., onsite RF transmitters, radar installations, or is older stone, masonry, or steel construction), an RF survey may be indicated. Vendors should be informed of these considerations when requesting estimates.

- Labor – cost should include the initial installation, training for network staff to maintain the solution, and training for the Security Operations team to utilize the solution.

- Physical and Virtual Infrastructure – equipment and service costs to support the solution should include: physical or virtual server costs, network infrastructure costs, network cabling, and power cabling.

- Maintenance and Support – costs include warranty, software support, and licensing costs that are part of the ongoing operations and maintenance of the solution.

Table 1 shows budgetary estimate example details for WIDS/WIPS solutions.

Table 1 Budgetary Estimate Example for WIDS/WIPS Solutions

Item Description	Purpose	Estimated Costs ($)	Unit
Predictive RF Survey	Utilizes facility plans to estimate coverage needs for sensors and APs		sq. ft.
Onsite Support	Utilized for training, system tuning, and configuration services, as well as an onsite RF spectrum survey, if desired		Per day
Sensor	Dual band 802.11AC sensor unit		Per sensor
Cell Sensor Option	Additional radio for detection of US cell phone signals by the 802.11 AC sensor		Per sensor

Item Description	Purpose	Estimated Costs ($)	Unit
Management Server Virtual Machine (VM)	A VM image for the management server that can support up to 50 sensors Cloud-based, physical appliances, and other license models are available depending on business needs and goals		Per VM or appliance
Service and Support	Support costs for each component varies depending on response time and level of service desired		Per device or license

8. Bluetooth Security Considerations

Bluetooth technologies (IEEE 802.15) in mobile devices present additional risks for the loss of data and the potential to eavesdrop on conversations. This exposes D/As to a higher risk for loss of confidentiality on D/A-managed devices and networks when Bluetooth is utilized while conducting D/A business. Any device – including laptops, cell phones, and tablets–that has this capability is subject to this risk.

Bluetooth technologies are designed to create a personal area network (PAN) that supports the connection of devices such as audio, keyboard, mice, or data storage devices to a system. All versions of the Bluetooth specification include unsecured modes of connection, and these are typically the easiest connections to establish. Bluetooth signals have been exploited at distances of several hundred feet, and this should be taken into consideration when evaluating the risks and establishing policies around its usage.

Mitigation methods for Bluetooth risks should include the development of a Bluetooth usage policy, enforcement of configuration management for D/A-managed devices based on this policy, and user awareness training that informs users of the risks associated with using Bluetooth. More detailed information on threats and mitigations for Bluetooth technologies can be found in NIST SP 800-121 rev 1.

Appendix A: Authorities and References

Listed below are the technical authorities, references, standards, and publications used in the creation of this guide.

Authorities and References	Description
CIO Council Mobile Security (Baseline, Framework, and Reference Architecture)	CIO Council's government mobile and wireless security baseline of standard security requirements *https://cio.gov/wp-content/uploads/downloads/2013/05/Federal-Mobile-Security-Baseline.pdf*
DHS 4300A	DHS Sensitive System Policy *https://www.dhs.gov/xlibrary/assets/foia/mgmt_directive_4300a_policy_v8.pdf*
CSC 12 Boundary Defense	The CIS Critical Security Controls for Effective Cyber Defense *https://www.cisecurity.org/critical-controls/*
CSC 15 Wireless Access Control	The CIS Critical Security Controls for Effective Cyber Defense *https://www.cisecurity.org/critical-controls/*
DoD Directive 8100.02	Use of Commercial Wireless Devices, Services, and Technologies in the Department of Defense (DoD) Global Information Grid (GIG) *http://www.dtic.mil/whs/directives/corres/pdf/810002p.pdf*
DoD Instruction 8420.01	Commercial Wireless Local Area Network Devices, Systems, and Technologies *http://www.dtic.mil/whs/directives/corres/pdf/842001p.pdf*
NIST SP 800-160	NIST SP 800-160 Systems Security Engineering: Considerations for a Multidisciplinary Approach in the Engineering of Trustworthy Secure Systems (While not specifically related to this topic, this publication provides guidance on security engineering applicable to all systems.) *http://nvlpubs.nist.gov/nistpubs/SpecialPublications/NIST.SP.800-160.pdf*
FIPS 140-2	Security Requirements for Cryptographic Modules *http://csrc.nist.gov/groups/STM/cmvp/standards.html*
GAO 11-43	GAO Report to Congressional Committees: Federal Agencies Have Taken Steps to Secure Wireless Networks, but Further Actions Can Mitigate Risk *http://www.gao.gov/new.items/d1143.pdf*
Gartner, Inc.	Next Generation Threats and Vulnerabilities, June 2009
HSPD-12	Policies for a Common Identification Standard for Federal Employees and Contractors *https://www.dhs.gov/homeland-security-presidential-directive-12*
NIST 800-153	Guidelines for Securing Wireless Local Networks (WLANs) *http://nvlpubs.nist.gov/nistpubs/Legacy/SP/nistspecialpublication800-153.pdf*

Authorities and References	Description
NIST 800-53 rev 4	Security and Privacy Controls for Federal Information Systems and Organizations *http://nvlpubs.nist.gov/nistpubs/SpecialPublications/NIST.SP.800-53r4.pdf*
NIST SP 800-121 rev 1	Guide to Bluetooth Security *http://nvlpubs.nist.gov/nistpubs/Legacy/SP/nistspecialpublication 800-121r1.pdf*
SANS CIS Critical Security Controls	The SANS CIS Critical Security Controls are a recommended set of actions for cyber defense that provide specific and actionable ways to stop today's most pervasive and dangerous attacks. *http://www.sans.org/critical-security-controls/*

Appendix B: Acronyms and Abbreviations

Acronym	Definition
AES	Advanced Encryption Standard
AP	access point
CAM	content addressable memory
CBC-MAC	cipher block chaining message authentication code
CCMP	Counter mode CBC-MAC protocol
CIO	Chief Information Officer
CSC	Critical Security Control
D/A	department and agency
DHS	Department of Homeland Security
DoD	Department of Defense
DoS	denial of service
EAP-TLS	Extensible Authentication Protocol-Transport Layer Security
FAR	Federal Acquisition Regulations
FCC	Federal Communications Commission
FIPS	Federal Information Processing Standards
GAO	Government Accounting Office
GIG	Global Information Grid

Acronym	Definition
HSPD	Homeland Security Presidential Directive
IEEE	Institute of Electrical and Electronics Engineers
IPS	intrusion prevention system
MAC	media access control
MiFi	mobile Wi-Fi
NIC	network interface card
NIST	National Institute of Standards and Technology
OMB	Office of Management and Budget
PAN	personal area network
PIV	personal identification verification
RF	radio frequency
SOC	security operations center
SNR	signal-to-noise ratio
SP	Special Publication
SSID	service set identifier
VM	virtual machine
WAP	wireless access point
WIDS	wireless intrusion detection system
WIPS	wireless intrusion prevention system
WLAN	wireless local area network

ABBREVIATIONS

4A	Anywhere, Anytime, by Anyone and Anything
ABS	Anti-lock Braking System
ADC	Analog to Digital Converter
AODV	Ad-hoc On Demand Distance Vector Routing
API	Application Programming Interface
ASCII	American Standard Code for Information Interchange
ASIC	Application Specific Integrated Circuit
BLE	Bluetooth Low Energy
BRAM	Block RAM
CDMA	Code Division Multiple Access
CNC	Computer Numerically Controller
CoAP	Constrained Application Protocol
CPLD	Complex Programmable Logic Device
CPU	Central Processing Unit
DCM	Digital Clock Manager
DMA	Direct Memory Access
DRAM	Dynamic RAM
DSN	Distributed Sensor Network
DSP	Digital Signal Processing/Processor
ETSI	European Telecommunications Standards Institute
FCC	Federal Communications Commission
FDDI	Fiber Distributed Data Interface
FDM	Frequency Division Multiplexing
FDMA	Frequency Division Multiple Access
FFD	Full Function Device

FFT	Fast Fourier Transform
FH	Frequency Hopping
FHSS	Frequency Hopping Spread Spectrum
FPGA	Field Programmble Gate Array
GHG	Green House Gases
GPP	General Purpose Processor
GPRS	General Packet Radio Service
GPS	Global Positioning System
GPS	Global Positioning by Satellite
GSM	Global System for Mobile communications
HVAC	Heating Ventilation Air Conditioning
ICT	Information and Communication Technologies
ID	Identity or Identification
IEEE	Institute of Electrical and Electronics Engineers
IETF	Internet Engineering Task Force
IoT	Internet of Things
IP	Intellectual Property
IP	Internet Protocol
IPv6	Internet Protocol Version 6
ISM	Industrial, Scientific and Medical Radio Bands
ISO	International Organization for Standardization
ITS	Intelligent Transport Systems
ITU-T	International Telecommunication Union/ Telecommunication Standardization Sector
LAN	Local Area Network
LoWPAN	IPv6 over Low power Wireless Personal Area Networks
MAC	Media Access Control
MEMS	Micro Electro Mechanical Systems
MOS	Metal Oxide Semiconductor
MSG	Message
NIST	National Institute of Standards and Technology
OFDM	Orthogonal Frequency Division Multiplexing
PAN	Personal Area Network
PC	Personal Computer

PDA	Personal Digital Assistant
PHY	Physical layer
PPP	Point to Point Protocol
QoS	Quality of Service
R&D	Research and Development
RAM	Random Access Memory
RF	Radio Frequency
RFD	Reduced Function Device
RFID	Radio Frequency Identification
RISC	Reduced Instruction Set Computer
SCM	Supply Chain Management
SDO	Standards Development Organisation
SOA	Service Oriented Architecture
SOAP	Service Oriented Architecture Protocol
SoC	System-on-Chip
SPI	Serial Peripheral Interface
SRAM	Static RAM
STM	Smart Transducer interface Module
TCP	Transmission Control Protocol
TDMA	Time Division Multiple Access
TEDS	Transducer Electronic Data Sheet
TG	Task Group
TIES	Telecom Information Exchanges Services
TSAG	Telecommunication Standardization Advisory Group (of ITU-T)
TSMP	Time Synchronized Mesh Protocol
U	Ubiquitous, as in "uHealth" or "uCity"
UCC	Urban Consolidation Centre
UCI	Universal Context Identifiers
UDP	User Datagram Protocol
USN	Ubiquitous Sensor Network
UWB	UltraWideBand
VHDL	Very High Speed Integrated Circuit (VHSIC) Hardware Description Language
W3C	World Wide Web Consortium
WiMAX	Wireless Interoperability for Microwave Access

WISA	Wireless Interface for Sensors and Actuators
WLAN	Wireless Local Area Network
WMAN	Wireless Metropolitan Area Network
WPAN	Wireless Personal Area Network
WSAN	Wireless Sensor and Actuator Network.
WSIS	World Summit on the Information Society
WSN	Wireless Sensor Network
WWAN	Wireless Wide Area Network
ZDO	ZigBee Device Object
ZED	ZigBee Extended Device

INDEX